T0317677

Advanced SiC/SiC Ceramic Composites: Developments and Applications in Energy Systems

Related titles published by The American Ceramic Society

Materials for Electrochemical Energy Conversion and Storage (Ceramic Transactions Volume 127)
Edited by A. Manthiram, P. Kumta, S.K. Sundaram, and G. Ceder
©2002, ISBN 1-57498-135-8

Perovskite Oxides for Electronic Energy Conversion, and Energy Efficiency Applications (Ceramic Transactions Volume 104)
Edited by W. Wong-Ng, T. Holesinger, G. Riley, and R. Guo
©2000, ISBN 1-57498-091-2

The Magic of Ceramics
By David W. Richerson
©2000, ISBN 1-57498-050-5

Ceramic Innovations in the 20th Century
Edited by John B. Wachtman, Jr.
©1999, ISBN 1-57498-093-9

Silicon-Based Structural Ceramics (Ceramic Transactions Volume 42)
Edited by B. Sheldon and S. Danforth
©1994, ISBN 0-944904-76-9

26th Annual Conference on Composites, Advanced Ceramics, Materials, and Structures: A and B (Ceramic Engineering and Science Proceedings Volume 23, Issues 3 and 4)
Edited by Hua-Tay Lin and Mrityunjay Singh
©2002

Advances in Ceramic Matrix Composites VI (Ceramic Transactions Volume 124)
Edited by J.P. Singh, Narottam P. Bansal, and Ersan Ustundag.
©2001, ISBN 1-57498-123-4

Ceramic Material Systems with Composite Structures—Towards Optimum Interface Control and Design (Ceramic Transactions Volume 99)
Edited by Nobuo Takeda, Laurel M. Sheppard, and Jun-ichi Kon
©1997, ISBN 1-57498-065-3

For information on ordering titles published by The American Ceramic Society, or to request a publications catalog, please contact our Customer Service Department at 614-794-5890 (phone), 614-794-5892 (fax), <customersrvc@acers.org> (e-mail), or write to Customer Service Department, 735 Ceramic Place, Westerville, OH 43081, USA.

Visit our on-line book catalog at <www.ceramics.org>.

C*eramic*
T*ransactions*

Volume 144

Advanced SiC/SiC Ceramic Composites: Developments and Applications in Energy Systems

Edited by

Akira Kohyama
Institute of Advanced Energy
Kyoto University

Mrityunjay Singh
NASA–Glenn Research Center

Hua-Tay Lin
Oak Ridge National Laboratory

Yutai Katoh
Institute of Advanced Energy
Kyoto University

Published by
The American Ceramic Society
735 Ceramic Place
Westerville, Ohio 43081
www.ceramics.org

Proceedings of the CREST International Symposium on SiC/SiC Composite Materials Research and Development and Its Application to Advanced Energy Systems, May 20–22, 2002, Kyoto, Japan.

COVER PHOTO: "Multilayered interphase infiltrated by P-CVI" is courtesy of Roger Naslain, Francis Langlais, René Pailler, and Gérard Vignoles, and appears as figure 8(b) in their paper "Processing of SiC/SiC Fibrous Composites According to CVI-Techniques," which begins on page 19.

For information on ordering titles published by The American Ceramic Society, or to request a publications catalog, please call 614-794-5890.

Contents

Processing for SiC/SiC Composite Constituent

Characterization of Thermomechanical Performance

Joining Technologies and Advanced Energy Applications

Preface

Most of the modern scientific and technological developments are interdisciplinary in nature. The field of composite materials is a good example. There have been many advances and new developments in the field of composite materials that have benefited society. The interdisciplinary nature of this field has attracted an enormous number of scientists, engineers, and researchers from different backgrounds. The quest for new materials and technologies in the field of ceramic matrix composites is enhancing and spreading in many areas, but it has not been an easy task. Many hurdles still remain in this field in spite of great advances and achievements over the last twenty-five years.

SiC/SiC composites are being viewed as high-risk but high-payoff candidates for structural materials in advanced energy systems for the 21st century, such as advanced gas turbines, high-temperature heat exchangers, and nuclear energy and hybrid fuel cell/turbine systems. This book, *Advanced SiC/SiC Ceramic Composites: Developments and Applications in Energy Systems*, is the proceedings of the CREST-International Symposium on SiC/SiC Composite Materials R & D and Its Application to Advanced Energy Systems, held May 20–22, 2002, in Kyoto, Japan. The major objective of the symposium was to bring together the international SiC/SiC composite materials R & D and design communities to exchange information, identify the design-related critical issues, discuss the latest material R & D accomplishments, and provide guidelines to focus future efforts.

It was our goal that the high payoff would clearly emerge from the design-related presentations and the material-related presentations would identify the current state-of-the-art and need for future critical technology developments. With both design goals and materials development status summarized, the discussion sessions focused on developing a materials R & D roadmap to address the key challenges during the course of SiC/SiC development. It is believed that readers from various technical disciplines will gain understanding of this area and will find these papers technically stimulating.

We would like to thank Ms. Yasuko Kimura for her hard work and diligence during the symposium and in the manuscript review process. Special thanks to Ms. Mary Cassells and Mr. Greg Geiger from The American Ceramic Society for their help in timely publication of this special volume.

Akira Kohyama

Mrityunjay Singh

Hua-Tay Lin

Yutai Katoh

Processing for SiC/SiC Composites

OVERVIEW OF CREST-ACE PROGRAM FOR SiC/SiC CERAMIC COMPOSITES AND THEIR ENERGY SYSTEM APPLICATIONS

Akira Kohyama and Yutai Katoh
Institute of Advanced Energy, Kyoto University
Gokasho, Uji, Kyoto 611-0011, Japan

ABSTRACT

The CREST-ACE Program, to establish high efficiency and environment -conscious energy conversion systems, has been carried out since 1997. The emphasis of this program is on R&D of SiC/SiC and other refractory composites with their system studies to establish sound material life cycles. For this purpose, studies ranging from fundamental materials design and process developments to applications of those materials to advanced energy systems have been systematically performed.

Significant progresses have been achieved in development of SiC fibers, interface engineering and SiC/SiC fabrication processes, such as F-CVI, PIP, MI, LPS and their combined processes, in the course of this program. The most outstanding accomplishment is the development of 'NITE' (Nano-Infiltration and Transient Eutectic-phase) process. These accomplishments are briefly overviewed and the attractive features of these material systems for advanced energy systems are identified.

INTRODUCTION

Energy supply for the 21st century will strongly require balance among our increasing need for energy at reasonable prices, our commitment to a safer, healthier environment and the moderated dependence on potentially unreliable energy suppliers [1,2]. It is also important to have better flexibility and efficiency maximization in the way energy is transformed and used.

As a key technology to establish high efficiency and environment-conscious (low impact on environment or reduced / zero-emission systems) energy conversion systems, multi-functional (structural) materials R&D is emphasized in this program. The program name, CREST-ACE, stands for Core Research for Evolutional Science and Technology - Advanced Material Systems for

Conversion of Energy. For this purpose, studies starting from materials design, process developments and applications of those materials to advanced energy systems towards the end of their material cycles are systematically carried out. The final goal is to produce and demonstrate model components of high efficiency and environment-safe energy conversion systems, which will have to be incorporated to the major energy plants in the 21st century.

As the relatively near-term energy issues, the development of new technologies to better utilize fossil fuels, such as coal gasification, advanced combustion systems, advanced fuel cell-based hybrid systems and combined heat utilization are recognized as the important technical challenge. For these applications, high-strength, corrosion resistant, durable and reliable materials are needed, where improved ceramics and ceramic composites will play key roles. As important energy options for the 21st century, nuclear fission and nuclear fusion energy will be very important as well. In these material systems, nuclear reactions and transmutations by high-energy beams and particles such as neutrons and γ-rays, have strong impacts on environment through the production of radioactive isotopes and electromagnetic emissions. Therefore, reduced-activation materials R&D have been a major task in fusion and fission energy research[3-5].

To meet the program goal, high temperature ceramics composites, such as silicon carbide (SiC) fiber-reinforced SiC matrix composite materials (SiC/SiC) and refractory metal composites, such as tungsten (W) alloy fiber-reinforced W matrix composite materials (W/W) have been selected as the base materials. The major part of present paper is dedicated for description of the development of SiC/SiC composite processes. The scope of CREST-ACE program includes the process development of SiC/SiC composites' constituents, characterization of refractory composites, development of evaluation techniques, development of utilization technologies such as joining and protective coating, and evaluation of advanced energy systems which incorporate SiC/SiC composites as key materials. The descriptions in those areas are found in numbers of separate papers in other chapters of these proceedings.

SiC/SiC MATERIAL SYSTEMS

SiC/SiC composites are considered to be the attractive candidates as materials for advanced energy systems, such as high performance combustion systems, fuel-flexible gasification systems, fuel cell/turbine hybrid systems, nuclear fusion reactors and high temperature gas-cooled fission reactors, because of the advantages represented by (1) high specific strength, (2) high temperature strength, (3) high fracture toughness, (4) small electrical conductivity (prevent Joule loss), (5) essentially high and tailorable thermal conductivity, and (6) low induced radioactivity under nuclear environments[6]. These characteristics are beneficial for "efficiency maximization", with inherent safety for the cases of

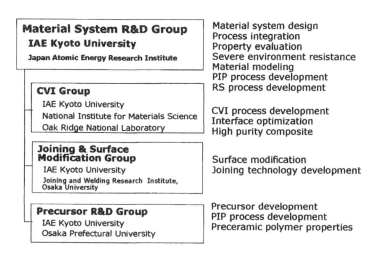

Material System R&D Group
IAE Kyoto University
Japan Atomic Energy Research Institute

Material system design
Process integration
Property evaluation
Severe environment resistance
Material modeling
PIP process development
RS process development

CVI Group
IAE Kyoto University
National Institute for Materials Science
Oak Ridge National Laboratory

CVI process development
Interface optimization
High purity composite

Joining & Surface Modification Group
IAE Kyoto University
Joining and Welding Research Institute, Osaka University

Surface modification
Joining technology development

Precursor R&D Group
IAE Kyoto University
Osaka Prefectural University

Precursor development
PIP process development
Preceramic polymer properties

Industrial partners include :
Ube Industries, Nippon Carbon, Toshiba, Kawasaki Heavy Industries,
Mitsubishi Heavy Industries

Fig. 1 – CREST-ACE R&D Organization

nuclear applications.

R&D of SiC/SiC composites in CREST-ACE program can be divided into three tasks; (1) process development of material production into composite material, (2) evaluation and prediction of materials performance, and (3) design and fabrication of multi-functional components for energy conversion systems. The first task consists of three sub-tasks; (a) improvement and innovation of SiC fibers, (b) process development of composite material production including matrix materials R&D, and (c) design and control of engineered interfaces for optimized material performances. The second task focuses on (a) mechanical properties, (b) thermal and electrical properties, (c) establishment of evaluation test methodology for SiC/SiC composite materials and fibers. In this task, studies on severe and complex environment effects are emphasized. For the third task, elements of energy conversion components for fusion reactor as well as and for high temperature gas-cooled reactor and elements for gas turbine system are designed and fabricated as the goal of this program which should verify the material performance and prove the system performance.

ORGANIZATION AND PROGRAM STRUCTURE [7]

The CREST-ACE program activity is supported by Japan Science and Technology Corporation (JST), and is operated as a JST activity with the

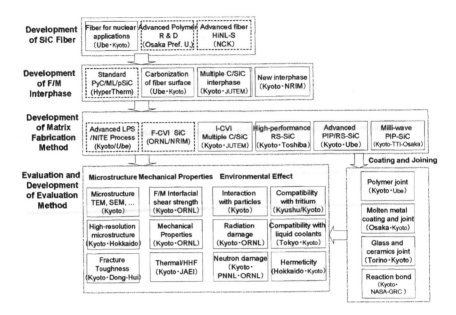

Fig. 2 – Research program structure of CREST-ACE.

participation of scientists, engineers and graduate students from universities, research institutes and industries to form a CREST-ACE team. The research organization is shown in Fig.1. Institute of Advanced Energy (IAE), Kyoto University is the central organization of this program and Joining and Welding Research Institute (JWRI), Osaka University and Osaka Prefecture University are the participants from universities. Nano-materials Laboratory (NML) at National Institute for Materials Science (NIMS), Japan Atomic Energy Research Institute (JAERI) and Osaka Industrial Research Institute (OIRI) are from national institutes and Ube Industries Ltd., Nippon Carbon Co., Toshiba, Mitsubishi Heavy Industries and Kawasaki Heavy Industries are participating from industries. Other collaborative efforts including international collaboration are listed in Fig.2, showing the program structure.

MAJOR ACCOMPLISHMENTS IN SiC/SiC R&D
(1) Progresses in Fabrication Process Development
At the beginning of the CREST-ACE program, the chemical vapor infiltration (CVI) process improvement activity was one of the three major tasks. Oak Ridge National Laboratory has been collaborating in this task. The research activity on melt-infiltration (MI) process (also called 'reaction-sintering' process) was initiated from the fundamental process survey as the second major task. The

Advanced SiC/SiC Ceramic Composites

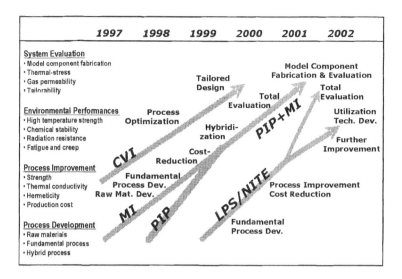

Fig. 3 – Progress in SiC/SiC fabrication process development in CREST-ACE

polymer-driven SiC fiber improvement efforts mostly carried out at Osaka Prefecture University and Ube Industries, Ltd., was the third one which is still continuing to produce new fibers with improved severe environment performance. A quick summary of the process development is shown in Fig.3. The polymer impregnation and pyrolysis (PIP) process development was started following the fiber R&D efforts and was finally developed into a hybrid process of near-stoichiometry PIP with MI post-treatment. Liquid phase sintering (LPS) process development was conducted as a potential matrix densification process but the improvement in high temperature process capability due to the improvement in SiC fibers brought to the drastic improvement in SiC/SiC composite fabrication. There is another important innovation from the LPS process development, which is Nano-Infiltration and Transient Eutectic-phase (NITE) process. There are on-going activities in PIP+MI combined process and NITE process as well as other process developments as back-ups.

(2) Advancement of CVI technique[8-16]

CVI process is already commonly used for commercial applications and is basically a well-established method. In our activities, CVI for multi-functional engineered interphase production and for high quality SiC matrix deposition have been carried out. Also, in order to characterize the fiber-matrix interphase features, varieties of mechanical testing were applied. In order to quantitatively evaluate mechanical properties of fibers, matrices and their interfaces in fiber reinforced

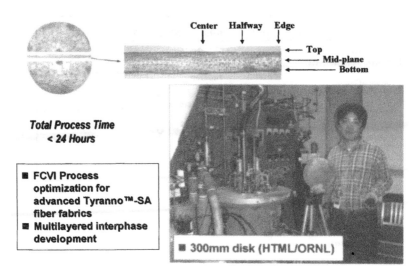

Center Halfway Edge

← Top
← Mid-plane
← Bottom

Total Process Time < 24 Hours

- FCVI Process optimization for advanced Tyranno™-SA fiber fabrics
- Multilayered interphase development

■ 300mm disk (HTML/ORNL)

Fig. 4 – Examples of advancement in FCVI processing.

SiC/SiC composites, nano-indentation tests have been carried out. Using the same technique, fiber push-out test was performed.

For studying the relationship between interfacial structures and macroscopic properties, a variety of mechanical property characterization was carried out on SiC/SiC composites with various interphases consisting of C and/or SiC. The influence of single C inter-layer thickness on mechanical properties was characterized in details for conventional Nicalon-CG/Hi-Nicalon and advanced Tyranno-SA reinforcements. It was found that the CTE mismatch, elastic properties and physical and chemical features of the fiber surfaces affects the dependency of composites' properties on inter-layer thickness. A SiC/C bi-layered interphase appeared to be effective in controlling the bonding strength and frictional stress at the interface. Multi-layered interphase composites demonstrated multiple fractures in tensile and flexural tests. These results showed that many aspects of composites' mechanical properties are tailorable through a structural control within the interphase. This shows, at the same time, that the optimization of CVI condition through the interface design can be quite complicated.

There have been many improvements in F(forced flow)-CVI process, performed at ORNL, in terms of interphase and density homogeneity, overall densification, strength, and neutron tolerance. Fig.4 indicates improvements in F-CVI process at ORNL, where 300mm diameter SiC/SiC disks were fabricated and presented encouraging results [17].

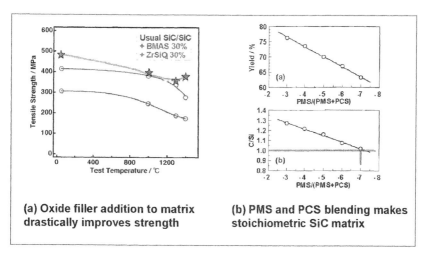

| (a) Oxide filler addition to matrix drastically improves strength | (b) PMS and PCS blending makes stoichiometric SiC matrix |

Fig. 5 – Improvement in polymer impregnation and pyrolysis process: strength and stoichiometry.

(3) PIP Process R&D by PVS and Other New Polymers [18-24]

As one of the most promising fabrication processes, there have been many efforts on polymer impregnation and pyrolysis (PIP) process R&D where improvements in performance and reliability of SiC/SiC composites for structural application are stressed. In this study, to reduce porosity in composites and to control microstructures of matrix and matrix/fiber interfaces are emphasized. The objective is to develop a fabrication scheme and process of SiC/SiC composites by PIP process with improved matrix integrity under variations of polymer-precursors and their blends. As is well known, the microstructure of ceramics derived from pre-ceramic polymer is very complicated and influenced by the polymer itself, heating scheme, environment and pressurization, in general.

In our study, polyvinylsilane (PVS) was selected as one of the promising novel precursor for the PIP process development. PVS is a pre-ceramic polymer with low viscosity at an ambient temperature. The polymer-to-ceramic conversion chemistry of PVS was studied by means of thermo-gravimetric analysis (TGA), differential thermal analysis (DTA), infrared spectroscopy, gas chromatography and X-ray powder diffraction. From these basic research activities, process optimization has been done. At the same time, in order to reduce the effect of volumetric loss of polymer by pyrolysis, an optimized scheme of SiC filler addition combined with a heating-pressurizing sequence was developed.

In general, interphase deposition through fiber coating is essential to secure

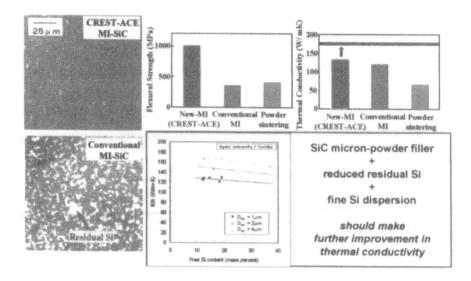

Fig. 6 – Improvement in reaction bonding process: strength and thermal conductivity of monolithic SiC.

pseudo-ductile fracture behavior of the composite products, before PIP processing using polycarbosilane (PCS) polymer. However, for the case of PVS utilization as matrix precursor, it was found that a sound fiber pull-out is achieved without applying a fiber coating.

Fig.5 presents other examples of improvements and development of PIP processing, where (a) effect of oxide filler addition to matrix on improvement in high temperature strength, and (b) designing polymer precursor for near-stoichiometry SiC through blending. Still the long and repeating PIP processing is necessary to make sufficiently dense material, which is the very critical remaining issue for PIP process to be used for more extensive applications.

(4) MI Process R & D [25-28]

Melt-infiltration (MI) process, also known as reaction-sintering (RS) process, is another processing technique for SiC and SiC/SiC composites, which has extensively been studied and is effective in producing dense and high thermal conductivity materials. Although this process is not very much process condition-sensitive nor costly, residual Si is detrimental for mechanical property and process induced damages to fiber and interfacial coating are additional technical issues.

Figure 6 shows representing summary of improvements in MI process from CREST-ACE program. Improvements in thermal conductivity and flexural

Advanced SiC/SiC Ceramic Composites

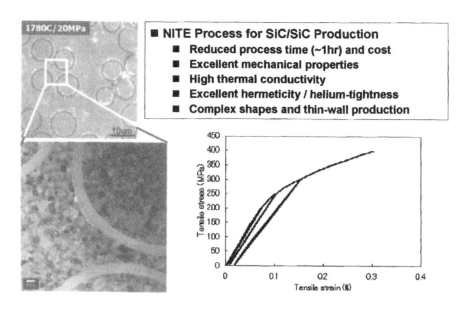

Fig. 7 – Advanced SiC/SiC by NITE process.

strength shown upper right are step wise from old powder sintering to conventional MI and followed by the new MI by CREST-ACE. As the result, monolithic SiC by this method reached over 1GPa in bending strength and over 180W/mK in thermal conductivity at room temperature. These were done by the addition of SiC micro-powder as filler, by reducing residual Si through process optimization, and by fine dispersion of residual Si. The third one is shown by SEM images at the left side of the figure. When the MI process is applied to SiC/SiC fabrication, C coating on SiC fiber is easily destroyed by reaction with molten Si and associated thermal spike and, therefore, to keep sufficient fiber/interface protection for MI process is not easy. Application of CVI multiple layer, stable nitride layer and very thick carbon layer can be solutions, but they are not always applicable. Thus this becomes the very critical issue for MI process.

Then an idea to combine the two processes, such as MI process and PIP process, came out from the CREST-ACE program. The concept of PIP and MI combined process is,

1st step: Stoichiometric PIP, which makes sufficient fiber/interface protection feasible for MI processing without detrimental effect.

2nd step: MI with fine-SiC powder as filler, which makes dense and high thermal conductivity SiC/SiC in a short time.

By this combined method, we were successful to produce 1/5-scale model

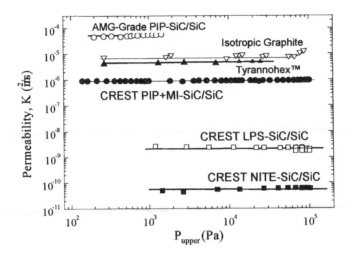

Fig. 8 –Improvement in hermeticity: helium permeability vs. upper stream pressure.

module of solid-breeding blanket cone for fusion reactor which is readily applicable to system thermomechanics interaction studies of fusion reactor blankets.

(5) LPS and NITE Process R & D [29-32]

Liquid phase sintering (LPS) process has been known as a technique to produce bulk ceramic materials from ceramic powders. This is a family of pressurized or unpressurized powder sintering process with utilizing liquid phases to densify materials under smaller pressures and at lower temperatures, compared to those required for solid-state sintering.

However, the LPS processes had not been attractive for producing SiC/SiC composites, since the temperature required for densification had been considered still far too high for composite processing. Conventional SiC-based fibers derived from polymer precursor have not been satisfactory in terms of heat resistance, even for the case of low oxygen fibers like Hi-Nicalon. The development of advanced SiC fibers with low-oxygen, well-crystallized microstructure and near-stoichiometric composition such as Hi-Nicalon Type-S and Tyranno-SA fibers, provide the potentiality to fabricate SiC/SiC composite at even high temperature. It is reported that the Si(-Al)-C Tyranno SA fiber has high tensile strength and modulus and shows very little degradation in strength or change in chemical composition after heat treatment at 1900C in an inert atmosphere.

As a trial to take advantage of these fiber characteristics, liquid phase

Advanced SiC/SiC Ceramic Composites

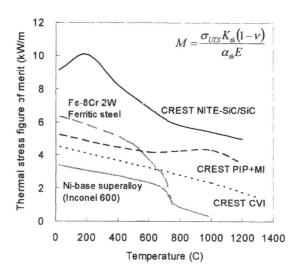

Fig. 9 – Temperature dependence of thermal stress figure of merit of advanced SiC/SiC composites and some of heat-resistant alloys.

sintering (LPS) process for SiC/SiC matrix densification was studied, where moderate matrix densification and fiber strength retention after sintering at 1780C and acceptable thermal properties under without a protective fiber coating were indicated.

Based on the encouraging results, the new process named Nano Infiltration and Transient Eutectic-phase (NITE) process has been developed. To minimize the process- induced fiber property degradation, Al_2O_3 and Y_2O_3 were selected as the sintering additives. By controlling the raw materials and process details, the amount of transient liquid phase and process time were minimized. The incorporation of nano-phase SiC powder is playing important roles in this effort and that is the basis to call this process NITE. Nano-infiltration means to utilize nano-powder-based slurry for sub-micron pore filling and utilization of nano-scale reaction domains for the matrix formation.

Figure 7 provides an example of tensile behavior at room temperature, where a proportional limit stress higher than 200MPa with very significant deformation strain before fracture, up to 0.3 %, is shown. SEM and TEM micrographs, on the left side of Fig. 7, show nearly no pores in matrix and fully crystallized SiC structure in fiber and matrix.

This process is attractive due to its very inexpensive process cost, time

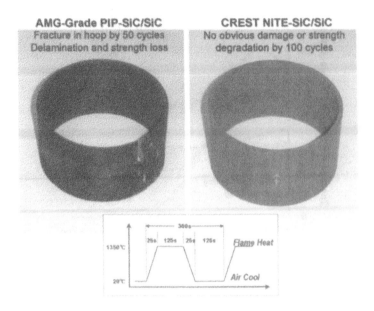

Fig. 10 – Result of combustion cycle test simulating gas turbine liner environment.

saving process characteristics and excellent flexibility in shape and size of products. From material property view point, this process is also very attractive because of the mechanical properties, high thermal conductivity, very exceptional hermeticity and radiation resistance anticipated.

Figure 8 clearly indicates the improvement in hermeticity by helium permeability measurement. Comparing with the conventional SiC/SiC by PIP process, CREST-NITE material showed almost 10^6 higher hermeticity. As materials for high temperature structural applications, thermal stress resistance is important and which can be evaluated by the thermal stress figure of merit. Figure 9 provides temperature dependence of thermal stress endurance of advanced SiC/SiC comparing with other high temperature structural materials, such as Inconel 600 and ferritic steels. For the case of CREST-NITE, thermal stress figure of merit is greater than others even from room temperature.

APPLICATIONS TO ADVANCED ENERGY SYSTEMS

In this research activity, SiC/SiC composites have been developed as attractive and promising materials for advanced energy systems, where superior high temperature and severe environment performances are essential needs. Although basic properties including major properties under neutron irradiation are

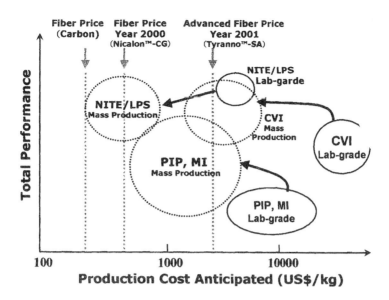

Fig. 11 – Cost trend estimation.

becoming acceptable level and many other technical issues looks within our sights to be solved, for the real application there still remains many issues.

Figure 10 shows the result from combustion cycle test using two different SiC/SiC materials with 10 cm diameter tubes at Mitsubishi Heavy Industries, Nagasaki Works. The test condition was, as shown in the figure, between 1350 C and 20 C by burning flame heating and air cooling with 300 sec. cyclic durations. The material by PIP process was fabricated along the process developed by the Japanese national project for advanced material Gas-generator (MITI project from 1991 to 2000). This project finished with a success to fabricate a real size combustor liner and test completed at 1200C. Another test material was fabricated, as the first time, by Ube Industries, Ltd., with NITE-SiC/SiC composite. The fabrication process was not optimized and shrinkage traces at the surface along axial direction of the tube were observed which were serious concerns as potential harmful surface notches during combustion cycle test. However, while the PIP sample was subjected to total fracture before 50 cycles, the CREST-NITE sample survived for 100 cycles and no macroscopic or mechanical damage was detected.

Limitations in size and shape of SiC/SiC products are important issue, where joining methods can reduce the limitation but may not be perfect solutions. In this meaning, other fabrication processes than LPS and NITE are not as attractive for making high quality structural component to be used for a large scale power generation plants. As mentioned, time-efficient processing technique such as

NITE are most eligible as the industrial fabrication processes.

Another important issue is the cost of materials and ultimately the cost of entire material related cycle. Figure 11 presents cost trend estimation based on the data available by the authors. As indicated, NITE process may provide inexpensive and high performance SiC/SiC materials for industrial usages in the near future, supposed that the price of advanced SiC fibers lowers with the increasing production scale.

CONCLUSIONS

The CREST-ACE program, conducted as planned since 1997, has been successful to make various improvements in SiC/SiC fabrication processes and even to invent a novel NITE process. The accomplishments with many breakthroughs in these years satisfy key requirements and indicate the way to solve remaining issues for utilization of SiC/SiC composites in advanced energy systems. Now, we are allowed to say with confidence that "SiC/SiC composites are not high risk/low pay-off materials any more".

ACKNOWLEDGMENTS

The authors would like to express their sincere appreciation to our colleagues working for the CREST-ACE Program in these five years. This work is performed as 'R&D of Composite Materials for Advanced Energy Systems' research project supported by Core Research for Evolutional Science and Technology (CREST) and partly supported by the Japan-USA collaborative program, JUPITER-II, and by the Research Program at IAE, Kyoto University..

REFERENCES
[1] National Energy Strategy, US DOE, DOE/S-0082P (1991)
[2] Y. Kaya, Proceedings of 10th Pacific Basin Nuclear Conference, 15-20(1996)
[3] A. Kohyama, H.Tezuka, N.Igata, Y.Imai, J. Nucl. Mater., 141-143 513-518(1986)
[4] A. Kohyama,, H. Matsui and A. Hishinuma, Proceedings of 10th Pacific Basin Nuclear Conference, 883-891(1996).
[5] A. Kohyama, A. Hishinuma, D.S. Gelles, R.L. Klueh, W. Dietz and K. Ehrlich, J. Nucl. Mater., 233-237 138-145(1996).
[6] L.L. Snead, R.H. Jones, P. Fenici and A. Kohyama, J. Nucl. Mater., 233-237 26-33 (1996).
[7] A. Kohyama, Y. Katoh, T. Hinoki, W. Zhang and M. Kotani, Proceedings of the Eighth European Conference on Composite Materials, Vol. 4, 15-22 (1998).
[8] T. Hinoki, W. Yang, T. Nozawa, T. Shibayama, Y. Katoh, and A. Kohyama, "Improvement of SiC/SiC Composites Mechanical Properties by Various Surface

Treatment of Fibers", J. Nucl. Mater., 289 23-39 (2001).

[9] W. Yang, H. Araki, T. Noda, J. Y. Park, Y. Katoh, T. Hinoki, J. Yu and A. Kohyama, "Bend Properties of CVI Hi-Nicalon™/SiC Composites with Various PyC-SiC Interlayers," Submitted to J. Am. Ceram. Soc.

[10] Y. Katoh, A. Kohyama, T. Hinoki, W. Yang and W. Zhang, "Mechanical Properties of Advanced SiC Fiber-Reinforced CVI-SiC Composites," Ceramic Engineering and Science Proceedings 21 [3], 399-406 (2000)

[11] H. Araki, W. Yang, Y. Shi, S. Sato, T. Noda and A. Kohyama, "Bending Properties of CVI SiC^f/SiC Composites at Elevated Temperatures," Ceramic Engineering and Science Proceedings 20 [4], 371-378 (1999)

[12] T. Noda, H. Araki, W. Yang and A. Kohyama, "Thermal Stabilities of CVI SiCf/SiC Composites," Ceramic Engineering and Science Proceedings 20 [4], 387-394 (1999).

[13] T. Hinoki, W. Yang, T. Nozawa, Y. Katoh and A. Kohyama, "Controlling Fiber-Matrix Interfacial Properties of SiC/SiC Composites by Simple Fiber Pre-Treatment," Proceedings of the Second Asian-Australasian Conference on Composite Materials, 739-744 (2000).

[14] W. Yang, H. Araki, J.Y. Park, T. Noda, A. Kohyama and J. Yu, "Effect of Multiple Coating Interfacial Structures on Bending Property of FCVI SiC^f/SiC Composites," Ceramic Engineering and Science Proceedings 21 [4], 259-266 (2000).

[15] T. Hinoki, W. Zhang, Y. Katoh, A. Kohyama and H. Tsunakawa, "Roles of Interfacial Microstructure on Interfacial Shear Strength of SiC/SiC," Proceeding of Eighth European Conference on Composites Materials, Vol. 4, 209-215 (1998).

[16] T. Hinoki, W. Zhang, A. Kohyama, S. Sato and T. Noda, "Effect of Fiber Coating on Interfacial Shear Strength of SiC/SiC by Nano-Indentation Technique," Journal of Nuclear Materials 258-263, 1567-1571 (1998).

[17] T. Hinoki, L.L. Snead, T. Taguchi, N. Igawa, W. Yang, T. Nozawa, Y. Katoh and A. Kohyama, "Optimization and Characterization of Chemical Vapor Infiltrated SiC/SiC Composites," in these proceedings.

[18] M. Kotani, A. Kohyama, Y. Katoh and K. Okamura, "PIP Process Optimization and Mechanical Properties of SiC/SiC Composites," Proceedings of the Second Asian-Australasian Conference on Composite Materials, 689-694 (2000).

[19] S.M. Dong, W. Zhang, T. Nozawa, Y. Katoh, A. Kohyama, S.T. Schwab and L.L. Snead, "Characterization of the Milliwave-PIP SiC/SiC Composites," Advances in Ceramic Composites IV, Ceramic Transactions (2000).

[20] M. Kotani, A. Kohyama, K. Okamura and T. Inoue, "Fabrication of High Performance SiC/SiC Composite by Polymer Impregnation and Pyrolysis Method," Ceramic Engineering and Science Proceedings 20 [4], 309-316 (1999).

[21] T. Nakayasu, M. Sato, T. Yamamura, K. Okamura, Y. Katoh and A.

Kohyama, "Recent Advancement of Tyranno/SiC Composites R & D," Ceramic Engineering and Science Proceedings 20 [4], 301-308 (1999).

[22]A. Idesaki, M. Narisawa, K. Okamura, M. Sugimoto, Y. Morita, T. Seguchi and M. Itoh, "SiC-Based Fibers Synthesized from Hybrid Polymer of Polycarbosilane and Polyvinylsilane," Key Engineering Materials 164-165, 39-42 (1999).

[23]M. Narisawa, A. Idesaki, S. Kitano, K. Okamura, M. Sugimoto, T. Seguchi and M. Itoh, "Use of Blended Precursors of Poly(vinylsilane) in Polycarbosilane for Silicon Carbide Fiber Synthesis with Radiation Curing," Journal of the American Ceramic Society, Vol. 82[4], 1045-1051 (1999).

[24]T. Iseki, M. Narisawa, K. Okamura, K. Oka and T. Dohmaru, "Highly Cross-Linked Precursors to Silicon Carbide," Ceramic Engineering and Science Proceedings 20 [4], 317-322 (1999).

[25]S.P. Lee, Y. Katoh and A. Kohyama, "Microstructure Analysis and Strength Evaluation of Reaction Sintered SiC/SiC Composites," Scripta Materialia 44, 153-157 (2001).

[26]S.P. Lee, Y. Katoh, J.S. Park, S.M. Dong, A. Kohyama and S. Suyama, "Microstructural and Mechanical Characteristics of SiC/SiC Composites with Modified-RS Process," Journal of Nuclear Materials 289, 30-36 (2001).

[27]S.P. Lee, Y. Katoh, T. Hinoki, M. Kotani, A. Kohyama, S. Suyama and Y. Ito, "Microstructure and Bending Properties of SiC/SiC Composites Fabricated by Reaction Sintering Process," Ceramic Engineering and Science Proceedings 21 [3], 339-346 (2000).

[28]S. Suyama, Y. Ito, S. Nakagawa, N. Tachikawa, A. Kohyama and Y. Katoh, "Effect of Residual Silicon Phase on Reaction-Sintered Silicon Carbide," Proceeding of the Third IEA International Energy Agency Workshop on SiC/SiC Ceramic Composites for Fusion Structural Applications, 108-112 (1999).

[29]A. Kohyama, S.M. Dong and Y. Katoh, "Development of SiC/SiC Composites by Nano-powder Infiltration and Transient Eutectoid Process," submitted to Ceramic Engineering & Science Proceedings.

[30]Katoh, A. Kohyama, D.M. Dong, T. Hinoki and J-J. Kai, "Microstructures and Properties of Liquid Phase Sintered SiC/SiC Composites," submitted to Ceramic Engineering & Science Proceedings.

[31]S.M. Dong, Y. Katoh and A. Kohyama, "Preparation of SiC/SiC Composites by Hot Pressing using Tyranno-SA fiber as Reinforcement," submitted to Journal of American Ceramic Society.

[32]T. Hino, T. Jinushi, Y. Hirohata, M. Hashiba, Y. Yamauchi, Y. Katoh and A. Kohyama, "Helium Gas Permeability of SiC/SiC Composite Developed for Blanket Component," J. Fusion Science and Technology, accepted.

PROCESSING OF SiC/SiC FIBROUS COMPOSITES ACCORDING TO CVI-TECHNIQUES

Roger Naslain, Francis Langlais, René Pailler and Gérard Vignoles
Laboratory for Thermostructural Composites, UMR-5801 (CNRS-SNECMA-CEA-UB1), University Bordeaux 1, 33600 Pessac, France

ABSTRACT

An overview of the CVI-processing for the fabrication of SiC/SiC composites from a fiber preform and a gaseous methyltrichlorosilane/H_2 precursor, is presented. Firstly, the chemistry of the process is discussed in terms of deposition mechanism and kinetics. The basic phenomena involved in the densification of a porous medium by CVI, i.e. the competition between mass transfers in the gas phase and surface chemical reactions, are discussed. Secondly, the basic isothermal/isobaric CVI processing of SiC/SiC composites is described, as well as modified versions designed to increase the SiC deposition rate or to deposit highly engineered interphases and matrices. Finally, the main characteristics of SiC/SiC CVI-composites are briefly presented.

1-   INTRODUCTION

Chemical vapor infiltration (CVI) is one of the main processes commonly used to fabricate SiC-matrix composites. Its success is related to a number of intrinsic advantages and the availability of different versions which have been specifically developed to improve the efficiency of the basic technique [1-3] or to produce materials with optimized interphases or/and matrices. In its basic form, the so-called isothermal/isobaric CVI (or I-CVI), it yields composites with a high quality SiC-matrix (which can be either stoichiometric or contain free carbon or silicon depending on deposition conditions), in terms of microstructure and mechanical properties, but with a relatively slow deposition rate [4-6]. The main advantage of this processing technique is its high flexibility.

The SiC deposition rate can be significantly improved when a temperature or/and pressure gradients are applied to the fiber preform to be densified. These modified versions of the basic CVI are referred to as TG-CVI, PG-CVI or F-CVI (T, P, F and G standing for temperature, pressure, forced and gradient, respectively) [7-9]. However, all of them are less flexible since some fixturing is necessary to generate the T/P gradients in the fiber preform. The so-called pressure-pulsed CVI (P-CVI) has been initially proposed as a way to increase the SiC-deposition rate [10, 11]. It now appears that P-CVI is much better suited to the deposition of highly engineered materials at the micro- or nano- scales. Examples in this field are the SiC/SiC composites with multilayered interphases (such as (PyC-SiC)$_n$) in which part of the conventional pyrocarbon (PyC) interphase is replaced by SiC [12] or/and multilayered self-healing SiC-based matrices [13] to improve the oxidation resistance and lifetime.

Finally, in the related calefaction process, which has been designed to still increase the deposition rate, the heated preform is directly immersed in a boiling liquid precursor. Although this process has been mainly used to produce C/C composites [14], it can be extended to other ceramic matrices [15] and could be utilized to fabricate SiC-matrix composites. Further, the CVI-process can also be used in combination with other processing techniques, such as a liquid phase route (PIP : polymer impregnation and pyrolysis or RMI : reactive melt impregnation) [10, 16]. As an example of the benefit which can be achieved through hybrid processes, the residual open porosity of the composites produced by I-CVI can be filled through and additional RMI step [17].

The aim of the present contribution is to give an overview of the CVI techniques in terms of basic phenomena, nature of fiber preforms and deposition conditions taking into account some of the properties of SiC/SiC composites required for applications in the field of advanced energy systems.

2- BASIC PHENOMENA

2.1- Chemistry of the process

CVD and CVI are processing techniques with the same chemical basis. In CVD, deposition is performed mainly on the external surface of a heated substrate to produce a coating whereas in CVI deposition occurs, in the pore network of a highly porous heated substrate, here a fiber preform, to fabricate a dense fiber-reinforced composite.

SiC can be deposited from a variety of gaseous precursors from the Si-C-H or Si-C-H-Cl systems. In the field of SiC/SiC composites, the most commonly used is a mixture of methyltrichlorosilane (MTS) and hydrogen. MTS is a liquid at room temperature (bp = 67.3°C at 1 atm) which can be easily purified by distillation. It is abundant and relatively cheap. Conversely, its decomposition on/in a heated substrate in a hydrogen atmosphere, which occurs according to the following overall equation :

$$CH_3SiCl_{3(g)} \xrightarrow{H_2} SiC_{(s)} + 3\ HCl_{(g)} \qquad (1)$$

yields significant amounts of corrosive HCl species. The actual chemical mechanism involved in the formation of SiC from MTS is much more complex and it still remains imperfectly established. Firstly and as shown in fig. 1, MTS undergoes a thermal homogeneous (i.e. in the gas phase) decomposition whose kinetics is second order (for P < 3 kPa) or first order (for P > 10 kPa) with respect to MTS with an apparent activation energy of 355-335 kJ/mol [18]. Hence, within the temperature range (900-1100°C) where the CVI-processing of SiC/SiC composites is performed, MTS is already partially decomposed in the hot zone of the CVI-furnace before reaching the hot fiber preform. It is generally accepted that MTS decomposes into CH_3 and $SiCl_3$ radicals which in turn react in the gas phase to yield CH_4, $SiCl_4$, $SiCl_2$ and $SiHCl_3$, tentatively according to the following scheme [18] :

$$CH_3SiCl_3^* \rightarrow CH_3 + SiCl_3 \quad (2)$$

$$CH_3 + H_2 \rightarrow CH_4 + H^· \quad (3)$$

$$SiCl_3 + SiCl_3 \rightarrow SiCl_2 + SiCl_4 \quad (4)$$

$$SiCl_3 + H^· \rightarrow SiHCl_3^* \quad (5)$$

$$SiCl_3^* + H^· \quad \underline{LP} \quad SiCl_2 + HCl \quad (6)$$

$$\underline{HP} \quad SiHCl_3 \quad (6')$$

where * and ˙ stand for an activated molecule and a free radical, respectively.

Hence, CH_3, $SiCl_3$ and $SiCl_2$, which are very reactive species, could be the main carbon and silicon *precursors* during SiC deposition.

The deposition of SiC from MTS has been studied through thermodynamic [6] and kinetic [18, 19] approaches. Since SiC-CVI is performed at relatively low temperatures, the chemical system is far from equilibrium with the result that kinetic considerations have a key role. This feature explains discrepancies between the results of thermodynamic calculations and experimental data. As an example, free silicon is often observed in SiC-deposits for conditions commonly used in CVD whereas its occurrence is not predicted for such conditions by thermodynamic calculations.

Figure 2 shows a *CVD-diagram* for the deposition of SiC from MTS/H_2 precursor, which has been drawn for $800 < T < 1100°C$; $0 < P < 30$ kPa and $\alpha = H_2/MTS$ ranging from 3 to 10 , from kinetic experiments [18]. In the *mass-transfer regime* (MTR) i.e. at high temperature and pressure, the deposit is highly crystalline and stoichiometric. It mainly consists of β-SiC (cubic 3C-polytype) with a columnar microstructure, a rough facetted surface and a preferred growth direction (fig. 3a). In all the other domains, referred to as CRR (for chemical reaction regimes), the deposition kinetics are rate-controlled by chemical reactions. In *CRR-1* (low T, P), the deposition kinetics is highly dependent on the MTS partial pressure (apparent reaction order n (MTS) = 2.5), with a very high apparent activation energy ($E_a > 500$ kJ/mol). It is supposed to be rate-limited by both the income of reactive species (and hence the MTS decomposition) and heterogeneous reactions. In *CRR-2*, the apparent activation energy is lower ($E_a =$

190 kJ/mol) and the apparent reaction order depends on the total pressure : n(MTS) = 0 for P ≈ 5 kPa and n(MTS) = 1 for P > 10 kPa. The rate-limiting step might be surface reactions (and no longer MTS decomposition). The SiC deposition kinetics does not depend on the partial pressure of hydrogen (n(H$_2$) = 0) in both CRR1 and CRR2. Finally, in CRR-3, the growth rate is assumed to be limited by Cl-bearing species (such as HCl) adsorption at the substrate surface. Hence, HCl could be regarded as a SiC-growth inhibitor [20].

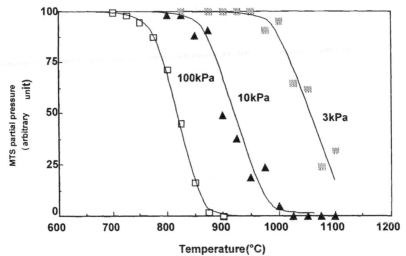

Figure 1 : Influence of temperature on the decomposition of MTS under different pressures (according to ref.18)

In the CRR-domains, the nature of the deposits can be different depending on CVD-conditions. In the CRR1/CRR2 domains and at low pressure (e.g. P ≈ 5 kPa), the deposit is *stoichiometric* (or close to stoichiometry), microcrystallized (200-700 nm β-SiC crystallites) with a relatively rough surface morphology. Conversely at high total pressure (P = 10-20 kPa), the deposit is very smooth, nanocrystallized (with a β-SiC grain size of a few nm) and it contains an *excess of silicon* (fig. 3b). As an example, for P = 17 kPa ; T = 890°C and α = 10, the Si/SiC ratio is equal to 7 and Si-nanocrystals are well apparent in the nanocrystallized SiC matrix.

Finally SiC can also be deposited with *free carbon*. Thermodynamic calculations [6] have shown that the co-deposition of SiC and free carbon, predicted to occur for low α-values (0.1 < α < 10) from MTS/H$_2$ precursor (for T = 1200 K and P = 100 kPa), is favored when argon is added to the gas phase. Its occurrence has also been established experimentally [21, 22]. Interestingly, such

Advanced SiC/SiC Ceramic Composites

CVD/CVI deposits display a lamellar microstructure thought to be related to a time oscillating reaction.

To conclude, the SiC-based deposits formed, under typical CVI-conditions, from the MTS-H_2 (Ar) precursor can be well-crystallized or almost amorphous, on the one hand, and stoichiometric, Si-rich or C-rich, on the other hand, depending on the T-P-α conditions, i.e. the kinetics of deposition. Kinetic laws, are known for some of the CRR-regimes.

2.2- Chemical vapor infiltration of SiC in a porous medium

The densification of a heated porous substrate by SiC, the so-called CVI-process, is much more difficult to control than its CVD-counterpart, the objective being here to fill as completely as possible the open porosity of the substrate, i.e. to avoid the early closing of the pores entrance. In CVI, two main phenomena are in competition, *chemical reactions* producing the SiC-based deposit on the pore wall and *mass transfers* in the pore network that fill the pore with gaseous reactants (MTS, its homogeneous decomposition products and hydrogen) and remove HCl and unreacted species.

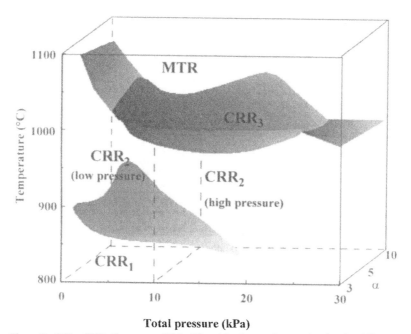

Total pressure (kPa)

Figure 2 : P-T-α CVD-diagram established from kinetic experiments showing the different kinetic regimes of SiC-deposition from flowing MTS/H_2 gaseous precursor (50 < D < 400 sccm gas flow rate) (according to ref.18)

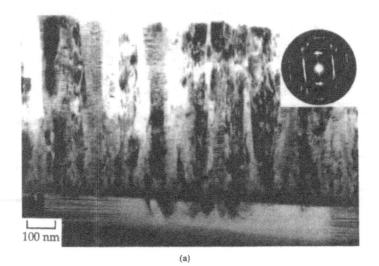

(a)

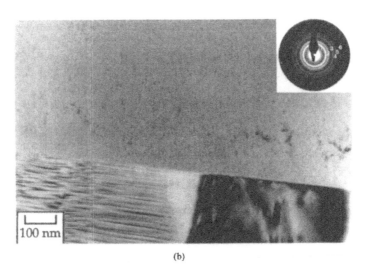

(b)

Figure 3 : Bright field TEM images of SiC-based deposits formed under different kinetic regimes from MTS/H$_2$: (a) MTR deposit ($\alpha = 5$; T = 1100°C ; P = 19 kPa ; Q = 200 sccm), (b) CRR2 deposit ($\alpha = 5$; T = 925°C ; P = 10 kPa ; Q = 200 sccm) (insets : selected area electron diffraction patterns) (according to ref.18)

Further, in the CVI of a porous fiber preform, the area of the inner surface on which SiC-deposition takes place is much larger than the external surface and it changes versus time, i.e. as densification proceeds. In CVI, the porous preform(s) "pump(s)" the reactant from the deposition chamber and reject(s) large amounts of product (HCl), particularly at the plant level where a large number of fiber preforms are treated simultaneously.

The competition between in-pore chemical reactions and mass transfer is often discussed on the basis of a dimensionless number, the Thiele modulus, first introduced for porous catalysts; and defined as :

$$\tau = \sqrt{\frac{kL^2}{D\phi_o}} \qquad \text{or} \qquad \tau = \frac{L}{\phi_o}\phi_o^{1/2}\sqrt{\frac{k}{D}} \qquad (7)$$

for a cylindrical pore, where $L/\emptyset_o$ is the aspect ratio of the pore, L its length and $\emptyset_o$ its diameter, k is the kinetic constant of the deposition reaction and D an overall diffusion coefficient of the gaseous species including both Knudsen and conventional diffusions. Densification is favored vs external surface coating when the Thiele modulus is small, i.e. when the deposition reaction kinetics are slow (low k value) and mass transfer fast (high D value) [23-28]. Under such conditions, which usually correspond to low temperature and pressure, the reactant species diffuse in the pore network far from the external surface before reacting (which favors densification) but the densification rate is relatively slow. Further even under such favorable conditions, SiC tends to be preferentially deposited near the external surface of the preform (where the reactant concentration is high) with the result that SiC/SiC composites may display a density gradient (the core being less densified than the skin), when processed according to conventional I-CVI. However, there are different ways to minimize such a gradient, that will be discussed later.

The effect of CVI-parameters (P, T, α) on the densification homogeneity can be worked out through experiments [4, 6] and modelling [25-31]. As an example, Figure 4 shows the influence of T and P on the calculated thickness profiles of SiC-deposit along a straight open-ended cylindrical pore. In the calculations, SiC was assumed to be deposited directly from MTS (equation (1), i.e. without formation of intermediate species) according to a MTS first order kinetic law [26]. Lowering both T and P results, as expected, in a more homogeneous densification of the pore. The pore geometry has also a strong influence on the densification homogeneity (equation (7)). For a given pore diameter, increasing the pore aspect ratio is detrimental to the densification homogeneity. Conversely, the chemical composition of the precursor, i.e. the α-ratio, has a relatively moderate influence on the SiC-thickness profile. Experiments performed on a model pore ($\emptyset_o$ = 33 μm, L = 3 to 10 mm) yield data which were only in partial agreement with the calculations, due to inappropriate hypotheses. As a matter of fact, a best fit was found when assuming the occurrence of an intermediate unknown species X (which could be one of the actual precursors discussed in section 2.1). More generally speaking, all modelling approaches of the I-CVI of SiC are strongly

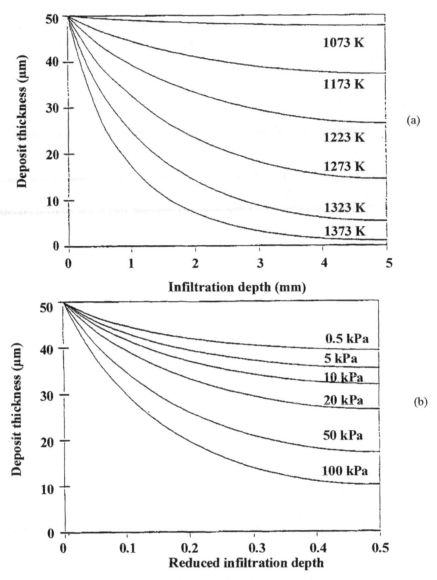

Figure 4 : Influence of temperature and total pressure on the calculated thickness profiles of a silicon carbide deposit (MTS/H$_2$ precursor with α = 10) along a straight open-ended cylindrical pore ($\varnothing$o = 100 μm ; L/$\varnothing$o = 100) for : (a) different temperatures at P = 20 kPa and (b) different total pressures at T = 1223 K. The pore being symmetrical, only half-profiles for 0 < z/L < 0.5 with z axial coordinate and L : pore length, are shown. The profiles are drawn at pore entrance closure (for t = t$_c$) (according to ref. 26)

Advanced SiC/SiC Ceramic Composites

dependent on pertinent deposition mechanisms and kinetic laws [27, 32, 33]. As an example, very few take into account the inhibiting effect of HCl (see domain CRR-3 in fig. 2 corresponding to many I-CVI experiments), although the HCl concentration is high at pore center (fig. 5) [20, 26].

Another weakness of many modelling lies in the fact that they assumed that the chemical composition at the fiber preform surface is the composition of the gas phase injected in the CVI-apparatus which is incorrect since MTS undergoes some decomposition in the gas chamber.

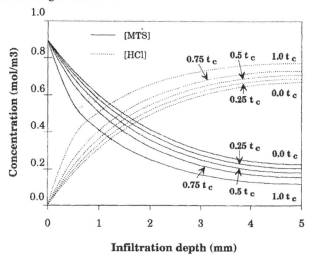

Figure 5 : MTS and HCl calculated concentration profiles along a straight open-ended cylindrical pore ($\varnothing o$ = 100 μm ; L/$\varnothing o$ = 100) during the deposition of SiC (MTS/H_2 precursor with α = 10), at various reduced infiltration times (t_c time at pore entrance closure), for T = 1223 K and P = 100 kPa (according to ref.26)

An example of a full modelling of the CVI (for the apparatus used to draw fig. 2) which includes the calculations of the MTS partial pressure outside and inside the fiber preform is shown in fig. 6 (calculations have also been performed for p_{H2} and p_{HCl} but the results are not shown here)[34]. The key point in such calculations is to take into account continuity conditions at the preform surface. In these calculations, the same deposition mechanism and kinetic law previously used in ref. 23, were considered. P_{MTS} decreases as the precursor flows through the hot zone due to mass transfer considerations and MTS depletion effects (fig. 6a). Further, a MTS gradient is observed inside the fiber preform which becomes steeper and steeper as densification proceeds (fig. 6b). It is worthy of note that the MTS gradient is steeper near the front surface of the substrate than near the back

MTS Partial Pressure (Pa)

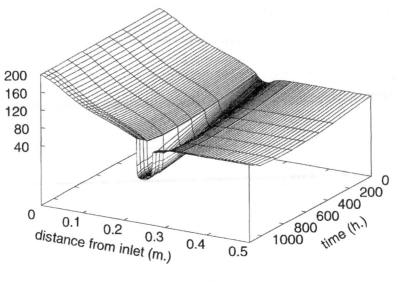

(a)

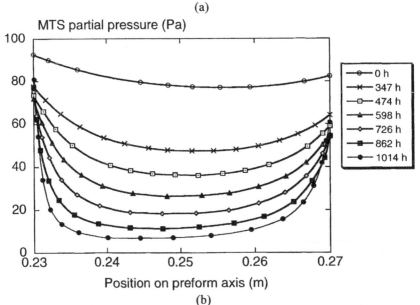

(b)

Figure 6 : Calculated P_{MTS} profiles inside an I-CVI apparatus (a) and inside a fiber prefrom set in the hot zone (b) during the deposition of SiC from MTS/H_2 with Q = 85 sccm (mean inlet velocity : 5 cms^{-1}) ; T = 1100 K ; P = 2 kPa, inlet gas compostiion : 10 mol.% MTS and 90 mol. % H_2 (according to ref.34)

Advanced SiC/SiC Ceramic Composites

face. An important point in the fabrication of a SiC/SiC composite by I-CVI is the nature of the preforms. The infiltrability of a preform depends on its fiber architecture. As shown by equation (7), the infiltrability of a pore is influenced by its diameter and shape parameter. It is well known that a unidirectional fiber preform is difficult to densify homogeneously by I-CVI. As long as densification proceeds, individual fibers are coated with SiC. As a result, the longitudinal pores are rapidly isolated from one another with limited connection. Hence, densification proceeds mainly in the axial direction, i.e. under very unfavorable conditions (high shape parameters) yielding an early pore entrance sealing by the SiC-deposit. Preforms with a 2D or 3D pore network, such as those produced from woven (or non-woven) fiber layers, comprising large size interconnected channels display a better infiltrability. The large channels are used to feed the core of the preform with reactants and to remove the reaction products. Conversely, these channels are too large to be completely filled in acceptable time, explaining that SiC/SiC composites often exhibit some residual porosity, which can eventually be eliminated by combining I-CVI with an appropriate liquid phase step.

3-   SiC-CVI PROCESS VERSIONS
3.1- SiC I-CVI process
   I-CVI is the reference technique for processing SiC/SiC composites at plant level. Fiber preforms are set inside a hot-wall infiltration chamber heated at 900-1100°C and maintained under reduced pressure with an appropriate pumping unit (fig. 7a). The preforms can be either self-standing (3D-preforms) or required some fixturing (2D-preforms), at least at the beginning of the densification process, to fix the fiber volume fraction value (typically 40-50%). The gaseous MTS/H_2 precursor flows through the infiltration chamber at a flow rate which has to be high enough to properly feed the preforms (whose high inner surface area "pumps" the reactant, as previously mentioned). The corrosive reaction products (HCl and different Cl-bearing species) should be eliminated from exhaust gas either with cold traps or with a scrubber. When the infiltration is performed at relatively high temperature and pressure (for a high densification rate), the SiC-deposit tends to occur near the external preform surface. Under such conditions, the pores entrance may experience an early sealing by the SiC-deposit and consequently should be reopened by an appropriate surface machining. Conversely, lowering both temperature and total pressure results in a more homogeneous densification (fig. 4) but a lower densification rate. Densification is usually stopped when residual overall porosity is 10-15% [1, 4, 30, 31, 33]. The main advantage of I-CVI is its *flexibility*. Since the reactants and products mass transfers through the preforms are mainly by diffusion and the furnace is run under CRR conditions, a large number of preforms can be treated simultaneously. Further, preforms can be different in shape and size. Although the densification overall time can be

relatively long, there is very few manipulation of the preforms and the process is clean. Finally, the densification cost per part is small compared to the present cost of SiC-based fibers.

3.2- TG-CVI process
In TG-CVI, a temperature gradient is applied to the fiber preform, the core being heated at a higher temperature than the skin (fig. 7b) with two important consequences. Firstly, densification starts in the core, i.e. near the susceptor, and moves from the core to the external surface, as the thermal conductivity of the preform and consequently its temperature progressively increase. Further, the core can be almost fully densified and the pores entrance remains open till the end of the densification process. Hence, there is here no need for a surface machining. Secondly, densification can be performed at higher temperature and total pressure, i.e. with a higher densification rate [8, 35]. Conversely, TG-CVI is less flexible than I-CVI. It can be only applied to fiber preforms of simple geometry in which a temperature gradient can be easily generated such as disks with an open hole at the center or tubes.

3.3- F-CVI process
In F-CVI, a temperature gradient is applied through the thickness of the fiber preform, the reactants being injected through the cold face of the preform by forced convection under pressure (fig. 7c). The hot face of the preform is rapidly sealed by SiC-deposit and the densification front moves again from the hot face towards the cold face, as the pores are being progressively filled and the thermal conductivity increasing, the reaction gaseous products leaving the preform through its lateral surface. The simultaneous use of temperature and pressure gradients through the preform allows the process to be run at higher temperatures and gas flow rates with consequently a shorter densification duration [7, 9, 32]. Conversely, F-CVI is still less flexible since a specific fixturing has to be used for each fiber preform in order to generate the TP-gradients.

3.4- The P-CVI process
Pressure-pulsed CVI has been first presented as a way to increase the densification rate and therefore to reduce the densification duration [36]. In P-CVI, the deposition chamber is periodically evacuated and then filled with reactants. As a consequence, the gaseous reaction products which may inhibit the SiC deposition (see CRR-3 domain in fig. 2), are periodically extracted from both the preform(s) and the deposition chamber and replaced by fresh reactants, enhancing thus the deposition rate. Further, the injection in a short time of cold reactants may cool to some extent the external surface of the fiber preform, creating some temperature gradient allowing thus the use of a higher inner temperature, a feature which also increases the deposition rate. Conversely,

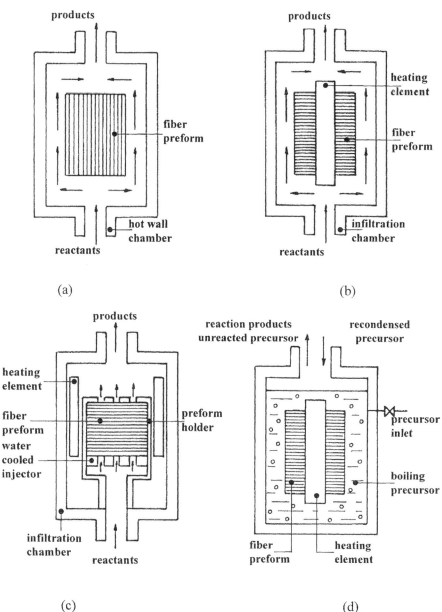

Figure 7 : Different versions of the CVI (and related) process : (a) I-CVI, (b) TG-CVI (c) F-CVI and (d) calefaction process

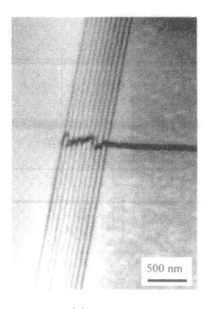 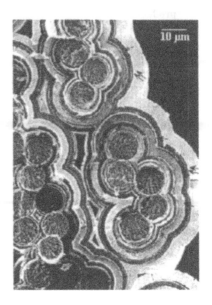

(a) (b)

Figure 8 : Multilayered interphase (a) and self-healing matrix (b) infitrated by P-CVI.
The multilayered $(PyC_{20}-SiC_{50})_{10}$ interphase was deposited in a 2D-Hi-Nicalon preform
further densified with SiC by I-CVI. The self-healing matrix, comprising mechanical
fuses and glass-former layers was infiltrated in a carbon tow (according to ref. 39 and 38,
respectively)

generating such pressure pulses, with a relatively short period, requires a much
more complex facility (specially for large size deposition chamber).

A unique advantage of P-CVI lies in the fact that it allows the design of highly
engineered interphases and matrices [37]. When the P-pulses are short enough i.e.
less than one second and the deposition rate slow (low T, P), the deposit can be
built almost atomic layer by atomic layer. Moreover, if the nature of the gaseous
precursor is periodically changed, multilayered ceramics can be deposited within a
fiber preform at the nanometer scale. Multilayered $(PyC-SiC)_n$ or $(BN-SiC)_n$
interphases, in which the elementary layer thickness is only a few nanometers with
n = 10 to 30, have been proposed as a way to improve the oxidation resistance of
SiC/SiC composites, the carbon or BN mechanical fuse being replaced by the SiC
glass former, in oxydizing atmospheres [12]. Lowering the amount of pyrocarbon
could also be useful in applications where the materials are submitted to
radiations, e.g. in nuclear fusion reactor. Finally, P-CVI has also been used to
produce composites with so-called self-healing matrices in which the SiC-matrix
is replaced by a multilayered matrix combining mechanical fuse layers and layers

that form fluid oxide phases when exposed to an oxydizing atmosphere (fig. 8) [13, 38].

3.5- The calefaction process

In the calefaction process, the fiber preform is directly immersed in a boiling liquid precursor. Hence, it has common features with CVI but also some noticeable differences. Firstly and as schematically shown in fig. 7d, the preform is usually heated with a heating element (such as a graphite susceptor) set in the core of the preform and which generates a *temperature gradient* between the hot core (whose temperature is close to ≈ 1000°C) and the external surface of the preform maintained at the boiling temperature of the precursor (67.3°C for MTS). Hence, densification proceeds from the core to the outer part of the preform, a common feature for most TG-processes, without pores entrance closing. Secondly, the calefaction process is usually run at the *atmospheric pressure*. It is actually the combination of high temperature and high pressure conditions which explains the very high densification rates. Although the calefaction process has been used mainly for the production of C/C composites [14], it could also be employed for that of SiC/SiC composites [15].

4- SiC/SiC COMPOSITES PROCESSED BY CVI

SiC/SiC composites produced by CVI, e.g. from 2D-preforms, display some residual porosity (fig. 9), which is higher in the core than near the external surface. The fiber volume fraction is ≈ 40-50% and that of the matrix ≈ 35-50%. The matrix can be deposited with a variety of composition : stoichiometric, Si or C-rich as well as multilayered. Further, it can be either almost fully amorphous, nanocrystalline or microcrystalline with, in this latter case, a preferred growth direction.

From a mechanical standpoint, SiC/SiC composites exhibit under tensile loading a *non-linear* stress-strain behavior and a *non-brittle* character when the fiber-matrix bonding has been properly optimized, i.e. when a fiber surface treatment has been applied and an interphase deposited on the fibers prior to the matrix infiltration. The most commonly used interphase, also deposited by CVI, is pyrocarbon. Other interphases, such as BN, $(PyC-SiC)_n$ and $(BN-SiC)_n$ can be used to enhance the oxidation resistance of the composites. All of them display a layered structure and are almost exclusively deposited by CVI. The fiber surface pre-treatment is used to increase the bonding between the fiber surface and the interphase in order that crack deflection takes place within the interphase and not at the fiber surface. It is specific to each fiber type. The occurrence of the residual porosity is generally not considered as detrimental to the mechanical behavior since SiC/SiC composites are not brittle. Conversely, it favors the in-depth diffusion of oxygen and its reaction with both the interphase and the fiber. The oxidation resistance is ameliorated when a self-healing matrix is used instead of pure SiC. For specific applications where the materials have to be gas tight, the residual open porosity of SiC/SiC composites produced by CVI, can be filled through an additional step, e.g. by RMI.

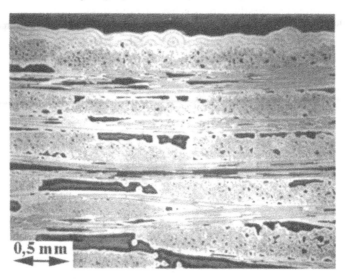

Figure 9 : Cross-section of a 2D-SiC/SiC composite with a self-healing matrix deposited by CVI, showing the fiber arrangement, the residual porosity and the multilayered character of the matrix (according to ref. 40) .

5- CONCLUSION

The CVI-process is well suited to the fabrication of advanced SiC/SiC composites. It is a clean process, performed at relatively low temperature and which yields near-net shape parts with a minimum of manpower. Further, the same technique can be used to deposit the interphase, the matrix and eventually an external coating. In its I-CVI version, the process is well adapted to volume production, the relatively slow deposition rate being balanced by the large number of parts treated simultaneously and by a high flexibility. The deposition rate can be improved with modified versions but at the expense of flexibility. Finally, it appears that the main factor which bridles the development of SiC/SiC composites is more the present cost of SiC fibers than that of the CVI-process by itself.

REFERENCES

[1] R. Naslain, F. Langlais, "CVD-processing of ceramic-ceramic composite materials", in "Tailoring Multiphase and Composite Ceramics" (R.E. Tressler et al., eds.), *Mater. Sci. Research*, **20** (1986) 145-164, Plenum Press, New York.

[2] F. Langlais, "Chemical vapor infiltration processing of ceramic matrix composites" in "*Comprehensive Composite Materials*" (A. Kelly, C. Zweben, eds.), **chap. 4-20** (2000) 611-644, Elsevier, Oxford.

[3] T.M. Besmann, B.W. Sheldon, R.A. Lowden, D.P. Stinton, "Vapor phase fabrication and properties of continuous-filament ceramic composites", *Science*, **253** (1991) 1104-1109.

[4] R. Naslain, J.Y. Rossignol, P. Hagenmuller, F. Christin, L. Héraud, J.J. Choury, "Synthesis and properties of new composite materials for high temperature applications based on carbon fibers and C-SiC or C-TiC hybrid matrices", *Revue de Chimie Minérale*, **18** (1981) 544-564.

[5] E. Fitzer, D. Hegen, H. Strohmeier, "Possibility of gas phase impregnation with silicon carbide", *Rev. Int. Hautes Temper. Refract.*, **17** (1980) 23-32.

[6] F. Christin, "Carbon-silicon carbide composites : a new family of materials for high temperature applications", *PhD Thesis*, **n° 641**, Univ. Bordeaux, December 13, 1979.

[7] D.P. Stinton, A.J. Caputo, R.A. Lowden, T.M. Besmann, "Improved fiber-reinforced SiC composites fabricated by Chemical Vapor Infiltration", *Ceram. Eng. Sci. Proc.*, **7**, (7-8) (1986) 983-989.

[8] I. Golecki, R.C. Morris, D. Narasimhan, N. Clements, "Rapid densification of carbon-carbon composites by thermal-gradient chemical vapor infiltration", *Ceram. Trans.*, **58** (1995) 231-236.

[9] W.M. Matlin, D.P. Stinton, T.M. Besmann, "Development of a two-step forced chemical vapor infiltration process", *Ceram. Trans.*, **58** (1995) 119-123.

[10] K. Suzuki, S. Kume, K. Nakano, Y. Kanno, "Pressure-pulsed chemical vapor infiltration of SiC to three-dimensional carbon fiber/SiC preform prepared by polymer impregnation and pyrolysis", *J. Ceram. Soc. Japan*, **110** [1] 44-50 (2002).

[11] P. Dupel, R. Pailler, X. Bourrat, R. Naslain, "Pulse chemical vapor deposition (P-CVD) and infiltration (P-CVI) of pyrocarbon in model pores with rectangular cross-sections : Part 2- Study of the Infiltration", *J. Mater. Sci.*, **29** (1994) 1056-1066.

[12] S. Bertrand, O. Boisron, R. Pailler, J. Lamon, R. Naslain, "(PyC-SiC)$_n$ and (BN/SiC)$_n$ nanoscale-multilayered interphases by pressure-pulsed CVI", *Key Eng. Mater.*, **164-165** (1999) 357-360.

[13] R.R. Naslain, R. Pailler, X. Bourrat, S. Bertrand, H. Heurtevent, P. Dupel, F. Lamouroux, "Synthesis of highly tailored ceramic matrix composites by pressure-pulsed CVI", *Solid State Ionics*, **141-142** (2001) 541-548.

[14]B. Narcy, F. Guillet, F. Ravel, P. David, "Characterization of carbon-carbon composites elaborated by rapid densification process", *Ceram. Trans.*, **58** (1995) 237-242.

[15]P. David, B. Narcy, J.D. Lulewicz, F. Ravel, S. Schulmann, "Elaboration of ceramic composites by rapid densification", in *Proc. ECCM-10* (A. Poursatip, K. Street, eds) (1995) 611-616, Woodhead Publishing, Cambridge, UK.

[16]S. Masaki, K. Moriya, M. Shibuya, H. Ohnabe, "Development of Si-Ti-C-O fiber reinforced SiC composites by chemical vapor infiltration and polymer impregnation and pyrolysis", *Ceram. Trans.*, **58** (1995) 187-192.

[17]C.A. Nannetti, A. Borello, D.A. de Pinto, "C fiber reinforced ceramic matrix composites by a combination of CVI, PIP and RB", in *High Temper. Ceram. Matrix Composites* (W. Krenkel et al., eds), pp. 368-374, Wiley-VCH, Weinheim, 2001.

[18]F. Loumagne, "CVD of silicon carbide from $CH_3SiCl_3H_2$: Homogeneous process and microstructural characterization", *PhD Thesis*, **N° 1044**, University Bordeaux 1, Dec. 17, 1993.

[19]S.V. Sotirchos, G.D. Papasouliotis, "On the homogeneous chemistry of the thermal decompostion of methyltrichlorosilane : Thermodynamic analysis and kinetic modelling", *J. Electrochem. Soc.*, **141** (1994) 1599-1627.

[20]T.M. Besmann, "CVI processing of ceramic matrix composites", *Ceram. Trans.*, **58** (1995) 1-12.

[21]C. Mallet, "On the co-deposition of pyrocarbon-silicon carbide by CVD : application to ceramed C/C composites for bone prostheses", *Master Thesis*, **n° 1811**, Univ. Bordeaux, Oct. 7, 1982.

[22]E. Fitzer, "Dynamische instabilitäten bei heterogenen gas/feststoff-reactionen" *Chemie-Ing. Techn.*, **41** [5-6] 331-339 (1968).

[23]E. Fitzer, W. Fritz, G. Schoch, "The chemical vapour impregnation of porous solids. Modelling of the CVI-process", *J. de Physique IV/C2* (1991) 143-150.

[24]E.W. Thiele, "Relation between catalytic activity and size of particles", *Ind. Eng. Chem.*, **31** [7] 916-920 (1939).

[25]R. Fédou, F. Langlais, R. Naslain, "A model for the isothermal isobaric chemical vapor infiltration in a straight cylindrical pore. 1 : description of the model", *J. Mater. Synth. and Process*, **1** [1] 43-52 (1993).

[26]R. Fédou, F. Langlais, R. Naslain, "A model for the isothermal isobaric chemical vapor infiltration in a straight cylindrical pore. 2 : application to the CVI of SiC", *J. Mater. Synth. and Process*, **1** [2] 61-74 (1993).

[27]S. Midleman, "The interaction of chemical kinetics and diffusion in the dynamics of chemical vapor infiltration", *J. Mater. Res.*, **4** (1989) 1515-1524.

[28]G.Q. Lu, "Modelling the densification of porous structures in CVI-ceramic composites processing", *J. Mater. Proc. Technol.*, **37** (1993) 487-498.

[29]S.V. Sotirchos, "Dynamic modelling of chemical vapor infiltration", *AIChE J.*, **37** (1991) 1365-1382.

[30]J.Y. Ofori, S.V. Sotirchos, "Optimal pressures and temperatures for isobaric isothermal CVI", *AIChE J.*, **42** (1996) 2028-2840.

[31]J.Y. Ofori, S.V. Sotirchos, "Multidimensional modelling of chemical vapor infiltration : application to isobaric CVI", *Ceram. Intern.*, **23** (1997) 119-126.

[32]N.H. Tai, T.W. Chou, G.C. Ma, "Effect of deposition mechanisms in the modelling of forced flow/temperature gradient chemical vapor infiltration", *J. Amer. Ceram. Soc.*, **77** (1994) 849-851.

[33]H.C. Chang, T.F. Morse, B.W. Sheldon, "Minimizing infiltration times during isothermal chemical vapor infiltration with methyltrichlorosilane", *J. Amer. Ceram. Soc.*, **80** (1997) 1805-1811.

[34]C. Descamps, "Modelling of isobaric-CVI : application to SiC and pyrocarbon", *PhD Thesis*, **n° 2166**, Univ. Bordeaux 1, Dec. 6, 1999.

[35]D. Gupta, J.W. Evans, "A mathematical model for CVI with microwave heating and external cooling", *J. Mater. Res.*, **6** (1991) 810-818.

[36]Y. Ohzawa, T. Sakurai, K. Sugiyama, "Preparation of fibrous SiC shape using pressure-pulsed chemical vapor infiltration and its properties as a high temperature filter", *J. Mater. Proc. Technology,* **96** (1999) 151-156.

[37]R. Naslain, R. Pailler, X. Bourrat, G. Vignoles, "Processing of ceramic matrix composites by pulsed-CVI and related techniques", *Key Engineering Mater.*, (H. Suzuki, K. Komaya, K. Uematsu, eds.), **159-160** (1999) 359-366, Trans. Tech. Publ., Zuerich.

[38]F. Lamouroux, S. Bertrand, R. Pailler, R. Naslain, M. Cataldi, "Oxidation-resistant carbon fibre reinforced ceramic-matrix composites", *Composites Sci. Technology*, **59** (1999) 1073-1085.

[39]S. Bertrand, "Improvement of the lifetime of SiC/SiC composites with (PyC/SiC)$_n$ or (BN/SiC)$_n$ nanostructured interphases", *PhD Thesis*, **n° 1927**, Univ. Bordeaux 1, Sept. 29, 1998.

[40]P. Forio, "Thermo-mechanical behavior and lifetime of a 2D-SiC/SiC composite with a multilayered self-healing matrix", *PhD Thesis*, **n° 2171**, Univ. Bordeaux 1, June 26, 2000.

RESEARCH ON THE HIGH PERFORMANCE SILICON CARBIDE CERAMICS AND SILICON CARBIDE BASED COMPOSITES

Dongliang Jiang
Shanghai Institute of Ceramics, Chinese Academy of Sciences (SICCAS)
1295 Ding-Xi Rd, Shanghai 200050, P. R. China

ABSTRACT

This is an overview of research work in silicon carbide field in this institute for the past 25 years. It will be divided into several parts to introduce the processing and properties of high performance SiC ceramics; surface modification technology, SiC based multiphase materials, and composites. Multilayer structure SiC composites prepared by aqueous tape casting technology and rheological properties of slurries were reviewed. The effects of composition and thickness of interface between layers on the properties of laminate composites were also discussed. The finite element method(FEM) was used to design the structure of TiC particle reinforced SiC layered composite.

INTRODUCTION

High performance ceramic is one of the most important materials except metal and polymer. Because of it's specially functions, such as high temperature, high strength, resistance to wear and corrosion, electric, magnetic, insulate, optic, absorb wave and so on, make it has received much attention by scientists and engineering since middle age of last century. Now, high performance ceramics have been widely applied in various industrial fields. Among of high performance ceramics, silicon carbide (SiC) with good resistance to high temperature, wear and corrosion is one of the most important high performance ceramics. But the brittleness is the main factor that prohibits it from wide applications

One effort to improve strength and toughness is through composites by particles, whiskers, and fibres reinforcement. The other one has been developed through macro or microstructure design and new processing methods by biomimetic analysis since later 1980,

This paper will briefly introduce the research work and some results of high performance silicon carbide ceramics and silicon carbide based composites in Shanghai Institute of Ceramics, Chinese academy of Sciences (SICCAS) for the past 25 years。

SILICON CARBIDE CERAMICS AND ITS COMPOSITES

1. Silicon carbide ceramics

Studies of SiC ceramics in SICCAS bagan since 1977[1,2,4,5]。 It is well known that sintering of SiC is very difficulty due to its covalent bond . In order to meet industrial requirements main effort is to get high density material through high purity (>99%) ; ultra-fine powder (<1μm) ; optimum additives (B+C, B_4C+C, Al_2O_3, Y_2O_3 and $Al_2O_3+Y_2O_3$) [3,5,7,14] and various sintering processing such as: Pressureless sintering (PLS) , Hot-pressing (HP) [3], Reaction bonded (RB) [6], and Hot-Isostatic-pressing (HIP) [11,13,18,19]。 Table1 shows some properties of SiC ceramics with various processing。In the table, composition of SW9, SY, SBC, and SA is $SiC+B_4C+C$, $SiC+Y_2O_3$, $SiC+B_4C+C$, and $SiC+Al_2O_3$, respectively.

Table 1. The properties comparison of SiC ceramics with various processing

Properties	PLS SiC		HP SiC		RB SiC
	SW9	SY	SBC	SSA	RB24
Density (g/cm3)	3.05-3.10	3.21	3.17	3.22	3.06-3.15
Bending Strength (MPa)	400-500	>700	>500	>710	>600
Hardness (HRA)	>93	93-95	>93	>93	>93
Fracture Toughness (MPa·m1/2)	3.5-4.0	>10.0	>4.0	>5.0	>4.3
Thermal Expansion	4.7	4.6	4.5	4.5	5.0

Recently SPS (Sparkling pulse sintering) technology was also applied to develop SiC ceramics with nano structure. Sintering temperature of most ceramics can be reduced 150 to 200°C by SPS as compared with HP sintering.

2. Surface modification of SiC ceramics by N_2-HIP processing

(15,20,21,22,23,26,31,32,33)

From chemical thermodynamic analysis, we found that most carbide will be transferred to nitride under the high temperature and high nitrogen pressure. It can be expressed by chemical equation(1-3) as below, figure 1 shows （A） representcative temperature and pressure of various carbides and （B） different stabilized region especially for TiC and SiC and shadow zone can be controlled for partial transfer of mixed TiC and SiC.

$$3/2\ SiC\ (s) + N_2\ (g) \rightarrow \!\!\!\!-\!\!\!\!- 1/2Si_3N_4 + 3/2C\ (s) \qquad (1)$$

$$2TiC\ (s) + N_2\ (g) \rightarrow \!\!\!\!-\!\!\!\!- 2TiN\ (s) + 2C\ (s) \qquad (2)$$

$$1/2B_4C\ (s) + N_2\ (g) \rightarrow \!\!\!\!-\!\!\!\!- 2BN\ (s) + 1/2C\ (s) \qquad (3)$$

SiC will transfer to Si_3N_4 with 17.6% volume expansion under 1850°C and 30 atm N_2 pressure and yielded surface residual compression stress can be calculating in

$$\sigma = (\alpha c - \alpha n)EnVn\Delta T/Vn(En/Ec-1)+1 \qquad (4)$$

where α and E are the linear thermal expansion coefficient and Young's modules of carbide(c) and nitride(n), Δ T is an effective temperature difference, and Vn is the volume fraction of nitride in carbide .

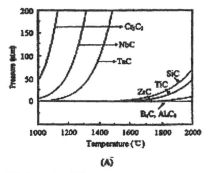
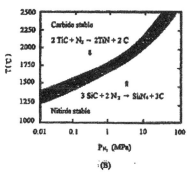

Fig1. Equilibrium partial pressures of N_2 at temperature from 1000 to 2000°C for nitridation of several carbides （A） special for SiC and TiC （B）

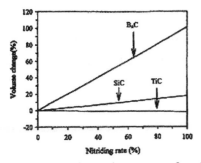

Fig.2 Relative volume changes as a function of transforming rates for nitridation of several carbides

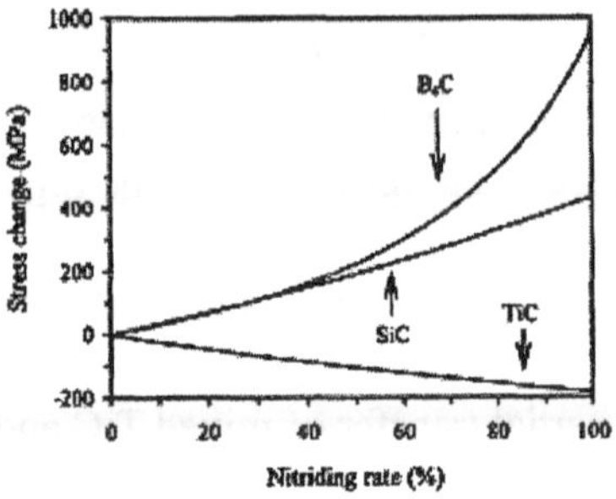

Fig.3 Residual stress as a function of transforming rate for several carbides

After N_2-HIP surface modification, mechanical properties of almost of all carbides or composites can be dramatically changed with almost 50%-100% increasing. These very successful results are listed in table.3.

Table2. Comparison of some properties of carbides before and after post-HIP nitridation

composition	Before	nitridation		After	nitridation	
	Density (gcm^{-3})	Bending strength (Mpa)	Fracture toughness $(Mpam^{1/2})$	Density (gcm^{-3})	Bending strength (Mpa)	Fracture toughness $(Mpam^{1/2})$
SiC	3.14	660	5.7	3.20	1000	8.4
SiC-25vol%TiC	3.45	354	6.5	3.60	690	7.9
SiC-5vol%SiCw	3.13	595	6.7	3.17	920	8.5
SiC-20vol%ß-Sialon	3.18	630		3.19	908	
Al₂O₃-20wt%SiCw-10wt%SiCp	3.70	705		3.75	1033	

3. Whiskers and particles reinforced SiC based multiphase composites[12,8,42]

In order to enhance toughness of SiC ceramic, one of the important way is through whiskers or particles reinforcement. In most situation, SiC whiskers were synthesized by carbon reduction method in Ar and with some catalysts existed。

$$SiO_2 + 2C \rightarrow \beta\text{-}SiC + CO_2 \qquad (5)$$

Properties of SiC whiskers by above method can be described as below; Diameter and length of whiskers is ϕ=0.5-1.0μm and L=30-50 μm, respectively. After acid washing and separation, final purity of SiC whisker can reach 99%. Main impurity of O_2 content is around 0.5-0.6%. SiC, Al_2O_3, Si_3N_4, and Mullite were reinforced by using these whiskers and some import whiskers(American Matrix Co.) in our studies[16,17,24,25,28].

Because of some toxicity of whiskers and cost high, we have pay much attention in particles reinforced SiC by using thermal expansion coefficient mismatch between second phase particle and SiC[12,28]. The strong compression residual stress between reinforced second phase particles and SiC matrix was the main factor to improve toughness and strength. TiC[8,34], ZrB_2[42], Sialon[35] Si_3N_4,[38], Al_2O_3[42], AlN[29,37,41,45,46], AlN+Y_2O_3[27,30,39],$TiSi_2$[43] and Al_2O_3+TiB_2[42] were selected to be second phase to reinforced SiC matrix. Some properties of particles and whiskers reinforced SiC were listed in Table 3

Table3 Properties of various SiC based composites with different processing

Composites	Processing	Strength (MPa)	Toughness (MPa·m$^{1/2}$)	Density (g/cm^3)
SiC	HIP	1000	8.4	3.20
SiC-TiC	HP	580	7.1	3.64
SiC-TiC- Al_2O_3	HP	700	6.5	3.65
β-SiC-TiC- Al_2O_3	HP	670	7.2	3.64
SiC-TiB_2- Al_2O_3	HP	888	8.7	3.55
SiC-ZrB_2	HP	560	6.5	3.59
SiC- Al_2O_3	HP	750	4.0	3.30
SiC-AlN	PLS	350-410	4.5	3.15
	HP	1128	6.6	3.2
SiC-YA	PLS	704	10.5	3.27
SiC-SiCw	HIP	920	8.5	3.17
SiC-β-Sialon	HIP	908	---	3.19
SiCw/mullite	SPS	570	4.5	3.13
Al2O3-SiCp	HP	580	5.4	3.72
Al2O3- SiCw-SiCp	HIP	1033	---	3.75

4. Some applications of silicon carbide ceramics and it's composite

Main application is for wear parts such as: mechanical sealing ring, sleeves, bearing ball, milling media, cutting tool, and armor. Another important application is using its good resistance to high temperature as for thermal resistance material: refractory, combustion liner, nozzle, heat exchange tube etc. Also, there are some special applications like semiconductor Si treat parts; atomic energy for structural parts; aeronautical and military application parts all of these application received much attention. Fig.4 Shows several wear parts and special application parts manufactured by SICCAS.

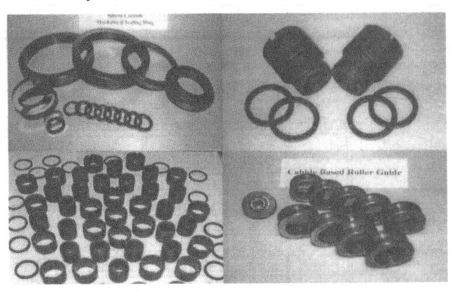

Fig 4. Some wear parts: mechanical sealing ring, sleeves, and roller guide

SILICON CARBIDE AQUEOUS TAPE CASTING AND LAMINATED COMPOSITES

Macro structure design of laminated composites is based on bio-mimetic point of view and environment friendly processing selected – aqueous tape casting[47,48] technology. Fig 2 shows flow chart of laminated composites by aqueous tape casting process.. Tape casting is the usual process for the preparation of laminated composites, in the past, organic solvents were commonly used to prepare concentrated suspensions due to good dispersion for binder and plasticizer and easy to remove. But over 50% organic material release to atmosphere will be

Advanced SiC/SiC Ceramic Composites

critical to our environment protection. In this case, aqueous tape casting should be taken even it still has lots of problem. The effects of PH value, Zeta potential, dispersion agent, solid content, binder, and plasticizer (selected suitable for aqueous tape casting) factors on the properties of slurry were studied.

1. Studies on the rheological properties of SiC aqueous slurries[49,50].

Commercial SiC (FCP-15, Norton Co.) powders with an average particle size of 0.9m and a special surface area as 9.03m^2/g were used in this work. Al$_2$O$_3$ and Y$_2$O$_3$ were selected as sintering additives. After a serious of tests, finally, It was found that well-dispersed SiC slurries could be obtained through surface treatment process of SiC raw materials by control PH value at alkaline region as well as the

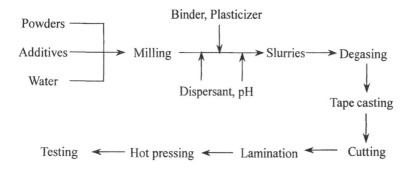

Fig 2 。 Flow chart for preparation of layered composite
through tape casting process

addition of PEI , PVA, and Glycerol as the dispersant, binder, and plasticizer. respectively. SiC slurries suitable for the aqueous tape casting process could be obtained by controlling apparent viscosity at 300-1000mpa •s region.. Fig.3 shows representative the relationship among of zeta potential, PH value and dispersants. Fig.4 shows that rheological properties of SiC slurries with binder and plasticizers.

2. Laminated composites without weak interface

A process of aqueous tape casting, laminating and hot pressing prepared laminated SiC samples. Between the layers there was no weak interface added. After sintering, it was difficult to detect the individual SiC layer and interface

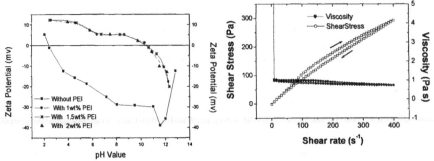

Fig 3 Zeta potential of 50vol% SiC slurries with and without dispersant

Fig. 4. Rheological properties of SiC slurries 24vol% with binder and plastizers

between adjacent layers. The microstructure of SiC laminates showed somewhat like a monolithic ones. The mechanical properties of laminated SiC samples were summarized in Table.4. The properties of monolithic samples were also listed for comparison. Table 4 shows the effects of the ratio of additives on the mechanical properties.

Table4. Mechanical properties of monolithic and laminated SiC samples

	Bulk SiC		Laminated SiC
Sintering process	Hot pressed	pressureless sintering	Hot pressed
Relative density(%)	99.5	98	99
Strength(Mpa)	754	650	724
Toughness(Mpam1/2)	4.5	6.8	9.5
Work of fracture(Jm1/2)		31.2	756

Table 5. The mechanical properties of laminated SiC samples without interface phase

$Al_2O_3:Y_2O_3$	Density (g·cm⁻³)	Porosity(%)	Strength (MPa)	Toughness (MPa·m¹ᐟ²)	Hardness (Hv)
5:3	3.24	0.25	724	9.5	2149
7:3	3.26	0.19	662	7.6	2133

Advanced SiC/SiC Ceramic Composites

3. Laminated composites with weak interface (SiC+C)

In order to increasing work of fracture, weak interface (SiC+C) was introduced in the layers. The optimized mechanical properties of SiC laminates with weak interface could be obtained by adjusting the thickness and composition of interfacial layers. Table6 shows the effects of thickness of interface on the mechanical properties. It seems to be that strength increased with decreasing thickness of interface, but for work of fracture was just opposite. For the optimized processing and composition of interface as 50vol%SiC+50vol%C, the strength, work of fracture and toughness properties of laminates composite can reach 580Mpa, 4282J/m^2 ,and 10.8Mpa $\cdot$ m$^{1/2}$, respectively.

Table 6. Influence of the interface thickness on the mechanical properties of SiC

	Aqueous tape casting	Screen printing	
Interfacial thickness (μm)	25	10	3
Strength (MPa)	429	603	687
Toughness (MPa·m$^{1/2}$)	6.7	8.8	4.7
Work of fracture (J·m^{-2})	1586	3449	797

4. Layered SiC/TiC composites designed by finite element method (FEM)[51,52]

Surface compressive stress can increase not only fracture toughness but also strength and weibull modulus. It is mainly caused by the coefficient of thermal expansion (CTE) mismatch between layers. In symmetrical layered composites, the design of residual stress is very important because the inner tensile stress should be as low as possible when the surface compressive stress is introduced. Also, thermal residual stress should be designed to vary from the surface compressive to central tensile gradually. FEM is proper for designing this structure. The optimized layered composite then was fabricated and its mechanical properties were listed for comparing with the FEM calculation results.

Fig.5 It is a model of symmetrical SiC/S10T structure The composition of S2T, S4T, S6T, S8T, and S10T was SiC+2wt%TiC, SiC+4wt%TiC, SiC+6wt%TiC, SiC+8wt%TiC, and SiC+10wt%TiC, respectively. The thickness of each layer was controlled in 220 μ m approximately by tape casting technology. Each composition contents 4 layers (symmetrical distributed both side) only S10T is 20 layers in center. The stacked layers were put in graphite die hot-pressed sintering at 1850°C, 35MPa for 30minuutes. Fig.6 shows the thermal residual stress distribution from surface to center. The mechanical properties of SiC/S10T

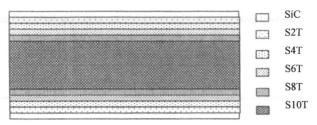

Fig.5 the gradient design of components for SiC/TiC laminates

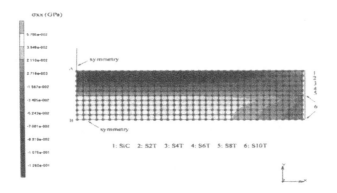

Fig.6 the stress distribution in gradient SiC/TiC laminates by FEM method

composites and comparison with SiC and S10T materials are listed in Table 7. The. toughness of SiC/S10T exhibits much more uniform than those of monolithic SiC and S10T. It also proved that compression–strengthened ceramics have a strong tendency to trap surface cracks in the outer layer. That also means flaw sensitivity of fracture of the layered structure with compressive stress is decreased. Tested results of SiC/S10T layered composite compared well with theoretical calculated

calculated by FEM method shown in Table.8.

Table 7. Mechanical properties of layered SiC, S10T and SiC/S10T

Materials	Bending strength (MPa)	Fracture toughness (MPa m$^{1/2}$)
SiC	714±86.3	6.66±0.97
S10T	663±38.7	7.86±0.72
iC/S10T	834±73.1	10.54±0.15

Table 8. Comparison of tested and theoretical strength of SiC/S10T layered ceramics

Material	Surface residual stress (MPa)		Strength (MPa)	
	Calculated	Tested	Theoretical	Tested
SiC				714
SiC/S10T	-126	-129	840	834

(-) Compression stress

SUMMARY

This paper provides a general theme of the research activities of SiC ceramics and its composites with related processing developed in Shanghai Institute of Ceramics, Chinese Academy of Sciences in the past 25 years. The more detail of each part could be found in related references.

ACKNOWLEDGEMENT:

Author would like to take this opportunity to thank CREST organized committee and chairmen of Professor Kohyama and also thank significant contributions of my colleagues in this research group in SICCAS.

REFERENCES

1. Dongliang   Jiang，" Silicon Carbide engineering ceramics" J. Chin. Ceram. Soc.，6 [2] 176. (1978)
2. T. K. Wu, H. C. Miao, T. C. Chiang, X. R. Fu, D.L. Jiang and X.X. Huang, "Studies on Nitride and Carbide Ceramics, International Symposium on

Factors in Densification and Sintering of oxide and Nonoxide Ceramics", Oct. 3-6, 1978, Hakone, Japan.

3. Dongliang, Jiang , Zengsu Pan, Yu. Lin. Li, "Study on the hot-pressing of SiC ceramics", J. Chin. Ceram. Soc., 9 [2] 133 (1981)

4. Dongliang, Jiang , "High performance silicon carbide engineering ceramics" , Mech. Eng. & Mater. , 9 [1] 23 (1985) (in Chinese.)

5. T.S. Yen, J.K. Guo ,X.R. Fu and D.L. Jiang, "High Temperature Structural Ceramics Development for Engines in the Chinese Academy of Sciences", 2nd International Symposium "Ceramic Materials and Components for Engines", Lubeck-travemunde, FRG, April 14-17, 1986

6. Dongliang, Jiang, Zensu Pan, Yu. Lin. Li,, "Reaction bonded SiC ceramics", J.Chin. Inorganic Material., 3 [2] 130 (1988)

7. Dongliang, Jiang, , Gao Kuang, Zensu Pan, S.H. Tan , "Oxidation and high temperature strength of SiC with difference additives", J. Chin. Inorganic Material., 3 [1] 47 (1988)

8. Dongliang, Jiang, J. H. Wang, Y. L. Li, "Study on the strengthening of Silicon carbide based multiphase ceramic. I. SiC-TiC system" Mater. Sci. Eng. A(109), (1989) , p401-406

9. Dongliang, Jiang, , " Surface strengthening and Toughening of ceramics", J. Inorganic Material., 4 [1] 97 (1989)

10. Dongliang, Jiang, , Y. L. Li,, J. H. Wang,, L. T. Ma, Xi Yang, "Study on the strengthening on SiC-ZrB$_2$ multiphase ceramics" , J. Chin. Ceram. Soc., 18 [2] 123 (1990)

11. J. H. She, D. L. Jiang and P. Greil, "Hot isostatic pressing of presintered silicon carbide ceramics", J. Euro. Ceram. Soc., 7 [ ]243-47 (1991)

12. Dongliang Jiang,, Jingkun Gao " Multiphase ceramics", J. Chin. Ceram. Soc., 19 [3] 258 (1991) .

13. J.H. She, D.L. Jiang, S. H. Tan and J. K. Guo, "Improvement of presintered silicon carbide ceramics by hot isostatic pressing", Mat. Res. Bull. 26 (1991) 1277-82

14. J. H. She, D. L. Jiang, S.H. Tan and J.K. Guo, "Microstructure and properties of hot isostatically processed SiC ceramics with an Al_2O_3 additive", Materials Letters, 14 (1992) 24.

15. D.L. Jiang, J.H. She, S.H. Tan and P. Greil, "Strengthing of silicon carbide ceramics by surface nitridation during hot isostatic pressing", J. Am. Ceram. Soc., 75 [9] 2586-89. (1992).

16. Zhenren Huang, Dongliang, Jiang, , S. H. Tan, "SiC whisker reinforced Al_2O_3-matrix composite", J. Chin. Ceram. Soc., 20 [4] 424−30 （1992）.

17. Zhenren Huang , Y. L. Li,, Dongliang, Jiang , S. H. Tan, "The Effect of dispersion procession on the properties of composites", J. Chin. Inorganic Material. 8 （1993） 214−20.

18. J. H. She, D. L. Jiang, and J. K. Guo, "Densification of silicon carbide ceramics by hot isostatic pressing", Scripta METALLURGICS et MATERIALLA, 28 (1993) 523-26.

19. J. H. She, J. K. Guo and D. L Jiang, "Hot isostatic pressing of α-silicon carbide ceramics", Ceramics International 19 (1993) 347-3

20. J. H. She, D.L. Jiang, S.H.Tan, "Surface toughening of liquid phase sintered silicon carbide by surface nitridation," J. Euro. Ceram. Soc., 13(1994) 159-66

21. Jiang Dongliang, She Jihong, Tan Shouhong, "Nitridation of SiC-TiC Composites By Hot Isostatic Pressing", p821, 3rd, Euro-Ceramic.3, Edited by P. Duran, J.F. Fernandez, 1993.

22. D. L. Jiang, J.H. She et al, "Hot Isostatic Pressing of Silicon Carbide Based Multiphase Composed Materials", Fall Meeting of the Materials Research Society, Boston, USA., 28 Nov. - 2 Dec., 1994..Proceeding of the 1994 Fall Meeting, v.365, p.159-164, 1995,

23. .P. Greil, H G. Bossemeyer, A. Kluner, D. L. Jiang and J. H. She, "Surface Toughening of Liquid Phase Sintered Silicon Carbide by Surface Nitridation", J. Europ. Ceram. Soc., 13 (1994) 159

24. Z.R. Huang, D.L. Jiang, S.H. Tan, "Colloidal Processing of Whisker Composites", J. Am. Ceram. Soc., 79 [9].2240-42. (1995)

25. J. H .She, D. L. Jiang, S.H. Tan and J.K. Guo, "Hot Isostatic Pressing of Paticule- and Whisker- Reinforced Silicon Carbide Matrix Composites" KEY ENGINEERING MATERIALS, vols. 108-110(1995)p.45-52, Trans Tech Publications, Switzerland.

26. J.H. She and D.L. Jiang, "Microstructural and Compositional Investigations of Surface-Nitrided Silicon Carbide Ceramics", Mater. Res. Bull., 30(1995) p.707-713.

27. Yu-bai Pan,, S. H. Tan, D. L. Jiang, "Oxidation behavior of SiC-AlN-Y_2O_3 multiphase ceramics", J. Chin. Inorganic Material., 10 [3] 359 (1995).

28. Jiang Dongliang, "A REVIEW: Silicon Carbide Based Composites" p,67-84, KEY ENGINEERING MATERIALS, Trans Tech Publications Ltd, vols. 108-110(1995).

29. Yu-bai Pan,, S. H. Tan, Dongliang, Jiang, "Microstructure of hot-pressed SiC-AlN multiphase ceramics", J. Chin. Inorganic Material., 10[4] 428-32 (1995)

30. Yu-bai Pan,, S. H. Tan, Dongliang, Jiang , " preparation and properties of SiC-AlN-Y_2O_3 multiphase ceramics", J. Chin. Inorganic Material., 10[2] 189-192 (1995).

31. D.L.Jiang, "Surface modification of cabides by HIP nitridation processing", Solid State Ionics 101-103 (1997) 1221-1227

32. J.H. She and D.L Jiang, "An Attractive way to Heal Surface defects of Some Carbide and Boride ceramics", Mater. Lett, 26 (1996) 313

33. W.M. Wang, R.R. Lu, J.Q. Zhu, J. H. She and D.L. Jiang, "Micro-analysis of SiC-Si$_3$N$_4$ ceramics made by hot isostatic pressing", Nuclear Instruments and Methods in Physics Research, B 108 (1996) 343

34. S.M. Dong, D.L. Jiang, S. H. Tan and J. K. Guo,"Mechanical properties of SiC/TiC composites by hot isostatic pressing", J. Mater. Sci. Lett, 15 (1996) 394

35. S.M. Dong, D.L. Jiang, S.H. Tan and J .K. Guo, "Hot isostatic pressing and post-hot isostatic pressing of SiC-β-Sialon composites", Mater. Lett, 29 (1996) 259.

36. Y.P. Zeng, D. L. Jiang, S. H. Tan and J. K. Guo, "Tape casting of TiB2 (Al$_2$O$_3$)/nano-TiC (Al$_2$O$_3$) multilayer composites", J. Mater. Sci. and Tech., v.13 No.4 (1997)p.324-326

37. Z.M. Chen, S.H. Tan, J. Tang and D. L .Jiang, "High performance SiC-AlN composites ", J. Mater. Sci. and Tech.,v.13 No.4 (1997)p.342-344

38. Dong Shaoming, Jiang Dongliang, Tan shouhong and Guo Jingkun, "Preparation and Characterization of Nano-Structured Monolithic SiC and Si$_3$N$_4$/SiC Composite by HIP", J. Mater. Sci. lett. 16(1997)1080.

39. S. H. Tan, Z. M. Cheng, D.L. Jiang, "Liquid sintering of SiC ceramics ", J. Chin. Ceram. Soc. 26 [2] 191-197. (1998)

40. D. L. Jiang, "Toughened Silicon Carbide Based Composites By Microstructure Design and Processing" High Temperature Ceramics Matrix Composites III, CSJ Series vol 3, p.179-184, Eds. K. Niihara, K. Nakano, T Sekino, E. Yasuda, Trans Tech Publication Ltd. Switzerland.1998-1999

41. Yu-bai Pan, Jian-hu Qiu, Mikio Morika, Shouhong Tan, Dongliang Jiang "The mechanical properties and microstructure of SiC-AlN particulate composite", J. Materials Sci. 33 (1998), 1233-37

42. D. L. Jiang and Z. R. Huang, " SiC Whiskers and Particles Reinforced Al$_2$O$_3$ Matrix Composites and N$_2$-HIP Post-Treatment",(**Invited**) Key Engineering Materials, vols 159-160 (1999)379-386. Trans Tech Publications, Switzerland.

43. Jianlin Li, <u>Dongliang Jiang</u>, Shouhong Tan, " Microstructure and Mechanical Properties of in situ Produced SiC/TiSi$_2$ Nanocomposite", J. Euro. Ceram. Soc., 20 (2000) 227-233

44. Zhengren Huang, Seong-jai Cho, <u>Dongliang Jiang</u>, Shouhong Tan," Surface nitridation of Al$_2$O$_3$ based composite by N$_2$-HIP post-treatment", J. Mater. Sci., 34 (1999) 2023-2027

45. Yu-bai Pan, Jian-hui Qiu, Makoto Kawagoe, Mikio Morita, Shou-hong Tan and Dongliang Jiang, "SiC-AlN Particulate Composite", J. Euro. Ceram. Soc., 19 (1999) 1789-1793.

46. Y. Pan, S. Tan and D. Jiang, J. Qiu, M. Kawagoe, M. Morita, "In-Situ Characterization of SiC-AlN Multiphase ceramics", J. Mater. Sci., 34 (1999) 5357-5360

47. Yu-Ping Zeng, Dong-Liang Jiang, Tadahiko Watanabe, "Fabrication and Properties of Tape-Cast Laminated and Functionally Gradient Almina-Titaanium Carbide Materials", J. Am. Ceraam. Soc., 83[12]2999-3003(2000)

48. Zeng YP, Jiang DL.," Fabrication and properties of Laminaated Al$_2$O$_3$/TiC ccomposites", Ceram. Int. 27 [5] 597-602 (2001)

49. J.X.Zhang, D.L. Jiang, S.H. Tan, L.H.Gui, M.L. Ruan, "Aqueous processing of SiC Green Sheets I. Dispersant", to be published in J. Mater. Res. 17[8] (2002).

50. J.X.Zhang, D.L. Jiang, S.H. Tan, L.H.Gui, M.L. Ruan, "Aqueous processing of SiC Green Sheets II. Binder and plastizer", to be published in J. Mater. Res. 17[8] (2002).

51. Dongliang Jiang, "The Structural Design and Processing Science of High-performance Ceramics", Rare Metal Mater.&Eng. 30[11]344-51(2001)

52. Shuyi Qin, Dongliang Jiang, Jingxian Zhang, "Evaluation of weak interface effect on the residual stresses in layered SiC/TiC composites by the finite element method and x-ray diffraction", J. Mater. Res. 17[5]1118-24(2002)

OPTIMIZATION AND CHARACTERIZATION OF
CHEMICAL VAPOR INFILTRATED SiC/SiC COMPOSITES

T. Hinoki and L.L. Snead
Oak Ridge National Laboratory
Oak Ridge, TN 37831, USA

T. Taguchi and N. Igawa
Japan Atomic Research Institute
Tokai, Ibaraki 319-1195, Japan

W. Yang, T. Nozawa, Y. Katoh and A. Kohyama
Institute of Advanced Energy, Kyoto University
Gokasho, Uji, Kyoto 611-0011, Japan
CREST, Japan Science and Technology Corporation
Kawaguchi, Saitama 332-0013, Japan

ABSTRACT

SiC/SiC composites were fabricated by the forced-flow, thermal gradient chemical vapor infiltration (FCVI) method at the Oak Ridge National Laboratory (ORNL) and by the iso-thermal chemical vapor infiltration (ICVI) method at the Japanese National Institute for Materials Science (NIMS). The FCVI approach can fabricate relatively large composites in relatively short time, while the ICVI has significant controllability of fiber/matrix interphase formation. Fiber types included the near stoichiometric Tyranno SA and Hi-Nicalon Type-S. SiC/SiC composites 12.5 mm in thickness with either 75 or 300 mm in diameter were fabricated at ORNL. SiC/SiC composites with 40 mm in diameter and 1.5~3.0 mm in thickness were fabricated at NIMS. The microstructure of these materials was studied using SEM with EDS and TEM while their mechanical properties were evaluated by tensile, flexural and single fiber push-out testing.

Density, the uniformity of fiber/matrix interphase and mechanical properties improved by increasing fiber volume fraction, optimizing processing conditions for both the FCVI and the ICVI processes. Porosity was decreased to approximately 15%. The effect of the interphase on mechanical properties and fracture behavior were studied. Tensile strength of 2D composites reinforced with Tyranno SA fibers and with optimized multilayer SiC/C interphase was approximately 300 MPa.

INTRODUCTION

CVI produces a stoichiometric, crystalline β-SiC. The major advantage of CVI over other processing routes is the low thermal and mechanical stress of the densification process owing in large part to the lower deposition temperature. In addition, the process imparts little mechanical stress to the preform. The excellent controllability of formation of fiber/matrix interphase, which affects mechanical properties significantly, is also advantage of the method. However it takes long time for CVI processing and the size was limited to relatively small fabric.

The forced-flow thermal-gradient chemical vapor infiltration (FCVI) developed at ORNL overcomes the problems of slow diffusion and restricted permeability [1,2] even in the large component with 300 mm in diameter and 15 mm in thickness. Composites have been fabricated using Nicalon™ fibers. However the optimum conditions using recent near stoichiometric high purity fibers such as Tyranno™ SA and Hi-Nicalon™ Type-S has not been established. The isothermal chemical vapor infiltration (ICVI) method at NIMS [3] can form precise uniform fiber/matrix interphase within a composite. The interphase is one of keys to improve mechanical properties of composites [4,5]. The optimization of the interphase for composites reinforced with high-purity fibers is required.

SiC-sintered fibers, which are near stoichiometric, highly crystalline and high elastic modulus, such as Sylramic™ [6] of Dow Corning, Hi-Nicalon™ Type-S [7] of Nippon Carbon and Tyranno™ SA [8,9] of UBE industries are now available. These SiC fibers have been reported to show superior thermal stability compared to low-oxygen fibers, since the oxidation of excess C in air into CO at high temperatures resulted in the formation of pores in the latter [9]. These fibers are also expected to be stable under neutron irradiation. Therefore development and evaluation of the SiC/SiC composites reinforced with the highly crystalline fibers is desired.

One of the objectives of this study is to optimize processing conditions of both ICVI and FCVI using high purity SiC fibers focusing to increase density and to obtain a uniform fiber/matrix interphase through the plate thickness. Another objective is to characterize the fiber/matrix interphase and improve mechanical properties of the composites. The effect of the interphase on mechanical properties was evaluated. Commonly advantages of ceramic matrix composite compared with monolithic ceramic are larger fracture toughness and a narrow distribution in failure strength. The trend of development of SiC/SiC composites using high elastic modulus fibers in this work is toward high modulus, high proportional limit stress and high strength with a narrow distribution rather than low strength with large fracture toughness.

EXPERIMENTAL

SiC/SiC composite disks with 40 mm in diameter and 1.5~3.0 mm thick were fabricated using plain woven Hi-Nicalon and Tyranno SA fibers by ICVI at NIMS. The fibers were stacked in [0°/90°] direction. 7 sheets of Hi-Nicalon fibers or 9 sheets of Tyranno SA fibers were used to fabricate composites with 1.7 mm thick. C and multilayers of C and SiC were applied as the interphase between fiber and matrix. The precursors were methane (CH_4) for C deposition and methyltrichlorosilane (MTS, CH_3SiCl_3) or ethyltrichlorosilane (ETS, $C_2H_5SiCl_3$) for SiC deposition. Hydrogen was used as carrier gas for MTS and ETS. The temperature (900~1100 °C) of the furnace and flow rate of gas were controlled. Each gas flows into the bottom of the furnace. Thickness control of interphase in this system is difficult due to changing conditions through the thickness of the preform. Temperature should preferably be constant within the fabric. Slower deposition rate is better to control the thickness of interphase. In order to obtain uniform thickness of interphase within a composite, effects of temperature, gas flow rate and position of a preform on microstructure were evaluated and those experimental conditions were optimized.

At ORNL, SiC/SiC composites were fabricated by FCVI using plain-woven high-purity SiC fibers, Tyranno SA and Hi-Nicalon Type-S. The fibers were stacked in [-30°/0°/30°] or [0°/90°] directions in a graphite holder. 45~55 sheets of Hi-Nicalon Type-S fibers or 55~60 sheets of Tyranno SA fibers were used. The fabric size for small furnace is 75 mm diameter and 12.5 mm thick and that for large furnace is 300 mm diameter and 12.5 mm thick. C, SiC/C or multilayer (SiC/C)[6]

Table I. Properties of SiC/SiC composites with 75 mm in diameter fabricated by FCVI

ID	Fiber	Orientation	F/M Interphase	Nominal thickness of Interphase [nm]	V_f [%]	Density [Mg/m^3]	Porosity [%]
1256	Tyranno SA	[-30/0/30]	PyC	150	37	2.76	15.1
1257	Hi-Nicalon Type-S	[-30/0/30]	PyC	150	33	2.39	23.5
1258	Hi-Nicalon Type-S	[-30/0/30]	PyC	75	36.1	2.7	-
1259	Hi-Nicalon Type-S	[-30/0/30]	PyC	300	35	2.58	13.9
1260	Tyranno SA	[-30/0/30]	PyC	300	30.2	2.28	24.2
1261	Tyranno SA	[-30/0/30]	PyC	75	33.3	2.54	20.4
1264	Tyranno SA	[0/90]	PyC	150	35.4	2.61	23.3
1265	Tyranno SA	[0/90]	PyC	300	35.3	2.72	18
1266	Tyranno SA	[0/90]	PyC	75	35.2	2.62	18.1
1267	Tyranno SA	[0/90]	SiC/C	100/150	38.8	2.74	15.7
1268	Tyranno SA	[0/90]	SiC/C	100/150	38.8	2.69	18
1269	Tyranno SA	[0/90]	SiC/C	100/300	38.8	2.71	17.2
1270	Tyranno SA	[0/90]	(SiC/C)^6	50/50/(200/50)^3/(500/50)^3	39.8	2.52	26.88
1271	Hi-Nicalon Type-S	[0/90]	PyC	150	Not evaluated		
1272	Tyranno SA	[0/90]	(SiC/C)^6	50/20/(200/20)^3/(500/20)^3			

was applied to fiber/matrix interphase followed by matrix SiC deposition. C was deposited by decomposition of propylene (C_3H_6) at 1100 °C. SiC was deposited by decomposition of MTS at 1000~1200 °C. A graphite coating chamber radiatively heats the fibrous preform exterior and its interior is cooled with a water-cooled line following deposition of the fiber/matrix interphase. The MTS carried in hydrogen is injected inside the preform. The gas infiltrates through the preform thickness and exhausts at atmospheric pressure. The properties of the composites fabricated in this study are summarized in table I. The ID number is the chronological order in which composites were fabricated. The fiber volume fraction (V_f) was estimated from the size and mass of the preform while the porosity values listed in the table I were estimated from cross-sectional scanning electron microscopy (SEM) images. The values of the thickness of interphase listed in Table I correspond to nominal values.

Mechanical properties of composites were evaluated by three-point flexural tests and tensile tests. Specimens with dimensions 30 mm (long) × 4 mm (wide) × 1.5 mm (thick) were used for flexural testing. More than three specimens were tested for each composite. The test span of the flexural tests was 18 mm. The flexural tests were conducted at a cross-head speed of 1.8 mm/min at ambient temperature. Tensile tests were conducted on the basis of ASTM C1275. The test specimens were edge-loaded with dimensions 41.3 mm (long) × 6.0 mm (wide) × 2.3 mm (thick). The dimensions of the gauge section were 15.0 mm (long) × 3.0 mm (wide) × 2.3 mm (thick). Details of the specimen and the tensile test are described elsewhere [10]. More than four valid test results were obtained for each composite. All tests were conducted at a cross-head speed of 0.5 mm/min at ambient temperature. The step-loading tests were performed for the precise evaluation of damage accumulation near proportional limit [11]. The mechanical properties of fiber/matrix interphase were evaluated by single-fiber push-out tests [12]. Specimens were sliced from composites normal to the fiber direction, which were mechanically polished to a final thickness of approximately 50 μm. For the tests the specimens were mounted on top of a holder containing a groove 50 μm wide. Isolated fibers with the fiber direction perpendicular to the holder surface on the groove were selected with a video microscope and were pushed out using a Berkovich-type pyramidal diamond indenter tip with maximum load capability of 1 N. Microstructure was observed by optical microscopy and field emission SEM (FE-SEM). The thickness of interphase was measured by FE-SEM at several regions in a specimen. Fracture surface after tensile tests was examined by SEM with EDS.

RESULTS

Optimization of ICVI

In previously fabricated composites, the deposition rate of C was so high that most of the C precursor was deposited at the upstream side. The thickness of C interphase was quite different between upstream side and downstream side. For example, the thickness of C interphase at upstream was more than 1 μm, whereas no C interphase was identified on the downstream side. However the uniformity of interphase thickness was significantly improved by optimizing the experimental conditions, temperature, gas flow rate and position of the sample in the furnace. The optimized deposition rates of C and SiC are approximately 2 μm/min and 20 μm/min respectively. The scatter of temperature within composites improved to less than 1 % of controlled temperature. Apparent difference of interfacial thickness between upstream side and downstream side was not identified in the optimized composites.

Density of SiC/SiC composites was also increased with optimization of CVI conditions, while CVI processing time was significantly shortened from 40~50 hour to 15~20 hour. In the case of previous composites, upstream side was deposited easily and sealed prior to infiltration within the sample. Then the composites were removed and reversed to infiltrate the opposite side. In contrast to previous composites, the deposition began inside of composites without sealing of SiC at upstream side in the optimized deposition conditions. Another key to increase the density was increasing fiber volume fraction. Fiber fabrics were well-aligned stacked and thickness of the preform was reduced from 2.0 mm to 1.7 mm in the case of Hi-Nicalon preform with 7 sheets. The density of the optimized composites reinforced with Hi-Nicalon was approximately 2.5~2.6 Mg/m^3, which corresponds to approximately 85% of the theoretical density. The density was 2.0~2.2 Mg/m^3 in the previous composites, which corresponds to approximately 70 % of the theoretical density.

Characterization of fiber/matrix interphase

To understand interfacial mechanical properties, single fiber push-out tests were carried out. The interfacial shear strength (ISS) (τ_{is}) was obtained from the 'push-out' load (P) in single fiber push-out testing and calculated from Eq. 1.

$$\tau_{is} = \frac{P}{\pi D t} \qquad (1)$$

where D is fiber diameter and t is specimen thickness. The resultant effect of thickness of C interphase on the interfacial shear strength of composites reinforced with Hi-Nicalon and composites reinforced with Tyranno SA is shown in Fig. 1. The interfacial shear strength drastically decreased with increasing C interphase thickness. Similar results using Nicalon CVI composites have been reported [13]. It was found that the interfacial shear strength of composites reinforced with Tyranno SA fibers was slightly larger than that of composites reinforced with Hi-Nicalon if the thickness of C interphase is same.

The effect of thickness of C interphase on the proportional limit stress (PLS) of

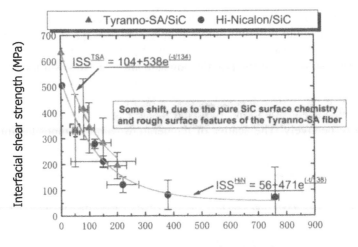

First PyC interphase thickness (nm)

Fig. 1. Effects of thickness of PyC interphase on interfacial shear strength

flexural test and flexural strength of composites reinforced with Hi-Nicalon (a) and composites reinforced with Tyranno SA (b) is shown in Fig. 2. The composites reinforced with Hi-Nicalon fibers showed the peak PLS and the flexural strength at about 150 nm thick of C interphase, while the composites reinforced with Tyranno SA showed the peak at about 100 nm. It was found that the mechanical properties of composites reinforced with Tyranno SA fibers are more insensitive to C interphase thickness than those of composites reinforced with Hi-Nicalon fibers.

SEM and TEM observation showed that the interfacial crack almost always propagated along the interface between fiber and interphase [12]. This behavior was not limited only to C interphase. Multilayer C/SiC and "porous" SiC interphase

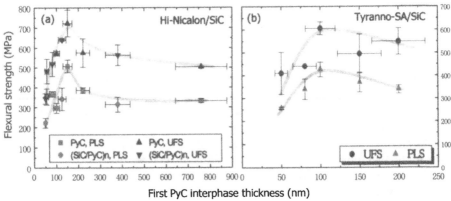

First PyC interphase thickness (nm)

Fig. 2. Effect of C interphase thickness on the Flexural Strength

Advanced SiC/SiC Ceramic Composites

showed the same behavior. Fracture surface of fibers was smooth. This led to low frictional stress of the debonded interface allowing easy crack propagation. This fracture behavior is attributed to a smooth fiber surface and weaker bond between the interphase and fiber than between the interphase and matrix SiC. In particular, the bond between fiber and C layer is weaker than the bond between C layer and SiC matrix. In order to improve interfacial fracture behavior and mechanical properties, PLS and maximum strength in particular, a SiC layer was formed on the fibers. Fiber and fiber bundle pull-out of composites with the SiC/C interphase reinforced with Hi-Nicalon fibers were shorter than those of composites without the first SiC layer. Both C and Si were detected from pull-out fiber surfaces of the composites without the SiC layer by EDS, and the atomic ratio of C to Si corresponded to that of Hi-Nicalon fiber [6]. In the case of composites with the first SiC layer, almost all species detected on pull-out fiber surface were C. The first SiC layer on Hi-Nicalon fiber induced strong bond between fiber and interphase and turned the crack path from between the fiber and the interphase to the inside of C interphase. Rough fracture fiber surface was seen, interfacial frictional stress was increased and mechanical properties of tensile tests were improved [14].

Optimization of FCVI

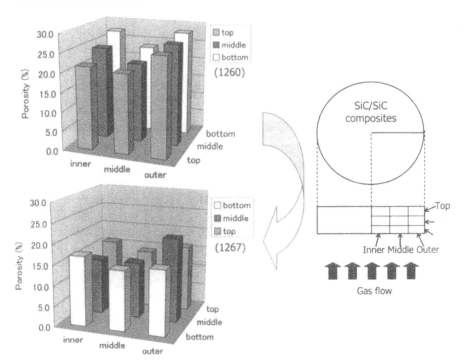

Fig. 3. Distribution of Porosity in Composites

To understand the distribution of thickness of interphase and porosity within a composite, thickness of interphase and porosity were measured at nine regions in a composite. The porosity of composites fabricated earlier was high and mostly more than 20% as shown in Table I. The bottom region, which is the upstream side of precursor gas, and the outer region tended to have higher porosity in the composites. It was found that porosity was significantly affected by fiber volume fraction. As the fiber volume fraction increased, the porosity decreased as shown in Table I. Fig. 3 compares the distribution of porosity of improved composites (1267) with that of previous composite (1260). In the improved composites porosity is independent of position, and the porosity trends seen in the previous composites was not observed.

The thickness of interphase varied thorough the plate thickness. The interphase of bottom region was thicker than that of top region as shown in Fig. 4 (1256). One of the solutions to decrease the scatter of the thickness was to change the upstream side and the downstream side of the preform in the middle of interphase deposition process. Fig. 4 shows the distribution of the thickness of interphase. The scatter of the thickness was significantly improved by changing the upstream side and the downstream side.

The thickness range of C interphase applied was about 20~300 nm. The average thickness of SiC layer of the SiC/C interphase was approximately 60 nm. Apparent effects of C interphase thickness and the first SiC layer on tensile properties were not seen. The multilayer interphase with new concept was developed in this study. Fig. 5 shows SiC/C multilayer interphase. The first thin SiC layer (50 nm) is to strengthen bond between fiber and interphase. The next three SiC layers (200 nm) are for multiple interfacial fractures of fibers. The next two thick SiC layers (500 nm) are for multiple fractures of bundles. Thin C layers (50 nm) are used just to separate SiC layers. The composites with the multilayer SiC/C interphase showed

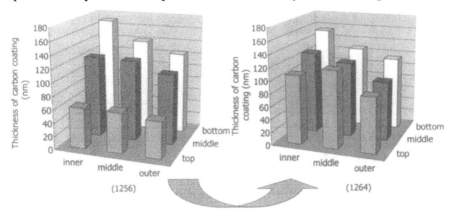

Reverse upstream side and downstream side in middle of deposition

Fig. 4. Improvement of Scatter of F/M Interphase Thickness

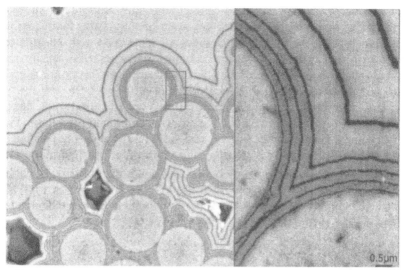

Fig. 5. SEM Images of Composites with SiC/C Multilayer Interphase

the multiple fracture of interphase. The tensile strength of the composites with the multilayer was nearly 30 % larger than the composites with C interphase and with SiC/C interphase as shown in Fig. 6. Elastic modulus and PLS was normalized by porosity as shown in Eq. 2.

$$\text{Normalized Value} = \frac{\text{Original Value}}{1 - \text{Porosity}} \qquad (2)$$

The elastic modulus of composites fabricated in this work is almost twice as

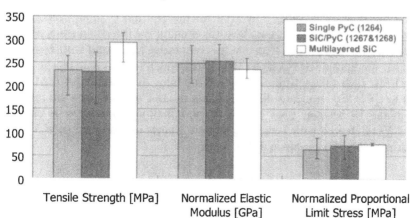

Fig. 6. The effect of fiber/matrix interphase on tensile properties

large as the reported elastic modulus of the composites reinforced with Nicalon fibers fabricated by CVI [15]. The fracture strain of composites fabricated in this work is less than half of that of the composites reinforced with Nicalon fibers. These properties are attributed to the intrinsic fiber properties. Both tensile strength and proportional limit stress of composites fabricated in this work are superior to those of composites reinforced with Nicalon fibers.

The large composite plates (300 mm diameter) are supposed to be fabricated using Tyranno SA fibers for irradiation experiments and round robin tests. Prior to using Tyranno SA fibers, the large composites were fabricated using Nicalon fibers to understand distribution of porosity and uniformity of interphase. Dense composites were fabricated in the first trial. Although the porosity of outermost regions was higher than that of the other, the average porosity was 14.4 %. The interphase was not formed uniformly. There was a gradient of the thickness, i.e. the interphase at the top side and the center was thicker.

DISCUSSION

Pores existing within fiber bundle are limited and they don't affect porosity significantly in case of CVI composites. It is difficult to fill large space at inter-bundle of fibers by CVI processing. The large pores existing at inter-bundle of fibers significantly affect the porosity. In this study, increasing fiber volume fraction decreased porosity. Increasing the fiber volume fraction decreased the porosity at inter-bundle of fibers and the total amount of large pores existing inter-bundle of fibers decreased.

The composites reinforced with Hi-Nicalon fibers showed the peak PLS and the flexural strength at about 150 nm in thickness of C interphase. The ISS of composites reinforced with Hi-Nicalon fibers was about 200 MPa for a 150 nm thick C interphase. The composites reinforced with Tyranno SA showed the peak at C interphase thickness of about 100 nm. The ISS of composites reinforced with Tyranno SA fibers was about 350 MPa for the 100 nm C interphase case. The optimum ISS to obtain the largest flexural strength for the composites reinforced with Hi-Nicalon fibers is less than that for composites reinforced with Tyranno SA. The elastic modulus of Hi-Nicalon (270 GPa) is much smaller than elastic modulus of CVD SiC (461 GPa), while the elastic modulus of Tyranno SA (381 GPa) is similar to CVD SiC. It means that the tensile stresses in the matrix are larger than those in the Hi-Nicalon fiber before crack initiation, whereas it is similar in composites reinforced with Tyranno SA fibers. Matrix cracking stress (σ_m) depends on fiber modulus (E_f) as shown in Eq. 3 [16].

$$\sigma_m = \left(\frac{6\tau G_m V_f^2 E_f E_{cl}^2}{(1 - V_f) E_m^2 r} \right)^{1/3} - \sigma_r \tag{3}$$

where σ_r is the residual stress, τ is the interfacial frictional stress, G_m is the critical mode I energy release rate, V_f is the volume fraction of fibers, E is the elastic modulus of the matrix $(_m)$ or composite $(_{cl})$ and r is fiber radius. One of the roles of the interphase is to deflect matrix cracks by interfacial debonding. It is assumed that the optimum ISS of the composites reinforced with Tyranno SA fibers is larger than that of composites reinforced with Hi-Nicalon fibers since the matrix cracking stress of the composites reinforced with Tyranno SA fibers is larger than that of the composites reinforced with Hi-Nicalon fibers. The gap between matrix cracking stress and maximum strength is relatively small in composites reinforced with Tyranno SA fibers. The effect of interphase on mechanical properties of composites reinforced with Tyranno SA fibers is smaller than that of composites reinforced with Hi-Nicalon.

The first SiC layer in the fiber/matrix interphase was very effective for composites reinforced with Hi-Nicalon fibers. The crack path in the interphase was turned from fiber/interphase interface to within the interphase. The interfacial frictional strength increased. The interfacial fracture behavior and mechanical properties were improved. However the apparent effect of the first SiC layer was not evident in the composites reinforced with Tyranno SA fibers. One of the reasons is due to fiber modulus as discussed in the previous paragraph. The matrix cracking stress of composites reinforced with Tyranno SA fibers should be larger than that of composites reinforced with Hi-Nicalon fibers. The effect of interfacial shear strength on mechanical properties of composites reinforced with Tyranno SA fibers is smaller than that of composites reinforced with Hi-Nicalon fibers. Another reason is the rougher surface of Tyranno SA fiber compared with that of Hi-Nicalon fiber. It is considered the interfacial frictional stress is sufficiently large without the first SiC layer. The first SiC layer is expected to be effective under severe environment in which the interfacial shear strength is reduced by oxidation or neutron irradiation. Further experiments under such severe environments are required to understand the necessity of the SiC layer.

The multilayer interphase with this new concept was developed in this study. It is considered that not only fiber pull-out but also fiber bundle pull-out plays an important role, so the concept of the bundle interphase was applied to the multilayer. Most composites with C/SiC multilayer interphase without the first SiC layer on fibers did not have multiple fracture of the interphase and the interfacial fracture behavior depended on the first C layer thickness. However both the multiple fracture of fibers and fiber bundles were attained in the composites with the multilayer SiC/C interphase. It is considered that the first SiC layer is the key for multiple fracture.

CONCLUSIONS

(1) The SiC/SiC composites fabricated by the ICVI system at NIMS were significantly improved by optimization of gas flow rate, temperature, the position of the preform inside the furnace, increasing fiber volume fraction and the precursor gas. The uniformity of the fiber/matrix interphase, the density and mechanical properties were significantly improved while the time for fabrication became less than half of that for previous runs.

(2) The interfacial shear strength drastically decreased with increasing C interphase thickness. It was found that the interfacial shear strength of composites reinforced with Tyranno SA fibers was slightly larger than that of composites reinforced with Hi-Nicalon if the thickness of C interphase is same.

(3) The composites reinforced with Hi-Nicalon fibers showed the peak PLS and the flexural strength at about 150 nm C interphase thickness, while the composites reinforced with Tyranno SA fibers showed the peak at about 100 nm thickness. It was found that the mechanical properties of composites reinforced with Tyranno SA fibers are more insensitive to C interphase thickness than those of composites reinforced with Hi-Nicalon fibers.

(4) The first SiC layer in the fiber/matrix interphase improved interfacial fracture behavior and mechanical properties significantly in composites reinforced with Hi-Nicalon fibers, while the apparent effect of the first SiC layer was not seen in composites reinforced with Tyranno SA fibers due to large modulus and rough feature of fiber surface.

(5) The density and the scattering on interphase of composites fabricated by FCVI was significantly improved by optimization of temperature and gas flow, increasing fiber volume fraction and reversing the up-stream side and down-stream side of the gas in the middle of deposition.

(6) Although thickness of C interphase and the first SiC layer didn't affect tensile properties of composites reinforced with Tyranno SA significantly, the tensile strength of the composites with the multilayer SiC/C interphase was improved.

(7) Large composites with 300 mm in diameter were successfully fabricated by FCVI with the porosity of 14.4 %, although the interphase was not formed uniformly.

ACKNOWLEDGEMENT

Authors would like to thank T. Noda and H. Araki of NIMS and J.C. McLaughlin of ORNL for the assistance of the fabrication of SiC/SiC composites. This work was supported by Core Research for Evolutional Science and Technology operated by the Japan Science and Technology Corporation, by Japan/USA Program of Integration of Technology and Engineering for Fusion Research (JUPITER II) and by the office of Fusion Energy Science, US DOE under contract DE-AC-05-00OR22725 with UT-Battelle, LLC.

REFFERENCES

1   T.M. Besmann, D.P. Stinton and R.A. Lowden, MRS Bull. XIII [11] (1988) 45.
2   T.M. Besmann, J.C. McLaughlin, Hua-Tay Lin, J. Nucl. Mater., 219 (1995) 31-35.
3   H. Araki, H Suzuki, W. Yang, S. Sato and T. Noda, J. Nucl. Mater., 258-263 (1998) 1540-1545.
4   E. Lara-Curzio, Comprehensive Composites Encyclopedia, Elsevier, (2000) 533-577.
5   E. Lara-Curzio and M. K. Ferber, Numerical Analysis and Modeling of Composite Materials, Ed. J. W. Bull, Blackie Academic & Professional (1995) 357-399.
6   R.E. Jones, D. Petrak, J. Rabe and A. Szweda, J. Nucl. Mater., 283-287 (2000) 556-559.
7   M. Takeda, A. Urano, J. Sakamoto and Y. Imai, J. Nucl. Mater., 258-263 (1998) 1594-1599.
8   T. Ishikawa, Y. Kohtoku, K. Kumagawa, T. Yamamura and T. Nagasawa, Nature, 391 (1998) 773-775.
9   T. Ishikawa, S. Kajii, T. Hisayuki, K. Matsunaga, T. Hogami and Y. Kohtoku, Key Eng. Mater., 164-165 (1999) 15-18.
10   T. Nozawa, E. Lara-Curzio, Y. Katoh, L.L. Snead and A. Kohyama, Fusion Materials Semi-Annual Progress Reports, DOE/ER-0313/31 (2001) 40-46.
11   M. Steen and J.L Valles, ASTM STP 1309, (1997) 49.
12   T. Hinoki, W. Zhang, A. Kohyama, S. Sato and T. Noda, J. Nucl. Mater., 258-263 (1998) 1567-1571.
13   E. Lara-Curzio, M. K. Ferber and R.A. Lowden, Ceram. Eng. and Sci., 15 [5] (1994) 989-1000.
14   T. Hinoki, W. Yang, T. Nozawa, T. Shibayama, Y. Katoh and A. Kohyama, J. Nucl. Mater., 289 (2001) 23-29.
15   P. Lipetzky, G.J. Dvorak and N.S. Stoloff, Mater. Sci. Eng., A216 (1996) 11-19.
16   W.A. Curtin, J. Am. Ceram. Soc., 74 [11] (1991) 2837-2845.

OPTIMIZING THE FABRICATION PROCESS FOR EXCELLENT MECHANICAL PROPERTIES IN STOICHIOMETIRIC SiC FIBER / FCVI SiC MATRIX COMPOSITES

T. Taguchi, N. Igawa and S. Jitsukawa
Japan Atomic Energy Research Institute
Tokai, Ibaraki 319-1195, Japan

T. Nozawa, Y. Katoh and A. Kohyama
Kyoto University
Gokasho,Uji, Kyoto 611-0011, Japan

L. L. Snead and J. C. McLaughlin
Oak Ridge National Laboratory
Oak Ridge, TN 37831, USA

ABSTRACT

Optimization of the fabrication for SiC composites with advanced stoichiometric SiC fibers (Hi-Nicalon Type S and Tyranno SA) by the forced thermal-gradient chemical vapor infiltration process was carried out. Porosity decreased to approximately 15% by increasing the fiber volume fraction and optimizing precursor gas flow rates. Uniformity of fiber / matrix interphase was improved by a preform flipped midway through the interphase deposition. The tensile strength slightly increased with the thickness of carbon interphase in the range of 75-300 nm. The effect of the fabric orientation on the tensile strength was also investigated. In the result, the tensile strength of the composite with the fabric orientation of [0°/90°] using Tyranno SA fiber is larger than that of [-30°/0°/+30°].

INTRODUCTION

Ceramic matrix composites show excellent high temperature mechanical properties and non-catastrophic failure behavior. These materials are, therefore, attractive candidates for structural applications at high temperature. For these reasons and their low residual radioactivity following the neutron irradiation, the continuous SiC-fiber reinforced SiC-matrix (SiC/SiC) composites are being actively investigated for the first wall and blanket components in power fusion reactors[1-4].

Among the various fabrication processes, the chemical vapor infiltration (CVI) is one of the best techniques to fabricate the SiC/SiC because of high purity SiC and minimal fiber damage during the composite fabrication[5]. Whereas the isothermal CVI (ICVI) can efficiently produce thin and complex shape parts, it requires a significantly long time to produce thicker materials. It takes high cost to produce thick materials by ICVI. On the other hand, the forced-thermal gradient CVI (FCVI) for a thicker part can be effected at sufficiently high rates to require only tens of hours or less[5]. It, therefore, can reduce the cost to produce thick parts.

Recently, high crystalline and stoichiometric SiC fibers have been produced. Typical examples are Hi-Nicalon Type S[6] and Tyranno SA[7], which possess superior mechanical and thermal properties, and a superior performance under neutron irradiation compared to their SiC-based predecessors[6,7]. Because the properties of these advanced fibers, e.g. thermal conductivity, waviness, tensile modulus and fiber diameter, are very different from their predecessors, optimization of composite fabrication and interphase improvement should be now focused.

Two kinds of advanced SiC fabrics were adopted as preforms: 2D-plain weave of Tyranno SA and Hi-Nicalon Type S. Though boron nitride shows excellent interphase for mechanical properties, it is undesirable for fusion applications. Because nitrogen transmutes into ^{14}C that has a very long half-life of â emitter by fusion neutrons, and boron simultaneously produces helium and enhances radiation damage through recoil interaction. Therefore, carbon, porous-SiC and multilayer-carbon/SiC as interphase materials are being generally studied for fusion applications[4,8]. Carbon interphase was selected in this study. The effects of the interphase and fabric orientation on the tensile strength were investigated.

EXPERIMENTAL PROCEDURE

Two kinds of advanced SiC fabric as performs was adopted: 2D-plain weave Tyranno SA (Ube Industries, Ube, Japan) and Hi-Nicalon Type S (Nippon Carbon Co., Ltd., Tokyo, Japan). The precursor for carbon deposition was 99% purity propylene (C_3H_6, Matheson, Morrow, GA, USA), and technical grade methyltrichlorosilane (MTS, CH_3SiCl_3, Gelest Inc., Tullytown, PA, USA) was used for SiC infiltration.

Table 1. Fiber fabric information of FCVI SiC/SiC composites

Specimen ID	Fiber	Number of fabric layers	Fiber orientation	Nominal carbon thickness (nm)
FCVI-1	Hi-Nicalon Type S	45	[-30°/0°/+30°]	75
FCVI-2	Hi-Nicalon Type S	55	[-30°/0°/+30°]	150
FCVI-3	Hi-Nicalon Type S	45	[-30°/0°/+30°]	300
FCVI-4	Tyranno SA	58	[-30°/0°/+30°]	75
FCVI-5	Tyranno SA	55	[-30°/0°/+30°]	150
FCVI-6	Tyranno SA	55	[-30°/0°/+30°]	300
FCVI-7	Tyranno SA	60	[0°/90°]	75
FCVI-8	Tyranno SA	60	[0°/90°]	150
FCVI-9	Tyranno SA	60	[0°/90°]	300

The SiC fabric layers with a fabric layer orientation of [-30°/0°/+30°] or [0°/90°] were restrained in a graphite fixture. The fiber volume fraction range was 30 to 39 vol% by changing the number of fabric layers. The size of preform was 75 mm in diameter and 12.5 mm thickness. The fiber fabric parameters in this study are shown in Table I. The carbon interphase was deposited at 1100 C and at 5 Pa in the condition of flow rates of 50 cm³/min C_3H_6 and 1000 cm³/min Argon. After

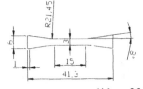

thickness: 2.3 mm

Fig. 1. Schematic illustration of tensile testing specimen.

interphase layer was deposited, the preforms were infiltrated at 1000-1200 C and at one atmospheric pressure. Methyltrichlorosilane flow rate was 0.17-0.50 g/min carried by 350-750 cm³/min of H_2. The back-pressure was monitored and the FCVI process was automatically finished after the back pressure reached to 1.7X10⁵ Pa from the initial pressure of 1.0 X10⁵ Pa, and the infiltration time was about 70 hours.

The density of each section was calculated from the dimensions and the mass of the plate to determine the distribution of the porosity in the composite. The microstructures of the composites were observed by optical microscopy and scanning electron microscopy (SEM).

Tensile testing was carried out at room temperature with a cross-head speed of 0.5 mm/min. The shape and size of the specimen were shown in Fig. 1. The strain of the specimen was measured with bonded strain gauges. The fracture surfaces were observed by SEM in order to investigate the effect of interphase on the mechanical properties.

RESULTS AND DISCUSSION

FCVI-SiC densification process

To determine distribution of the porosity and the carbon interphase thickness, the plate was cut into nine sections as shown in Fig. 2. Typical distribution of the porosity is shown in Fig. 2. The porosity in FCVI-2 specimen was relatively large (23.5 %) and depending on position: we found that the closer position to the inlet gas flow and outer position in the composite had the highest porosity. Much better uniform porosity in the composites (for example FCVI-3 specimen) was obtained

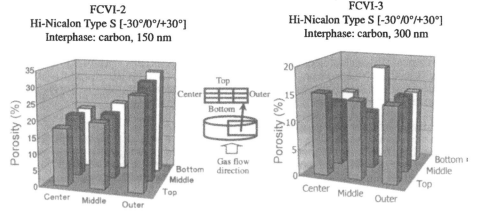

Fig.2. Typical distribution of the porosity of the fabricated SiC/SiC composite.

by decreasing the MTS and H_2 gas flow rates at the latter part of the FCVI process. Furthermore, the amount of porosity in the composite decreased. Because termination of the FCVI process was controlled by the back-pressure that decreased with decreasing the gas flow rate, the FCVI process was extended by the decrease in the flow rate.

The average porosity as a function of fiber volume fraction was shown in Fig. 3. High fiber volume fraction is very effective in decreasing the

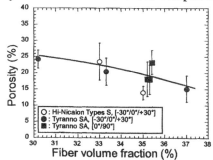

Fig.3. Effect of fiber volume fraction on the porosity.

porosity; the average porosity decreased to 15 % by increasing the fiber volume fraction to 39 vol%, but it seems that the porosity was independence of fiber type and infiltration condition.

The typical cross-sections of SiC/SiC are shown in Fig. 4. Even in the specimen that had smallest porosity (15.1 %), two kinds of pores could be observed; one is the large inter-bundle pore and the other is the small intra-bundle pore. The porosity in the composite mainly depends on the amount of inter-bundle pore because the size of inter-bundle pore is much larger than that of intra-bundle pore. From the cross-section micro-photographs, it was found that the size of inter-bundle pores was

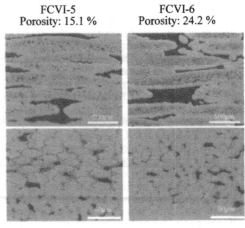

FCVI-5	FCVI-6
Porosity: 15.1 %	Porosity: 24.2 %

Fig.4. Cross-sections photographs of SiC/SiC composites.

smaller with a short distance between the fiber bundles in the specimen. The distance between fiber bundles is shorter when the fiber volume fraction is higher, which means that space between bundles with higher fiber volume fraction is smaller than those with lower fiber volume fraction; therefore the porosity of the composite with higher fiber volume fractions is smaller than that of lower fiber volume fraction.

Interphase fabrication process

Figure 5 shows the typical distribution of the thickness of carbon interphase between SiC fiber and SiC matrix in the composites estimated from cross-section SEM images. In FCVI-5 specimen, we found that the lower position in the composite had the thicker carbon interphase: the average thickness of carbon

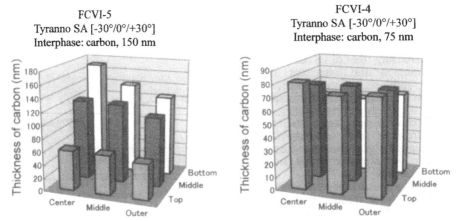

Fig.5. Typical distribution of the carbon thickness in the SiC/SiC composite.

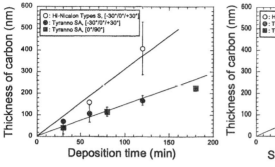
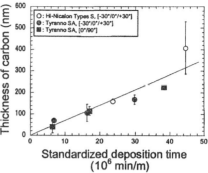

Fig.6. Relationship between deposition time and carbon thickness.

Fig.7. Relationship between standardized deposition time and carbon thickness.

interphase at bottom was about 145 nm whereas that at top it was 60 nm. This tendency is attributed to a concentration gradient in C_3H_6 gas. This gas traveled from the bottom to the top of the specimen, and the thickness of specimen was so large that the concentration of C_3H_6 gas was decreased due to its consumption at the lower part of the specimen. To mitigate this problem the preform was flipped over midway through the interphase infiltration. The distribution of carbon interphase thickness was remarkably improved and we obtained the specimens with near uniform through-thickness carbon interphase thickness (see FCVI-4 specimen in Fig.5).

The relationship between the average thickness of carbon interphase and deposition time is shown in Fig.6. The thickness of carbon interphase increased linearly with increasing deposition time. The carbon deposition rate of the composite using Hi-Nicalon Type S fiber, however, is larger than that of the composite using Tyranno SA fiber. At one constant fiber volume fraction, the total superficies of all fibers in the composite using Hi-Nicalon Type S is smaller than that in the composite using Tyranno SA fiber because the diameter of Hi-Nicalon Type S fiber is larger than that of Tyranno SA fiber. It is, therefore, considered that the carbon deposition rate of the composite using Hi-Nicalon Type S was larger than that of the composite using Tyranno SA. The relationship between the thickness of carbon and the standardized deposition time (the deposition time divided by the total superficies of fibers) is shown in Fig.7. The average thickness of carbon in all composites increased as one linear series with increasing the standardized deposition time. Therefore the carbon thickness could be estimate from the standardized deposition time, even if any kinds of fiber are used in the composite.

Mechanical properties

Typical tensile stress-strain curves are seen in Fig. 8. Each specimen was exhibited nonlinear stress-strain behavior, which means noncatastrophic failure behavior.

Each composite in this study has different density and fiber volume fraction. Relationship between the porosity of composite and the tensile strength is shown in Fig.9. No tendency can be, however, observed in this relationship. Because the load is mainly maintained by unfractured fibers and friction between fractured

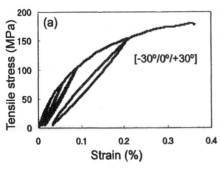
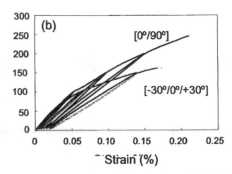

Fig. 8. Typical tensile behaviors of [-30°/0°/+30°] and [0°/90°] SiC/SiC composites using (a) Hi-Nicalon Type S and (b) Tyranno SA.

fiber and interphase above the proportional limit stress, the tensile strengths were, therefore, normalized by the fiber volume fraction of each composite in order to compare to them each other. Optimization of the interphase is one of the important factors to improve the mechanical properties of SiC/SiC composites. The effect of carbon thickness on normalized tensile strength of SiC/SiC composites was shown in Fig.10. For the bend strength, the optimum carbon interphase thickness is in the range of 170-1000 nm. In both [0°/90°] and [-30°/0°/+30°] composites using Tyranno SA fiber, it seems that normalized tensile strength is slightly increased with the thickness of carbon interphase in the range of 75-150 nm and then almost constant in the range of 150-300 nm. And in the composite using Hi-Nicalon Type S fiber, the normalized tensile strength was slightly increased with the thickness of carbon interphase in the range of 75-300 nm..

The tensile strength of the composite with the fabric orientation of [0°/90°] using Tyranno SA fiber is larger than that of [-30°/0°/+30°]. The tensile fracture surfaces of the composites were shown in Fig.11. For the excellent tensile strength, a large number of fibers retained at a

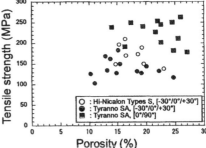

Fig.9. Effect of porosity on the tensile strength.

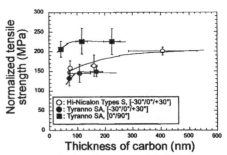

Fig.10. Effect of carbon thickness on the normalized tensile strength.

Advanced SiC/SiC Ceramic Composites

maximum load are required as well as the fiber strength itself[10]. In Fig.11 the pull-out of fiber was observed in the tensile fracture surface of all specimens. It seems that the pull-out length of the fibers canted to the tensile axis is shorter than that of the aligned fiber with the tensile axis in the [-30°/0°/+30°] composite. So the canted fibers might be unable to keep the enough strength. It is considered that the tensile strength depends on the number of the aligned fiber with the tensile axis efficiently. The aligned fiber volume fraction with the tensile axis in the [0°/90°] composite is higher than that of [-30°/0°/+30°] composite. The tensile strength of [0°/90°] composite was, therefore, larger than that of [-30°/0°/+30°] composite.

In the case of fabric orientation of [-30°/0°/+30°], the tensile strength of the composite using Hi-Nicalon Type S is larger than that of Tyranno SA. From the result of the fracture surface observation (see Fig.11), it is found that the pull-out length of the composite using Hi-Nicalon Type S is longer than that of Tyranno SA. Especially the canted fibers with respect to the tensile axis also caused the long pull-out as well as the aligned fibers. It is considered that the canted fibers in the composite using Hi-Nicalon Type S could retain their strength.

| FCVI-2 Hi-Nicalon Type S fiber [-30°/0°/+30°] | FCVI-5 Tyranno SA fiber [-30°/0°/+30°] | FCVI-8 Tyranno SA fiber [0°/90°] |

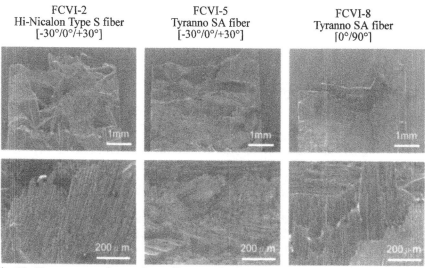

Fig.11. Fracture surface of tensile tested SiC/SiC composites with 150 nm carbon as an interphase layer.

CONCLUSION

Process optimization for FCVI SiC matrix composites with advanced SiC fibers, Hi-Nicalon Type S and Tyranno SA, was carried out. The SiC/SiC composites exhibited significant improvement for reduction in porosity (15.1%) and more uniform pore distribution by increasing the fiber volume fraction and by decreasing the MTS and H_2 gases flow rates at the latter part of the FCVI process. Uniform carbon interphase between advanced SiC fibers and FCVI-SiC matrix could be obtained by reversing the gas-flow direction mid-way through the coating process. The tensile strength was slightly increased with thickness of carbon interphase in the range of 75-300 nm. The effect of the fabric orientation on the tensile strength was also investigated. The tensile strength of the composite with the fabric orientation of [0°/90°] using Tyranno SA fiber is larger than that of [-30°/0°/+30°].

ACKNOWLEDGEMENTS

The authors would like to thank Dr. T. M. Besmann at ORNL for useful discussion. This study has been carried out under the US-DOE/ JAERI Collaborative Program on FWB Structural materials in Mixed-Spectrum Fission Reactors, Phase IV. This study is also supported by CREST-ACE (Core Research for Evolutional Science and Technology / Advanced material Systems for Energy Conversion) program sponsored by Japan Science and Technology Corporation.

REFERENCES

[1]L.L. Snead, R.H. Jones, A. Kohyama and P. Fenici, "Status of silicon carbide composites for fusion," *J. Nucl. Mater.*, **26**, 233-237(1996).

[2]R.H. Jones, D. Steiner, H.L. Heinisch, G.A. Newsome and H.M. Kerch, "Radiation resistant ceramic matrix composites," *J. Nucl. Mater.*, **245**, 87-107 (1997).

[3]P. Fenici, A.J. Frias Rebelo, R. H.Jones, A. Kohyama and L.L. Snead, "Current status of SiC/SiC composites R&D," *J. Nucl. Mater.*, **215**, 258-263(1998).

[4]A. Hasegawa, A. Kohyama, R.H. Jones, L.L. Snead, B. Riccardi and P. Fenici, "Critical issues and current status of SiC/SiC composites for fusion," *J. Nucl. Mater.*, **283-287**, 128-137 (2000).

[5]T.M. Besmann, J.C. McLaughlin and H-T. Lin, "Fabrication of ceramic composites: forced CVI," *J. Nucl. Mater.*, **219**, 31-35 (1995).

[6]M. Takeda, A. Urano, J. Sakamoto and Y. Imai, "Microstructure and oxidative degradation behavior of silicon carbide fiber Hi-Nicalon Type S," *J. Nucl. Mater.*, **258-263** 1594-1599 (1998).

[7]T.Ishikawa, Y. Kohtoku, K. Kumagawa, T Yamamura and T. Nagasawa, "High-strength alkali-resistant sintered SiC fiber stable to 2,200 C," *Nature* **391**, 773-775 (1998).

[8]L.L. Snead, M.C. Osborne, R.A. lowden, J. Strizak, R.J. Shinavski, K.L. More, W.S. Eatherly, J. Bailey and A.M. Williams, "Low dose irradiation performance of SiC interphase SiC/SiC composites," *J. Nucl. Mater.*, **253**, 20-30 (1998)

[9]W. Yang, H. Araki, J.Y. Park, T. Noda, A. Kohyama and J. Yu, "Effect of multiple coating interfacial structures on bending property of FCVI SiС$_f$/SIC composites," *Ceram. Eng. Sci. Proc.* **21**, 259-266 (2000).

[10]T. Hinoki, W. Yang, T. Nozawa, T. Shibayama, Y. Katoh and A. Kohyama, "Improvement of mechanical properties of SiC/SiC composites by various surface treatments of fibers," *J. Nucl. Mater.*, **289**, 23-29 (2001).

A NOVEL PROCESSING TECHNIQUE OF SILICON CARBIDE-BASED CERAMIC COMPOSITES FOR HIGH TEMPERATURE APPLICATIONS

Yutai Katoh, Shaoming Dong* and Akira Kohyama
Institute of Advanced Energy, Kyoto University
Gokasho, Uji, Kyoto 611-0011, Japan
*Presently at Shanghai Institute of Ceramics, CAS
Shanghai, China

ABSTRACT

It was demonstrated that silicon carbide (SiC) fiber reinforced SiC matrix composites (SiC/SiC composites) with outstanding performances could be produced through a transient eutectic phase route, by employing moderate matrix densification conditions and applying appropriate fiber and/or interphase protections, incorporating a technique named nano-infiltration. This paper provides an introduction of the nano-infiltration and transient eutectic phase (NITE) process for SiC/SiC composite processing, followed by the evaluation of thermo-mechanical properties including tensile strength, fracture behavior and thermal conductivity. The lab-grade unidirectionally-reinforced composites with fiber volume fraction of 20 percent exhibited proportional limit tensile stresses of >200MPa and ultimate tensile strength of ~400MPa. The thermal conductivity was in the range of 15~30W/m-K at 20~1200C.

INTRODUCTION

Production process of ceramics utilizing transient eutectic phases is attractive for matrix densification of ceramic-matrix composites (CMC), because of the quality of the ceramic products through that route. Generally, monolithic silicide ceramics produced by so-called liquid phase sintering (LPS) process that makes use of transient eutectoid phases possess dense and robust structures, reasonable thermal conductivity and gas tightness with substantially lower process costs compared to conventional processes, e.g., chemical vapor deposition[1,2]. Successful application of the transient eutectic phase process to CMC processing expectedly brings merits, such as unlimited product thickness, thin plate producibility, surface smoothness and applicability of existing (near) net-shaping

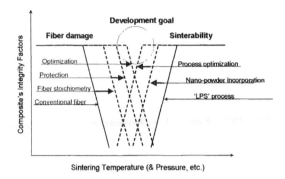

Fig. 1 – Directions of SiC/SiC process development through a transient eutectic phase route. 'Fiber damage' is an example of the composites' integrity factors

techniques, in addition to the inheritance of advantages from the monolithic ceramics.

However, such attempt is challenging at the same time for several reasons, as identified by previous works [3-5]. First, severe damages in reinforcing fibers and fiber-matrix (F-M) interfaces have to be avoided in a reactive environment during the pressurized processing at very high temperatures. Second, for the production of continuous fiber-reinforced ceramic composites with acceptable performances, submicron- to micron-scale openings inside fiber bundles need to be sufficiently filled with the matrix raw materials, i.e., mixed-powder slurry in the present case. Third, provided that the matrices are sufficiently densified, remnants of sintering additives have to be appropriately controlled in terms of both quantity and quality. Glassy or networked oxides should spoil the thermo-mechanical properties of the ceramic products [6].

In the present study, development of silicon carbide (SiC) fiber-reinforced SiC matrix ceramic composite (SiC/SiC composite) through the transient eutectic phase route was attempted. Major efforts are made to moderate the process conditions and to secure integrity of the composite constituents during densification. As shown in a conceptual illustration in Fig.1, there is no applicable condition for SiC/SiC composite processing using a combination of conventional SiC fibers and conventional LPS process. However, employing advanced near-stoichiometry SiC fibers and applying appropriate fiber/interface protection and matrix densification techniques, together with detailed process optimization, there proved to be an appropriate range of sintering condition that can produce

Table 1 – Approach in the present work and key factors for SiC/SiC process development through a transient eutectic phase route.

```
Approach
  ■ Lower process temperature
  ■ Apply appropriate fiber protection
  ■ Slurry infiltration to sub-micron pores
  ■ Minimize amount of sintering agent remnants
  ■ Eliminate glassy oxide remnants
  ■ Avoid mechanical damage to fibers
  ■ Control matrix grain size
Key factors
  ■ Near stoichiometry SiC fiber
    ► Heat resistance in inert environment ~1800C
  ■ Protective interphase (fiber coating)
    ► CVD-carbon
    ► Phenolic resin-derived carbon
    ► CVD-hBN
  ■ Nano powder-based matrix slurry
    ► Infiltrativity to intra-fiber bundle openings
    ► Large specific surface area
    ► Surface features (physical & chemical)
```

SiC/SiC composites with outstanding performance. The integrity factors other than fiber/interface damages include matrix porosity, process-induced damages (micro-cracking, etc.) and the phases, amount and geometrical distribution of the remnants of sintering additives. The most successful process was named nano-infiltration and transient eutectic phase process (NITE process) [7,8]. This paper provides overview of the process development effort. The approach and key factors in the process development are summarized in Table 1.

EXPERIMENTAL

The process development effort started from a study on sinterability of several commercially available SiC powders with various sintering additives. The SiC powders studied ranged from 'nano-phase' beta-SiC with nominal diameter of 30nm (Marketech International, Inc., Port Townsend, US) to crushed ones with average size of 270nm (Showa Denko K.K., Tokyo, Japan), in terms of the powder size. The sintering additives studied were mixed oxide systems consisting of at least two of alumina, calcia, magnesia, yttria, silica and carbon. Silica was assumed to be provided from the surface oxide layer of SiC raw materials. Only the results from alumina-yttria-silica system are discussed in this paper.

As the reinforcement, three commercial near-stoichiometry SiC continuous fibers, i.e., Tyranno™-SA Grade-3 (Ube Industries, Ltd., Ube, Japan), Hi-Nicalon™ Type-S (Nippon Carbon Co., Tokyo, Japan) and Sylramic™ Fiber (Dow Corning Co., Midland, US) were considered for use. The Tyranno™-SA was selected for this particular study because of the availability and the proven mechanical property retention after heat treatment at 1800C in an inert

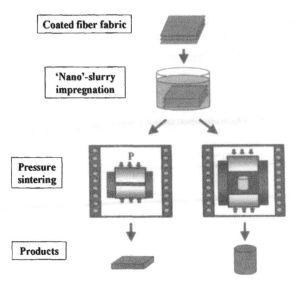

Fig.2 – Schematic illustration of the lab-grade composite production process.

environment [9].

A coating was applied to the fiber surface prior to the composite processing. The roles of coating are to produce a reasonably weak interphase layer between the fiber and the matrix, to prevent fibers from being sintered with the matrix raw materials, and optionally to provide an additional sintering additive for enhanced matrix densification in fiber bundle interiors where a sufficient pressurization may not occur during a pressure-sintering process. As the fiber coating, chemically vapor deposited carbon (CVD-carbon), phenolic resin-derived carbon and chemically vapor deposited hexagonal boron nitride were attempted. We show the result using the CVD-carbon.

Composites were prepared as schematically illustrated in Fig.2. The coated fiber bundles aligned unidirectionally were dipped in the matrix slurry, consisting of SiC raw powders and up to 10 mass percent sintering additives dispersed in organic solvent. Small amount of polycarbosilane (PCS) was added to the matrix slurry in some of the early study [10]. The slurry-impregnated prepreg was stacked and dried followed by pressure-sintering in an argon flow. Typical fiber volume fractions of the final products were 20 percent. The technique to prepare the optimized nano-phase powder-based slurry and infiltrate it into fiber fabric openings even with very small sizes without hurting the fibers and interphase coatings was named 'nano-infiltration.' Other techniques that do not fall in the nano-infiltration category will be addressed as 'LPS' hereafter. The sintering

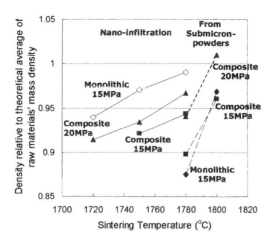

Fig.3 – Influences of sintering temperature and pressure on mass density of the final product of monolithic SiC and SiC/SiC composites from two different matrix raw powders.

temperature and pressure were 1720~1820C and 5~25MPa, respectively. The trial products then underwent evaluations of apparent density and scanning electron microscopy. Selected materials were subjected to extensive characterization by transmission electron microscopy, tensile and/or flexural mechanical tests, thermal conductivity measurement and X-ray diffraction.

RESULTS AND DISCUSSION

The sinterability of 'nano-phase' SiC powder is compared to that of sub-micron crushed SiC powder in Fig.3. The former powder is spherical and has specific surface area of ~140m²/g, while the latter is irregular-shaped typical of crushed powders with specific surface area of approximately 20m²/g. The surface chemistry is silica in both cases. The crushed powder requires sintering temperature >1800C for production of sufficiently compact monolithic SiC at the pressure of 15MPa and exhibits strong temperature dependence of densification behavior. On the other hand, the nano-phase powder could be sintered at 1750C or higher and exhibited moderate temperature dependence. The composites exhibited densification behavior apparently similar to that of the monolithic ceramics.

Fig.4 compares scanning electron micrographs of the submicron SiC powder (submicron LPS) and nano-infiltrated composites produced at 1780C and 15MPa. In the former composite, a significantly higher matrix porosity and severe fiber deformation are observed. Therefore, the primary mechanism of increased density

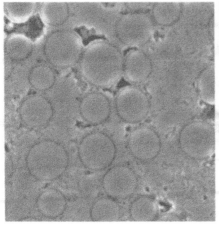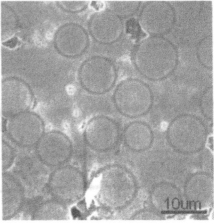

LPS (from submicron-powder)   **NITE (nano-infiltrated)**

Fig.4 – Scanning electron micrographs of the polished cross-sections of Tyranno™-SA / SiC composites produced from two different matrix raw materials but in identical sintering conditions.

with increased applied pressure in this case is fiber deformation and it leads to sacrifices in the fiber strength. On the other hand, no apparent fiber deformation is observed in the nano-infiltrated composites and the significantly lower matrix porosity indicates an enhanced matrix fluidity during the pressurizing.

The effect of sintering temperature on matrix densification at a pressure of 20MPa is shown in Fig.5, in which a set of scanning electron micrographs taken on the polished cross-sectional samples is presented. The entire cross-sectional areas of the composites could be classified into the apparently porous fiber-bundle regions and the less porous fiber-free regions. The micrographs are showing the appearance of fiber-bundle regions of the nano-infiltrated composites. The sizes of intra-bundle pores are likely be determined by those of openings inside the fiber agglomeration. The number of those pores clearly reduced with the increasing sintering temperature. The sintering pressure increase was not as effective for the porosity reduction.

The appearance of intra-bundle pore surfaces was typical of the polymer-derived matrix pores, for these composites had been densified with preceramic polymers added to the matrix slurry. The matrix close to those pores was a mixture of glassy carbon and very fine SiC crystallites, as identified by TEM, which evidenced that it had been produced by the pyrolysis of PCS. This observation suggests that the inclusion of PCS hampers the infiltration of matrix

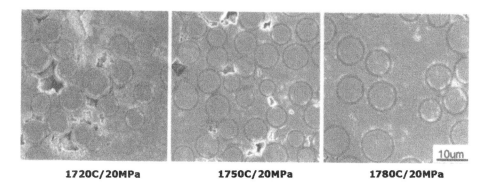

1720C/20MPa	1750C/20MPa	1780C/20MPa

Fig.5 – Effect of sintering temperature on matrix densification of nano-infiltrated Tyranno™-SA composites at a sintering pressure of 20MPa.

raw materials into the fiber bundles since it hardens well before the temperature reaches to allow formation of the transient liquid phases.

Deformation of SiC fibers was not observed even at the sintering condition of 1780°C and 20MPa. Reduction in the carbon interphases thickness with the increasing of sintering temperature from 1720 to 1780°C was not identified. This suggests that the higher temperatures and increased pressure might be allowed for property tailoring toward the enhanced thermal conductivity and hermeticity.

In Fig.6, the stress-strain relationship of nano-infiltrated composites processed at 1780C and 20MPa during a tensile test is presented. A significant stress increase beyond the proportional limit stress was observed for all the nano-infiltrated composites. The average proportional limit stress, ultimate tensile stress and tensile modulus of this composite were 215MPa, 383MPa and 288GPa, respectively. Both the ultimate tensile stress and the proportional limit stress increased with the increasing process temperature up to 1780C. All the nano-infiltrated composites sintered below 1780C exhibited similar fracture strain of about 0.3 percent, which is a typical value for chemically vapor infiltrated near-stoichiometry SiC fiber composites [11,12]. This shows that the enhanced matrix densification through a sufficient matrix sintering is beneficial at least in terms of fast fracture properties and significant damage is not occurring to the fibers even at the sintering temperature of 1780C.

The pseudo-ductility of submicron LPS composites was not as encouraging as of the nano-infiltrated composites probably due to the fiber damage during processing, since their fracture occurred shortly after the proportional limits well before reaching 0.3 percent. The fractography presented in Fig.7 supports this interpretation, where a brittle fracture mode was dominating for the cases of

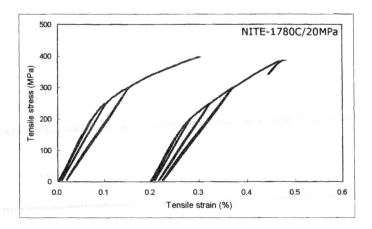

Fig.6 – Tensile stress – strain relationship of nano-infiltrated Tyranno™-SA composite produced at 1780C and 20MPa. The reinforcement is uni-directional and the fiber volume fraction is approximately 20 percent.

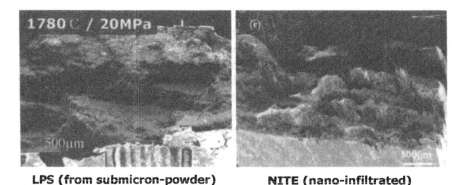

LPS (from submicron-powder)   **NITE (nano-infiltrated)**

Fig.7 – Tensile fracture surfaces of LPS (left) and NITE (right) composites produced at 1780C and 20MPa.

sub-micron LPS composites after sintering at 1780C, while the nano-infiltrated composite exhibited a sufficient extent of fiber pull-out on the fracture surfaces.

The thermal conductivity calculated from the measured thermal diffusivity, specific heat, room temperature mass density and linear coefficient of thermal expansion (CTE) is plotted against temperature in Fig.8. The measurement of thermal diffusivity and specific heat was made at each temperature. CTE was measured along the fiber length direction. Data from a monolithic liquid

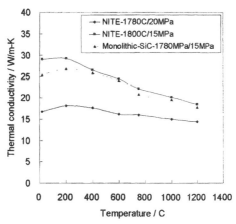

Fig.8 – Temperature dependence of thermal conductivity of liquid phase sintered monolithic SiC and NITE SiC/SiC composites.

phase-sintered SiC sample produced from the same raw material composition are plotted together for comparison. The nominal thermal conductivity of Tyranno™-SA fiber is about 65W/m-K at 20°C. The significantly reduced thermal conductivity in the composite compared to the monolith sintered at the same temperature, in spite of the inclusion of high thermal conductivity fibers, indicates that the presence of intra-bundle pores and/or polymer-derived porous intra-bundle matrix is hampering the composite's thermal conductivity. The thermal conductivity of the monolithic sample was insensitive to temperature, which suggests that the grain boundaries, scattered second phases or other lattice defects are the primary thermal resistances.

CONCLUSIONS

SiC/SiC composites with outstanding thermo-mechanical properties were successfully produced through a transient eutectic phase route, by employing moderate matrix densification conditions and applying appropriate fiber and/or interphase protections, incorporating a nano-infiltration technique. The Tyranno™-SA SiC fiber composites with pyrolytic carbon interphase and the nano-infiltrated SiC matrix with alumina-yttria sintering additive system exhibited excellent performances at the process temperatures of 1750~1800C. The lab-grade composites with fiber volume fraction of 20 percent exhibited proportional limit tensile stresses of >200MPa and ultimate tensile strength of ~400MPa. The thermal conductivity was in the range of 15~30W/m-K at 20~1200C. The non-NITE grade liquid phase-sintered composites were less appealing in terms of mechanical properties but might be   suitable for less severe applications.

Future work will be performed toward the further improved matrix quality and advanced control of matrix and interphase structures for improved high temperature and environmental performances and tailorable thermal properties.

ACKNOWLEDGMENTS

This work was performed as a part of Core Research for Evolutional Science and Technology (CREST) program operated by Japan Science and Technology Corporation (JST).

REFERENCES

[1] F.F. Lange, *Journal of Materials Science*, 10, 314-320 (1975).

[2] H. Suzuki, *Materials Chemistry and Physics*, 42, 1-5 (1995).

[3] K. Nakano, A. Kamiya, M. Iwata and K. Oshima, "Fabrication of Fiber Reinforced Ceramic Composites," A.R. Bunsell, et al., Eds., Development in the Science and Technology of Composite Materials, Elsevier Science (1989).

[4] K. Park and T. Vasilos, Journal of Materials Science, 32, 295-300 (1997).

[5] T. Yano, K. Budiyanto, K. Yoshida and T. Iseki, Fusion Engineering and Design, 41, 157-163 (1998).

[6] A. Gallardo-Lopez, A. Munoz, J. Martinez-Fernandez and A. Dominguez-Rodriguez, Acta Materialia, 47, 2185-2195 (1999)

[7] A. Kohyama, S.M. Dong and Y. Katoh, "Development of SiC/SiC Composites by Nano-powder Infiltration and Transient Eutectoid Process," submitted to Ceramic Engineering & Science Proceedings.

[8] Y. Katoh, A. Kohyama, D.M. Dong, T. Hinoki and J-J. Kai, "Microstructures and Properties of Liquid Phase Sintered SiC/SiC Composites," submitted to Ceramic Engineering & Science Proceedings.

[9] T. Ishikawa, Y. Kohtoku, K. Kumagawa, T. Yamamura and T. Nagasawa, "High–Strength Alkali-Resistant Sintered SiC Fibre Stable to 2200°C", Nature, 391, 773-775 (1998).

[10] S.M. Dong, Y. Katoh and A. Kohyama, "Preparation of SiC/SiC Composites by Hot Pressing using Tyranno-SA fiber as Reinforcement," submitted to Journal of American Ceramic Society.

[11] K. Hironaka, T. Nozawa, T. Hinoki, N.Igawa, Y. Katoh, L.L. Snead, and A. Kohyama, "High-temperature Tensile Strength of Near-stoichiometric SiC/SiC Composites," Journal of Nuclear Materials, in printing.

[12] W. Yang, A. Kohyama, Y. Katoh, H. Araki, J. Yu and T. Noda, "Performances of Tyranno-SA Fiber Reinforced SiC/SiC Composites under Unloading-Reloading Tensile Tests," submitted to Journal of American Ceramic Society.

FACILE FABRICATION OF SiC MATRIX COMPOSITES USING NOVEL
PRECERAMIC POLYMERS

Leo Macdonald
Starfire Systems Inc.
877 25th St
Watervliet NY 12189

1. ABSTRACT

Starfire Systems, Inc. offers a family of unique, patented, polysilylene-methylene polymers for the manufacture of advanced silicon carbide (SiC) ceramic materials with unequalled performance, safety, environmental and cost advantages. The resultant advanced ceramic materials provide superior resistance to wear, corrosion and elevated temperatures, when compared to many other materials. These new precursors greatly simplify the formation of SiC ceramic materials, making possible new applications in aerospace, power materials, microelectronics, and other 21st century industries. Typical applications are for increasing energy efficiency, pollution control, improved capital effectiveness and increased productivity in many industrial sectors.

Starfire has performed experimental research on the composites generated using Starfire Precursors. These composites have been formed via Polymer Infiltration Process (PIP), Ceramic Molding operation, Paint On process, and Chemical Vapor Deposition/Infiltration (CVD/I). These four applications are discussed. First, our precursor "SP-Matrix Polymer" is used in the Polymer Infiltration Process (PIP) to infiltrate fibrous performs and form a SiC matrix. The ceramic matrix is formed by a pyrolysis at 850°C. Typically the PIP process is repeated a number of times as each step results in a higher density and lower porosity. Second, our SP-Matrix Polymer can be mixed with powders and used to bond SiC pieces to SiC pieces and other substrates. It acts as a high temperature SiC glue. This enables the forming of complex structures with joints of silicon carbide. Third, we have manufactured a SiC ceramic molding compound which can be formed into a wide variety of shapes. This moldable compound yields a very hard, dense part after just one pyrolysis cycle. And fourth, we have used our SP-4000 and SP-2000 low molecular weight precursors for CVD and CVI reactions to yield SiC coated and SiC matrix infiltrated parts. The process is done at a low temperature (850°C) and under vacuum/inert gas atmosphere. The CVD coatings are very dense and uniform in thickness. The CVI process infiltrates a variety of fiber tow thicknesses and can be tailored to meet density and porosity needs. All of the above precursors have been demonstrated to adhere (as SiC) to a variety of substrates including graphite, silicon carbide, silica, alumina, diamond, zirconia, and some refractory metals. Typical SiC yields from the precursor at high temperatures are 85-90% by wt.

These preceramic polymers offer SiC manufacturers a greater design freedom than most other methods of SiC processing. The temperatures of application are much lower (typically 850°C), the process equipment required is much simpler and more cost effective, and these precursors have much lower environmental, personnel and usage hazards. These polymers are currently offered as SiC precursors for the industrial, academic, and governmental markets around the world.

2. INTRODUCTION

Starfire Systems, Inc. offers a family of unique, patented, polysilylene-methylene polymers for the manufacture of silicon carbide (SiC) ceramic materials with superior performance, safety, environmental and cost advantages. The resultant advanced ceramic materials, SiC composites, are very strong, rigid, inert and hard making them highly resistant to wear, corrosion and elevated temperatures, when compared to many other materials. Starfire's new precursors greatly simplify the formation of SiC ceramic materials, making possible new applications in aerospace, power materials, microelectronics, and other 21st century industries. Typical applications are for parts in harsh environments such as rocket and combustion engines, power plants, high temperature furnaces, and pollution control devices. Additionally SiC is useful for the electronics industry for a variety of on chip and manufacturing hardware applications.

Starfire Systems has performed experimental research to demonstrate the manufacturing process for a variety of SiC ceramics and composites using Starfire Precursors. These composites have been formed via the Polymer Infiltration and Pyrolysis (PIP) and Chemical Vapor Deposition/Infiltration(CVD/CVI) processes. Additionally our precursors have been used for a molding compound, a ceramic adhesive, a coating and other processes.

3. EXPERIMENTAL PROCEDURE

3.1 The Polymer Infiltration and Pyrolysis (PIP) process has been demonstrated using our "SP-Matrix Polymer". This chemical is used for the infiltration of fiber performs, forming a SiC matrix upon pyrolysis at 850°C under N_2 gas. Typically the PIP process is repeated a number of times as each step increases the preform density and decreases the porosity. Typically the PIP cycle takes 24 hrs for any given part. The sample is machined to shape typically after the third cycle. After a second to last cycle, the sample is often fired to 1650°C to crystallize and then finish with a final PIP to yield a high quality SiC composite. Preforms demonstrated include carbon fabric, felt and fibers, silicon carbide fabric, fibers, and alumina (Al_2O_3) felt and fabric. The PIP process has also been demonstrated to work as a binder for particulates to form joints and coatings.

3.2 The Chemical Vapor Deposition/Infiltration (CVD/CVI) process has been demonstrated using our SiC precursors, the "SP-4000" oligomer mixture and the "SP-2000" pure chemical. These two products are both readily evaporable and react via heat to form SiC solid on a surface (CVD) or in a matrix (CVI). Typical reaction conditions are 0.1 KPa (1 mm Hg) to 26.6KPa (200 mm Hg) vacuum with N_2 or Ar gas environment with substrate temperatures from 600°C to 1000°C. The precursor is delivered using a heated bubbler evaporator with a carrier gas (N_2 or Ar) and a delivery tube with valve. The clean and concentrated nature of these precursors results in very efficient conversion to SiC solid. The few byproducts are relatively mild and benign when compared to the copious corrosive toxic byproducts of other halide based SiC precursors (containing Cl or F atoms).

4. RESULTS AND DISCUSSION

4.1 A wide variety of parts have been prototyped by Starfire Systems using the PIP process based on the "SP-Matrix Polymer". These parts encompass many fields of manufacturing and technology. They range from heat exchangers and structural ceramics to abrasive parts and friction devices.

The Polymer Infiltration and Pyrolysis (PIP) process is simple and efficient. The typical process cycle is:
1. Assembly/preparation of preform in stainless steel holder.
2. Infiltration of preform with SP-Matrix Polymer (vacuum assist optional).
3. Pyrolysis at 850°C under inert gas (N_2). Ramp rate to 850°C is 3°C/min.
4. Cool down, remove part and clean off loose material.
5. Machine to net shape (after 3rd cycle).
6. Measure density and porosity via nondestructive immersion method.
7. Repeat Cycle.

The PIP cycle is repeated a number of times to get to increasing density and decreasing porosity. The desired density determines the number of cycles required. During the pyrolysis, the polymer precursor changes density to ~1/3 its original density. This is due to the increased bonding between adjacent atoms to form a strong multi-bond lattice. The repeating cycles fill in the successively smaller and smaller pores in the matrix. A plot of the increasing density and decreasing porosity vs the PIP cycle number is shown in figure 1.

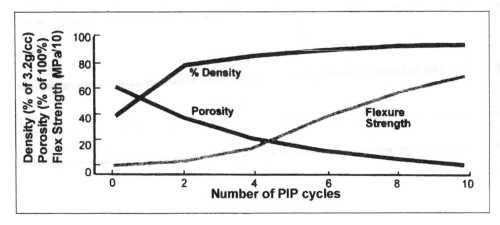

Figure 1. Graph showing typical properties of a SiC fiber, SP-Matrix Polymer ceramic composite as a function of number of PIP cycles. The Y-axis units are Density (% of 3.2 g/cc) & Porosity (% of total volume) & Flexure Strength (MPa/10)

4.1.1. Ceramic Parts Descriptions

Ceramic fiber preforms yield continuous fiber ceramic composites (CFCCs), while whiskers or powder preforms yield particle reinforced ceramic composites. All of the ceramics can be machined to a near net shape, typically after the third PIP cycle. The machining is relatively easy and fast using diamond tooling. Tool life is longer than usual because the low cycle SiC has porosity which enables easy material removal. After the rest of the cycles (typically 5 more), the part is very hard and dense and extremely difficult to machine. Diamond tool life is short and cutting speeds are very slow. These fiber and particle reinforced SiC ceramics have been made into a wide variety of parts for industrial uses. Structural ceramics have been made using carbon felt that are up to 4 cm thick x 40 cm diameter. Fabric preforms of SiC or carbon have been used for flat panels, disks, and joined panels to form T shapes and I shapes. Some of the disks have been machined to final shape for brakes (aircraft, automotive, motorcycle) and some are undergoing testing. Other shapes manufactured are tubes and heat exchangers. The tubes are up to 6.3 cm OD with 1-5 mm thick walls and lengths of up to 30 cm. The heat exchangers have been manufactured in parallel and cross flow designs. The passages are 1 mm to 6 mm ID and overall sizes are 15 cm x 15 cm by 2 to 8 layers thick. These heat exchangers have very good thermal conductivity and high temperature resistance to a variety of chemicals.

Advanced SiC/SiC Ceramic Composites

Larger or smaller parts can be made easily through the selection of appropriately larger or smaller tooling. Additionally the use of SP-Matrix Polymer has been demonstrated as a low loss, high temperature binder for bonding various particulates.

4.2 A number of coatings and matrix infiltrations have been demonstrated at Starfire Systems using the CVD/CVI processes based on "SP-4000" and "SP-2000" precursors. The SP-4000 chemical is a oligomeric precursor composed of several molecules of varying atomic weights. The molecules are all stoichiometric SiC and yield a stoichiometric SiC deposition. The evaporation temperature is a range from 30°C to 150°C. The SP-2000 is a single component chemical that is high in carbon content and so yields a carbon rich SiC based deposit. The boiling temperature of this chemical is 56°C. Both of these precursors have been used for both coatings on and infiltrations of samples. The typical CVD process for using these precursors is as follows:

1. Assemble and prepare sample (substrate or preform)
2. Load substrate into vacuum CVD equipment.
3. Heat substrate to 850°C (typical) under vacuum of 0.1 KPa (1 mm Hg)
4. Open precursor delivery valve and heat precursor source (bubbler, evaporator, etc) to cause vaporization.
5. A carrier gas is used to effect vapor transfer from the bubbler to the furnace and sample.
6. After coating is complete (time/thickness), shut down and remove sample.
7. Analyze and evaluate.

This CVD process is quite standard and can be carried out using simple and inexpensive equipment. Typical equipment costs are 50 to 100 times less than equipment sold for other silicon carbide precursors such as methyltrichlorosilane (MTS). The film deposition yields few byproducts which are rather benign when compared to the byproducts from the halide based precursors. The byproducts were analyzed by gas chromatography. The major byproduct of SP-4000 is hydrogen gas which is flammable but can be safely diluted with air and vented. The SP-2000 also has some methane byproduct which is flammable as well. The total byproducts are typically 8% for SP-4000 and 30% for SP-2000, both by weight of the total precursor used, while the MTS byproducts are ~68% by weight. The halide byproducts are hydrochloric acid (HCl) or hydrofluoric acid (HF), both of which are quite corrosive and are very harmful to people and the environment.

4.2.1. The CVD process has been demonstrated to form coatings on various substrates using SP-4000 and SP-2000 precursors. These SiC based coatings have been demonstrated on many different substrates including carbon, silicon carbide, aluminum oxide, silicon nitride, quartz, cordierite, as well as various fibers. Typical samples are piece 2.54 cm x 2.54 cm and 3.17 mm thick. The CVD coating seals most sub micron pores and provides a uniform surface matching the underlying substrate. The SP-4000 has been deposited at rates from 100 nm/hr to 75 μm/hr. The rate is proportional to the amount of porosity in the coatings because the precursor is outgassing hydrogen as it changes to SiC ceramic. The slow and moderate rates allow all the H_2 to escape, while the fast rates trap pockets of H_2 within the coating. Additionally higher substrate temperatures speed the decomposition and give lower porosity coatings at higher rates. The SP-2000 has been deposited at rates from 1 μm/hr to 70 μm/hr. It also has a rate dependent porosity in the coating. One advantage of the rate dependence of the porosity is a custom tailorable porosity in manufactured coatings. Sometimes it is advantageous to have a fully dense coating and other times it is advantageous to have a specific pore size. With the SP-4000 and SP-2000, both can be accomplished during the same run by just changing temperature or flow rate.

The SiC based coatings were analyzed for uniformity by cross sectional examination through metalography, and for composition by electron microprobe. The microprobe was used to analyze for amounts of Si, C, O, and N. The SP-4000 yields a stoichiometric SiC deposit (1:1 Si:C), while the SP-2000 yields a SiC with excess Carbon (1-15% depending on temperature). Typical SiC yields from the precursor at high temperatures are 85-90% by wt. In addition a heat treatment monitored by x-ray diffraction was used to look for crystallinity of SiC. The samples were CVD coated at 850°C and then heat treated at 1000, 1200, 1400, and 1600°C for 30 min in an argon environment. The subsequent x-ray showed the evolution of the β-SiC phase with a complete match of peaks to the standard. The crystallization temperature onset was 1200°C, and the sample was completely crystalline at 1600°C. See Figure 2.

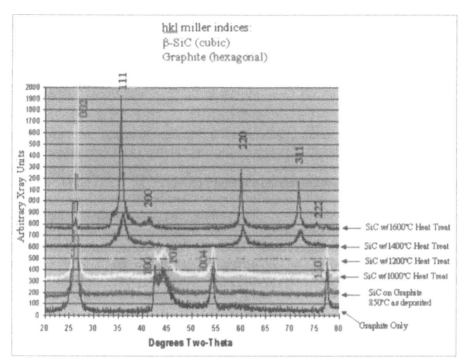

Figure 2. Demonstration of deposition and crystallization of silicon carbide on graphite using SP-4000 precursor as measured by X-ray diffraction.

4.2.2. The use of the SP-4000 and SP-2000 for CVI has been demonstrated successfully on several types of fiber preforms. Preform shapes are typically 2.54 cm x 2.54 cm x 2.54 cm cubes. Preforms which have been demonstrated include SiC fabrics and lay-ups, carbon felts, fabrics, and fibers, aluminum oxide fibers and mats, and fiberglass (silicon dioxide) mats. The CVI process is quite similar to the CVD process except for a few details. The precursor is flowed into the furnace along with a large dilution of bypass gas (typically N_2). This bypass gas serves a number of functions. It dilutes the precursor concentration in the gas phase to prevent gas phase particle nucleation of SiC particles. And it increases the flow speed of the gas through the reactor thereby assisting the precursor molecules in deeply penetrating into the substrate for a very uniform deposition. The temperature used is lower (at 700°C) to prevent sealing the outer surfaces of the preform. The runs are typically much longer, a few days rather than minutes or hours because of the large volumes of deposit required to fill the preform.

The uniformity of the infiltrated matrix was measured using cross sectional metalography and optical and electron microscopes. It was found to be uniform with small pores (10-50 nm dia). All of the fibers were encased in the SiC matrix.

5. CONCLUSIONS

The silicon carbide forming precursors by Starfire Systems Inc have been demonstrated in manufacturing processes for SiC ceramics and composites . The use of the Starfire polymers results in greater design freedom, a simple/easy to use process and superior performance in the end product. The PIP process with the SP-Matrix Polymer is used regularly here at Starfire Systems with excellent repeatability and reproducibility. The tooling and equipment is simple and inexpensive due to the decreased processing temperatures and highly pure and concentrated nature of the precursor. The CVD/CVI processes using the SP-4000 and SP-2000 precursors are also in use regularly at Starfire and have proven to be simple, straightforward and low capital requirements (typically US$5000-10,000 for a 6 to 20 cm diameter by 1 m length reactor).The deposits obtained using this method are high quality, high purity SiC or SiC/C

Overall these preceramic polymers offer SiC manufacturers significant benefits. This extremely versatile technology has capabilities to be used on a wide variety of substrates and fiber preforms for manufacturing SiC based products for a wide range of applications. Additionally the manufacturer will face much lower environmental, personnel and usage hazards than competing SiC manufacturers.

These polymer SiC precursors are currently offered to industry, universities, and government funded programs around the world.

ACKNOWLEDGEMENTS

The author would like to thank the RPI metalographic and analytical staff as well as the other Starfire employees who helped make this research successful.

References

[1] C. Whitmarsh and L.V. Interrante, "Synthesis and structure of a highly branched polycarbosilane derived from chloromethyltrichlorosilane", *Organometallics*, 10, 1336-1344 (1991); C.W. Whitmarsh and L.V. Interrante, U.S. Patent No. No. 5,153,295, "Carbosilane polymer precursors to silicon carbide ceramics", 10/6/92.

[2] Ph.D. thesis, "The Densification, Crystallization and Mechanical Properties of Allylhydridopolycarbiosilane Derived Silicon Carbide", Kevin Moraes, Department of Materials Engineering, RPI, 2000

PROCESS DESIGN FOR SiC/SiC COMPOSITE WITH POLYMERIC PRECURSOR

Masaki Kotani
National Space Development Agency of Japan
Tsukuba Space Center, 2-1-1 Sengen, Tsukuba, Ibaraki 305-8505, Japan

Yutai Katoh and Akira Kohyama
Institute of Advanced Energy, Kyoto University and CREST-ACE
Gokasho, Uji, Kyoto 611-0011, Japan

ABSTRACT

SiC fiber reinforced SiC matrix (SiC/SiC) composites have been developed through the polymer impregnation and pyrolysis (PIP) process. The process conditions to make a consolidated body, including curing temperature, pressure and SiC particle content, were systematically optimized for reducing porosity and improving fiber distribution. As a consequence, composites of more than 600MPa flexural strength were successfully fabricated, showing non-catastrophic failure mode with extensive fiber pullout. The present study reveals that microstructure improvement in the first PIP cycle was the most important aspect for success with the PIP process.

INTRODUCTION

Since silicon carbide possesses such superior properties as thermal resistance, oxidation resistance, strength and low activation, there have been many efforts in R&D of SiC/SiC composite for use in future fusion reactor and aerospace components [1, 2]. Among potential fabrication processes of ceramics-based composites, one of the most promising methods is the PIP method. Though the PIP method has advantages with regard to large-scale fabrication with complicated shapes, microstructural control and fabrication cost, it is difficult, by this process, to fabricate a composite of high density and fine/uniform fiber distribution due to volume shrinkage and gas evolution of the polymer precursor. Also, microstructure of a polymer-derived composite is very dependent on the precursor polymer and process conditions [3-7]. Therefore, it is essential to select appropriate polymers with small volumetric shrinkage and near stoichiometric composition (in terms of products) and to optimize process parameters. Here, small shrinkage provides higher potential of making high-density ceramics, and a stoichiometric product composition provides excellent radiation resistance with high strength at elevated temperature.

Polyvinylsilane (PVS), which is a SiC precursor having many functional Si-H bonds, was employed for the matrix precursor. Because of low viscosity and excellent wetting properties against SiC_{PCS}, PVS has advantages in designing fabrication processes and controlling microstructure [8, 9]. Its thermosetting property can be advantageous for reducing porosity by pressurization. The objectives in this work are to establish a fabrication process for high performance SiC/SiC composites, together with a precise knowledge about the evolution of the polymeric precursor during the process.

EXPERIMENTAL

PVS was employed as the matrix precursor. It was obtained as a transparent viscous liquid. As the reinforcement, Hi-Nicalon[TM] (Nippon Carbon Co., Ltd.) was utilized. Fine SiC particles were added to the polymer as a filler material. The PIP process in the present work was composed of the following four steps,

1. To make compound sheets, unidirectional Hi-Nicalon[TM] sheets were dipped into a slurry in ambient environment,
2. To make green sheets, the compound sheets were heated up to curing temperatures in inert gas,
3. To make a consolidated body, the green sheets were stacked and heated under pressure in inert gas up to 1473 K,
4. To densify the material, the consolidated body was subjected to repetitive impregnation and pyrolysis.

Process optimizations were conducted for (1) curing temperature and pressure, and (2) filler content. Since thermosetting in PVS occurred at temperatures around 600 K, the curing temperature was optimized between 583 K and 663 K. Detailed evaluations were performed for the specimens obtained after the third infiltration/pyrolysis step. Densities were measured by the Archimedes method. The sample IDs were set to reflect the processing conditions (curing temperature / K, pressure / MPa).

Mechanical properties of the composites obtained after 6 cycles of PIP processing were evaluated by 3-point flexural test at room temperature. Span length and crosshead speed were 25 mm and 0.5 mm/min, respectively. Dimensions of the specimens were 4x1x25 mm. Ultimate strength, σ_u, was determined from the peak load in the load-displacement curve. Work-of-fracture was calculated from the area of load-displacement chart. Specimen fracture surfaces were observed using optical microscopy and SEM.

RESULTS AND DISCUSSION
Optimization of process conditions
Curing temperature and pressure
Figure 1 shows the relative and apparent densities of the consolidated bodies produced under various curing temperatures and pressures. These results indicate that pressure during consolidation was indispensable. Relative densities of 5 and 10 MPa showed similar trends with temperature. Remarkable high relative densities such as 68 % and 70 % were obtained in (603, 5) and (623, 10). These results demonstrated the importance of the consolidation condition for densification. The apparent density provided useful information to estimate the volume fraction of fiber and matrix in consolidated bodies. It was valid only under the assumption that all porosity was open. Highest apparent densities were found for the same process temperature for which highest relative densities were obtained at each consolidation pressure. Based on the density of the fibers (2.74 Mg/m^3) and polymer-derived matrix (more than 2.9 Mg/m^3), it was considered that increasing the amount of matrix in a consolidated body was very important for reducing porosity. To evaluate microstructure of the consolidated bodies, SEM observation was carried out. A very uniform fiber distribution was found for the composite of (603, 5) though many fine pores were present between fibers. Based on these good results, the condition (603, 5) was applied to the next series of experiments in process optimization.

Mass fraction of the filler material in the matrix precursor

(a)   (b)

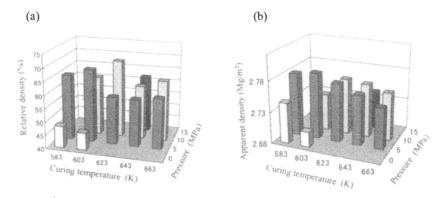

Figure 1. Relative density and apparent density of the consolidated bodies; (a) relative density, (b) apparent density

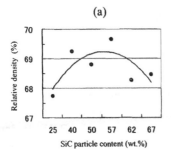
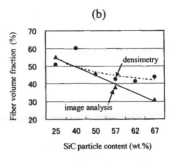

Figure 2. Relative density and fiber volume fractions of the consolidated bodies; (a) relative density, (b) fiber volume fraction

Figure 2 shows relative density and fiber volume fraction of the consolidated bodies. Those were fabricated with polymeric slurries of various filler contents. Highest relative density appeared at 57 %, even if effective volume yield increased with increasing SiC particle content. This indicated that high filler loading had some negative influence on densification of composites. Fiber volume fractions measured by apparent densities (circle) were higher than those estimated by image analysis (triangle). These differences in measurements increased with increasing filler content. Because the density of Hi-Nicalon[TM] was known to be smaller than that of dense materials derived from polymeric slurries, these results meant that the porosity of the slurry-derived matrix increased as filler content increased.

Figure 3 shows SEM micrographs of the consolidated bodies at various filler contents. By the low magnification pictures (upper line), large pores can be identified in (a) and (b) but very scarce in (c). By the high magnification pictures (lower line), micro-pores between fibers can be seen many in (a), a few in (b) and almost no pores in (c). These indicated that (c) was capable to be higher in density. However, the results by the Archimedean method showed that the density of (b) was larger than that of (c). Potential explanation for this discrepancy was the porosity of matrix as suggested above. Further precise investigation of this issue is under investigating. Also it was demonstrated that filler loading was one of influential factors to change fiber volume fraction.

Mechanical Properties

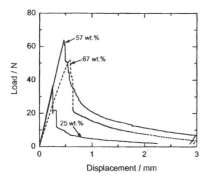

Figure 4. Load-displacement
curves of the composites

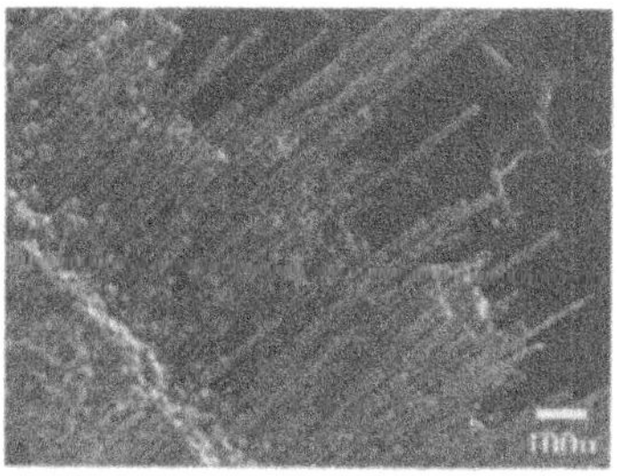

Figure 5. SEM fractograph of the
composite at the filler content of
57%

In order to investigate the effects of consolidated microstructure on
mechanical properties, flexural tests were performed on specimens machined from
the composites. Figure 4 shows load-displacement curves of the composite test
specimens. These composites exhibited non-catastrophic fracture behavior, with
one (67 mass%) showing an initial failure by delamination, and the other (57

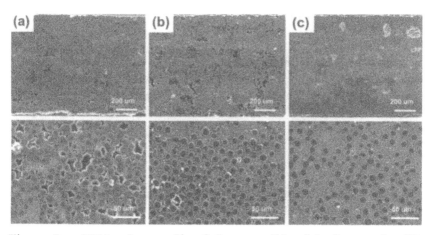

Figure 3. SEM micrographs of the consolidated bodies at the filler
contents of (a) 25 %, (b) 57 %, (c) 67 %.

mass%) showing a gradual failure through the specimen. The average flexural strength and work-of-fracture were (602 MPa, 5.1 kJ/m^2) and (575 MPa, 4.2 kJ/m^2), respectively. Figure 5 shows a SEM micrograph of the fracture surface of the stronger and tougher specimen. Though interfacial coating barriers like C, BN and other ceramics used elsewhere were not applied to SiC fibers, there was extensive fiber pullout of significant length, with smooth fiber surfaces and no detectable damages on the surfaces of the fibers. This type of failure mode, which typically leads to good mechanical properties, is another great advantage of the use of PVS as the matrix precursor.

SUMMARY

In order to develop an improved fabrication process for SiC/SiC composites, PVS, a new type of liquid SiC precursor, was adopted as matrix precursor. Based on the precise analysis of the pyrolysis behavior of PVS, systematic optimization was performed of the process conditions including curing temperature, pressure and filler content to make a consolidated body. By optimizing consolidation conditions, relative density was improved up to 70 %, and a good, uniform fiber distribution was formed. Due to reductions in inter-fiber pores and improved fiber distribution, a composite material was produced that has flexural strength of more than 600 MPa and work-of-fracture of about 6 kJ/m^2, and shows non-catastrophic flexure behavior. These accomplishments are encouraging for future development of the PIP process for production of high performance SiC/SiC composite materials.

ACKNOWLEDGEMENTS

This work was performed as a part of 'R&D of Composite Materials for Advanced Energy Systems' research project, supported by Core Research for Evolutional Science and Technology (CREST). The author is also grateful to Dr. M. Itoh (main researcher of Mitsui chemicals Co., Ltd.) for providing the polymer precursor.

REFERENCES

[1]A. Kohyama, Y. Katoh, T. Hinoki, W. Zhang and M. Kotani, "Progress in the Development of SiC/SiC Composite for Advanced Energy Systems: CREST-ACE Program," Proceedings of 8th European Conference on Composite Materials, 4, 15-22 (1998).

[2]L. Snead, R. Jones, A. Kohyama and P. Fenici, "Status of silicon carbide composites for fusion," Journal of Nuclear Materials, 233-237, 26-36 (1996).

[3]T. Tanaka, N. Tamari, I. Kondoh and M. Iwasa, "Fabrication and Evaluation of 3-Dimensional Tyranno Fiber reinforced SiC composites by Repeated

Infiltration of Polycarbosilane," Journal of the Ceramic Society of Japan, 103 [1], 1-5 (1995).

[4]T. Tanaka, N. Tamari, I. Kondoh and M. Iwasa, "Fabrication and Mechanical Properties of 3-Dimensional Tyranno Fiber Reinforced Infiltration of Polycarbosilane," Journal of the Ceramic Society of Japan, 104, 454-457 (1996).

[5]D. W. Shin and H. Tanaka, "Low-Temperature Processing of Ceramic Woven Fabric/Ceramic Matrix Composites," Journal of American Ceramic Society, 77 [1], 97-104 (1994).

[6]J. Jamet, J. R. Spann, R. W. Rice, D. Lewis and W. S. Coblenz, "Ceramic-Fiber Composite Processing via Polymer-Filler Matrices," Ceramic Engineering Science and Processing, 677 [5], 7-8 (1984).

[7]D. Suttor, T. Erny, P. Greil, H. Goedeke and T. Hung, "Fiber-Reinforced CMC with Polymer/Filler-Derived Matrix"; pp. 211 in Ceramic Transactions, Vol. 51, Ceramic Science and Processing Technology. Edited by H. Hausner, S.-I. Hirano and G. L. Messing. American Ceramic Society, Westerville, OH, 1995.

[8]W. R. Schmidt, L. V. Interrante and R. H. Doremus, "Pyrolysis Chemistry of an Organometallic Precursor to Silicon Carbide," Chemical Material, 3, 257-267 (1991).

[9]A. Boury, R. J. P. Corriu and W. E. Douglas, "Poly(carbosilane) Precursors of Silicon Carbide: The Effect of Cross-Linking on Ceramic Residue," Chemical Material, 3, 487-489 (1991).

MATRIX FILLING BEHAVIOR OF SiCf/SiC COMPOSITE BY WHISKERING AND CVI PROCESS

Ji Yeon Park, Ho Soo Hwang, Weon-Ju Kim,
Nuclear Materials Technology Development, Korea Atomic Energy Research
Institute, 150 Dukjin-dong, Yusong-gu, Daejon, 305-353, Korea

Ji Hye Son and Doo Jin Choi
Dept. of Ceramic Engineering, Yonsei University,
134 Shinchon-dong, Sodaemoon-gu, Seoul 120-749, Korea

ABSTRACT

In-situ whisker growing and then matrix filling (called whiskering process) was applied to reduce the porosity in make SiCf/SiC composite that is mainly due to a canning. Whiskers grew well in the voids both between fibers and between bundles. The morphology of the grown whisker was dependent upon the reaction temperature. The grown whiskers served to divide the large voids and then the matrix filling was performed inside the modified pore structure. MTS (CH_3SiCl_3) and H_2 were used as source and diluent gases, respectively. Density and flexural strength of SiCf/SiC composites fabricated by the whiskering process were higher than those by a conventional CVI process. These results suggest that improved SiCf/SiC composites can be fabricated using this process

INTRODUCTION

Silicon carbide has the potential advantages for structural applications in fusion reactor due to its unique properties such as good irradiation resistance and thermo-mechanical properties, reduction in the high Z (atomic number) impurities transported into the plasma from the first wall, less severe waste generation due to neutron activation and improved plant conversion efficiencies by higher operating temperatures [1]. The use of SiC monolith in high-temperature structural applications has been severely limited by their low fracture toughness and lack of predictable service life. Any flaw in SiC monolith can lead to a catastrophic failure. However, by using continuous long fibers to reinforce SiC matrix, fracture toughness has been significantly improved [1-5].

Several fabrication processes for SiC_f/SiC composites, such as chemical vapor infiltration (CVI) process [6], polymer impregnation and pyrolysis (PIP) process [7-8], reaction sintering (RS) process [9], etc., have been under investigation for more than 10 years. The CVI process is an effective method for fabricating SiC_f/SiC matrix composites, but this process is slow with an inherent drawback of substantial residual porosity [1]. As the infiltration process proceeds, the CVI grown matrix obstructs infiltration of vapor reagents, decreasing further infiltration efficiency [6, 10]. Because of this canning phenomenon, large closed pores are left in the interior of fiber preform. In the previous works [11, 12], we proposed a new method, so called the whiskering process, to obtain a dense composite in the C_f/SiC and SiC_f/SiC systems. This process consists of two steps: a whisker growing step and a matrix filling step. In the first step, the grown SiC whiskers may serve to divide large natural pores into smaller ones and modify the void structure. In addition, the grown whiskers offer new deposition sites for the CVI process. Therefore, matrix filling may be efficiently performed inside the whisker grown composites in the next step.

In this study, the matrix filling behavior by the whiskering process was investigated in preforms of 2D plain weave Nicalon^{TM} fiber cloth. Density and flexural strength of SiC_f/SiC composite by this process compared to those by a conventional CVI process.

EXPERIMENTAL PROCEDURE

For the CVI-SiC densification, acetylene (C_2H_2) and methyltrichlorosilane (CH_3SiCl_3, MTS) were chosen as source precursors of pyrolytic carbon and SiC, respectively. Hydrogen was used as a carrier gas to transfer the source precursor through the bubbler to the main reactor and as a diluent gas to regulate the concentration of the mixture involving MTS vapor and carrier gas. Diluent gas and carrier gas containing MTS vapor were mixed together before being introduced into the reactor. The flow rate of MTS vapor was controlled by adjusting the bubbler pressure and the flow rate of the carrier gas, maintaining the temperature of the bubbler containing liquid MTS at 0°C. The pressure in the reactor was monitored with a capacitance manometer and controlled with a throttle valve located between the reactor and the mechanical pump. Details of the reactor system are shown in the previous report [11].

Fibrous preforms were fabricated by depositing a thin pyrolytic carbon layer of about 0.1 μm thick in a stack of 2D plain-weave Nicalon^{TM} cloth (up to 10 layers). The whisker growing step in the whiskering process was performed at 1100°C for up to 6 h and the matrix filling step was carried out at 1000°C for up to 24 h. The input gas ratio of H_2 to MTS, α ($= F_{(diluent+carrier\ gas)}/F_{MTS}$) was 15 and the total flow rate was 500 sccm. Crystalline phases were detected and characterized by the X-ray diffraction (XRD) method. Microstructures and the constituent elements of SiC$_f$/SiC composites were observed using scanning electron microscopy (SEM ; Model JS-6300, Jeol, Japan) equipped with EDS. Bulk densities were determined by the Archimedes method. Three-point bending strengths were determined using a fixture with a span of 30 mm. For each material, 5 specimens with dimensions of 45^l x 4^{w}x 3^t mm were tested. The samples were loaded with a constant cross-head speed of 0.05 mm/s at room temperature.

RESULTS AND DISCUSSION

The whisker growing process was performed to observe the growing behavior of SiC whiskers at the input gas ratio of 15 and in the reaction temperature range of 1060 to 1150 °C. Fig. 1 shows microstructures of SiC deposits on SiC fibers produced at (a) 1060°C, (b) 1080 °C, (c) 1100 °C and (d) 1150 °C, respectively.

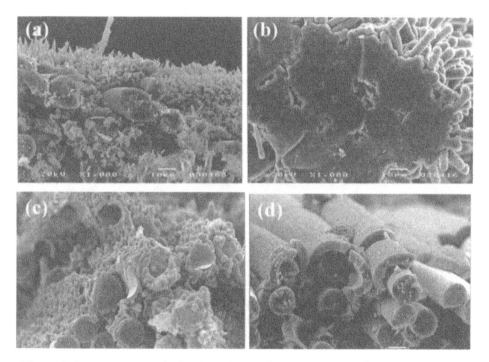

Fig. 1. Microstructures of SiC deposits produced at (a) 1060°C, (b) 1080 °C, (c) 1100 °C and (d) 1150 °C.

Different morphologies of SiC deposits are observed. While SiC whiskers grew at 1080°C and 1100°C, circumferential deposition occurred at 1150°C and spike like SiC deposits with a small aspect ratio grew at 1060°C. On the other hand, Ahn reported that the input gas ratio and the total reaction pressure in the H_2/MTS system were important parameters to produce SiC whiskers on graphite [13]. As the input gas ratio increased from 10 to 40, the deposition rate and the mean diameter of SiC whisker decreased. In addition, only in the total reaction pressure range of 4 to 6 torr, SiC whiskers grew at 1100°C. Considering the deposition rate and morphology of SiC whisker, SiC whiskers on SiC fibers can properly grow at the input gas ratio of 15 and 1100°C.

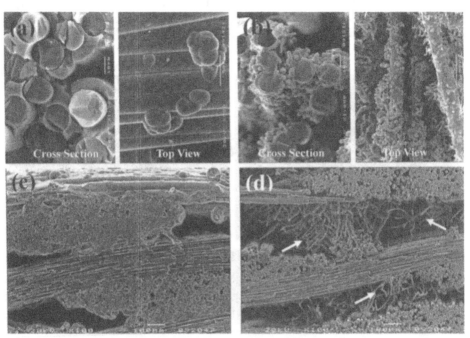

Fig. 2. Microstructures of top and cross section of SiC_f/SiC composites infiltrated (a) at 1000°C for 5 h by conventional CVI and (b) at 1100°C for 4 h by whisker growing step. (c) and (d) are cross sections of composites after further matrix filling of (a) and (b) at 1000 °C for 5 h, respectively.

To compare the matrix filling behaviors of both specimens by a conventional CVI process and by the whiskering process, partially matrix-filled specimens were prepared. Fig. 2 (a) shows microstructures of the top and cross sections of SiC_f/SiC composite infiltrated at 1000°C only for 5 h by a conventional CVI process. Generally, SiC was circumferentially deposited on fibers in a conventional CVI process. As the circumferential deposition of SiC proceeded, the grown SiC deposits met each other and a further deposition into the closed pores would be stopped. On the other hand, as shown in Fig. 2 (b), if the whisker growing step was applied to make SiC_f/SiC composite at 1100°C for 4 h, whiskers grew in the small voids between fibers. These whiskers divided the voids between fibers and left the open channels for matrix filling. Additionally, these grown whiskers were expected to act as new sites for the deposition of SiC. Therefore, the CVI process using the whisker growing process suggests the possibility of an effective matrix filling. To observe the intermediate stage of the matrix filling, the specimens shown in Figs. 2(a) and (b) were further infiltrated at 1000°C for 5 h.

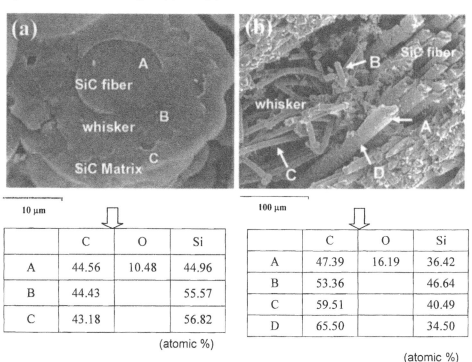

	C	O	Si
A	44.56	10.48	44.96
B	44.43		55.57
C	43.18		56.82

(atomic %)

	C	O	Si
A	47.39	16.19	36.42
B	53.36		46.64
C	59.51		40.49
D	65.50		34.50

(atomic %)

Fig. 3. Microstructures and EDS results of SiC_f/SiC composites prepared by 4 h whiskering and 5 h matrix filling

As shown in Fig. 2 (c), large voids between fiber bundles can be observed in the infiltrated specimen using only a conventional CVI specimen. On the contrary, many whiskers (arrow marks) are grown in the voids between fiber bundles and some filling of SiC between whiskers is observed in the whisker grown specimen (Fig. 2 (d)). The whiskers grown between the fiber bundles provide a number of new deposition sites of SiC and offer more paths to the reactant gases.

The constituent elements and the deposition behavior of grown whiskers after the partial matrix filling were observed by SEM images with EDS analysis. As shown in Fig. 3(a), whiskers radially grew on fibers and then, matrix filling occurred through filling the inter-voids between whiskers and depositing on whiskers. Whiskers and deposits from matrix filling consisted of only Si and C without oxygen (marked B and C). Oxygen was only detected from fiber, which is resulted from original Nicalon™ fibers (marked A). The EDS results suggest that excess Si or C exists in whiskers and deposits. Because quantitative analysis using EDS with fracture surface of samples has a limitation when detecting the exact composition, another method is required to exactly confirm the composition of whiskers and deposits. The constituent elements of the whiskers grown in the voids between SiC fiber bundles were also analyzed with the same method (Fig. 3(b)). Whiskers (marked B and C) also consisted of only Si and C. Oxygen was detected only in original fibers.

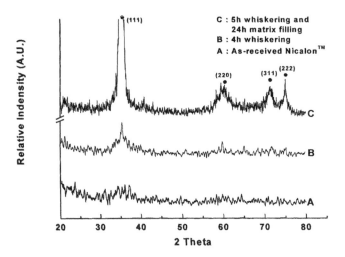

Fig. 4. X-ray diffraction patterns of as-received Nicalon™ fabric and SiC$_f$/SiC composites infiltrated with the whisker growing step at 1100°C for 4 h and matrix filling step at 1000°C for 24 h.

The phases of whiskers and deposits by matrix filling were detected by XRD. Fig. 4 shows XRD patterns of as-received Nicalon[TM] cloth, SiC$_f$/SiC composite after 4 h whisker growing and that after 5 h whisker growing and 24 h matrix filling. Crystalline phases were not clearly detected in the as-received Nicalon [TM] cloth but the crystalline β-SiC was observed in the specimens subjected to whisker growing. Considering the results of XRD and EDS analysis, whiskers and deposits by the whiskering process were β-SiC.

Further evaluation of density and flexural strength of SiC$_f$/SiC composites, using two types of processing, 50 mm diameter by 3 mm thickness specimens were prepared

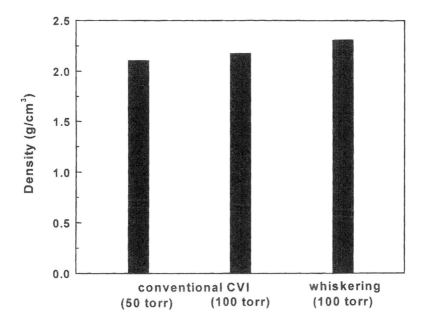

Fig. 5. Density of SiC$_f$/SiC composites prepared by (5h + 24h) conventional CVI and 5 h whiskering +24 h matrix filling

by the different processes; one is a process of 5 h whiskering and 24 h matrix filling and the other is that of 5 h and 24 h conventional CVI. Densities of both specimens are shown in Fig. 5. Density of the specimen by the whiskering process was 2.31 g/cm³ and those by a conventional CVI process were 2.18 g/cm³ at 100 torr and 2.10 g/cm³ at 50 torr, respectively. By applying the whiskering process for making the SiC$_f$/SiC composite, a specimen with the higher density could be obtained at the same processing time. These specimens were cut into 5 bars with dimensions of 45 x 4 x 3 mm to measure the flexural strength. As shown in Fig. 6, the flexural strength of the specimen by the whiskering process is approximately 25% higher than that by a conventional CVI process.

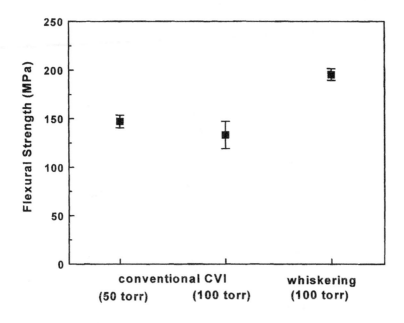

Fig. 6. Three-point flexural strength of SiC$_f$/SiC composites prepared by (5h + 24h) conventional CVI and 5 h whiskering +24 h matrix filling.

SUMMARY

the input gas ratio and the total pressure are important parameters to make SiC whiskers in the SiC$_f$/SiC system. In this study, SiC whiskers on SiC fibers could be radially grown at the input gas ratio of 15 and at around 1100°C. All whiskers grown on fibers and in the voids between SiC fiber bundles were β-SiC, which consisted of Si and C without oxygen. These whiskers are expected to modify the pore structure of SiC$_f$/SiC composite and act as new SiC deposition sites for further matrix filling. Density and flexural strength of the specimen fabricated by the whiskering process were higher than those by a conventional CVI process

ACKNOWLEDGEMENTS
This work was financially supported by Ministry of Science and Technology (MOST) through the Nuclear R&D program

REFERENCES
[1]A. Hasegawa, A. Kohyama, R.H. Jones, L.L. Snead, B. Riccardi, and P. Fenici, "Critical issues and current status of SiC/SiC composites for fusion," *Journal of the Nuclear Materials*, **283-287**, 128-137 (2000).

[2]N. Miriyala, P.K. Liaw, C.J. McHargue and L.L. Snead, "The mechanical behavior of a Nicalon/SiC composite at room temperature and 1000°C," *Journal of the Nuclear Materials*, **253**, 1-9 (1998).

[3]R. H. Jones, L. L. Snead, A. Kohyama, and P. Fenici, "Recent advances in the development of SiC/SiC as a fusion structural material," *Fusion Engineering and Design*, **41**, 15-24 (1998).

[4]A.G. Evans, "Perspective on the development of high toughness ceramics," *Journal of the American Ceramic Society*, **73**[2] 187-206 (1990) .

[5]D.W. Freitag and D.W. Richerson, "Opportunities for Advanced Ceramics to Meet the Needs of the Industries of the Future," DOE/ORO 2076 (1998).

[6]T. M. Besmann, B. W. Sheldon, R. A. Lowden, and D. P. Stinton, , "Vapor-Phase Fabrication and Properties of Continuous-Filament Ceramic Composites," *Science*, **253**, 1104-1109 (1991).

[7]R. Jones, A. Szweda, D. Petrak, "Polymer derived ceramic matrix composites." *Composites part A*, **30**, 569-575 (1999).

[8]A. Kohyama, M. Kotani, Y. Katoh, T. Nakayasu, M. Sato, T. Yamamura, K. Okamura, "High-performance SiC/SiC composites by improved PIP processing with new precursor polymers," *Journal of the Nuclear Materials*, **283-287**, 565 - 569 (2000).

[9]A. Sayano, C. Sutoh, S. Suyama, Y. Itoh, S. Nakagawa, "Development of a reaction-sintered silicon carbide matrix composite," *Journal of the Nuclear Materials*, **271-272**, 467-471 (1999).

[10]N. H. Tai and T. W. Chou, "Modeling of an Improved Chemical Vapor Infiltration Process for Ceramic Composites Fabrication," *Journal of the American Ceramic Society*, **73**[6], 1489-1498 (1990).

[11]B.J. Oh, Y.J. Lee, D.J. Choi, G.W. Hong, J.Y. Park, and W.J. Kim, "Fabrication of Carbon/Silicon Carbide Composites by Isothermal Chemical Vapor Infiltration, Using the In Situ Whisker-Growing and Matrix-Filling Process," *Journal of the*

American Ceramic Society, **84**[1], 245-247 (2001).

[12]J.Y. Park, H.S. Hwang, W.-J. Kim, J.I. Kim, J.H. Son, B.J. Oh and D.J. Choi, "Fabrication and characterization of SiC$_f$/SiC composite by CVI using the whiskering process," *Journal of the Nuclear Materials*, accepted for publication, (2002).

[13]H.S. Ahn and D.J. Choi, "Fabrication of silicon carbide whiskers and whisker-containing composite coatings without using a metallic catalyst," *Surface and Coatings Technology*, **154**, 276-281 (2002).

DEVELOPMENT OF SiC$_f$/SiC COMPOSITES BY MELT INFILTRATION PROCESS

S. P. Lee
Dept. of Mechanical Engineering, Dong-Eui University, Gaya-Dong 24, Busanjin-Gu, Busan 614-714, Korea

Y. Katoh and A. Kohyama
CREST-ACE and IAE, Kyoto University, Gokasho, Uji, Kyoto 611-0011, Japan

ABSTRACT

The present study deals with the fiber preform preparation route for the fabrication of high performance MI-SiC$_f$/SiC composites. Especially, the efficiency of matrix slurry infiltration process, in which the infiltration pressure and the cold pressing are combined for the preparation of fiber preform, is examined. The compatibility among SiC fibers, interfacial protection layer and molten silicon is also examined. The microstructures and the mechanical properties of MI-SiC$_f$/SiC composites are examined through means of SEM, TEM, EDS and three point bending test. Based on mechanical property-microstructure correlation, the process optimization is also discussed.

INTRODUCTION

SiC fiber reinforced SiC matrix composites (SiC$_f$/SiC) have been considered as a promising material for components in fusion energy systems such as first wall or divertor coolant channel, and advanced gas turbine engines in aerospace vehicles [1, 2]. It has excellent high temperature strength, remarkable dimensional stability and low induced radioactivity under severe nuclear environments. SiC$_f$/SiC composites have been fabricated by chemical vapor infiltration (CVI), polymer impregnation and pyrolysis (PIP), hot pressing (HP), and melt infiltration (MI) [3-

6]. The recent development of low oxygen SiC fibers such as Hi-Nicalon type S and Tyranno SA greatly extends the applicability of MI and HP processes [7,8]. The MI process can be recognized as an attractive technique, because it offers a high density and a good thermal conductivity, compared to those of PIP and CVI processes [6]. However, the mechanical properties of MI-SiC$_f$/SiC composites will be closely affected by the preparation methods of fiber preform, which associated with the infiltration techniques of matrix slurry into SiC fabrics, the blending technique of complex SiC/C matrix slurry, and the size of starting SiC fillers. Especially, the interconnected residual Si phase in the matrix must be reduced to avoid the reduction of creep resistance and the undesirable irradiation behavior. The functional interphases including SiC, C, C/SiC, BN/SiC layers must be also examined for a pseudo-ductile fracture of MI-SiC$_f$/SiC composites.

The purpose of the present study is to investigate the preparation routes of fiber perform related with matrix slurry infiltration methods, staring SiC powder sizes, and fabric structures for MI-SiC$_f$/SiC composites. The detailed analysis of composite microstructure and mechanical properties has also been carried out.

EXPERIMENTAL PROCEDURES

The reinforcing materials were unidirectional (UD) Tyranno SA fiber, plain-woven (P/W) Tyranno SA fiber, three-dimensional (3D) Tyranno ZMI fiber, and braided Hi-Nicalon fiber. The coating layers of BN/SiC, SiC and C were deposited on the surface of Hi-Nicalon fiber, Tyranno ZMI fiber, and UD-Tyranno SA fiber, respectively. The matrix slurry was a mixture of SiC powder, C powder, and water. Four kinds of SiC particles with average sizes of 0.03, 0.3, 0.1 and 4.0 μm, were utilized. The average size of C particle was 85 nm. The matrix slurry of SiC (4.0 μm) and C powders were injected into the preform of Hi-Nicalon fiber under a high pressure impregnation of 6.0 MPa. The preform of Tyranno ZMI fiber was prepared by the matrix slurry of SiC (0.3 μm) and C powders under a low impregnation pressure of 0.9 MPa. The fiber preforms of UD-Tyranno SA fiber and P/W-Tyranno SA fiber were prepared by the slurry infiltration process, in which the fiber was compacted by different magnitudes of cold pressure (3.5, 7.0, 10.5 MPa), after injecting the matrix slurry of SiC (0.03, 0.3, 0.1 μm) and C powders under a low impregnation pressure of 0.9 MPa. MI-SiC$_f$/SiC composites were fabricated by infiltrating molten Si into each preform under a vacuum atmosphere. The melt infiltration temperature and holding time of all composites were 1450 ℃ and 2 hours, respectively.

The density of composites was determined by the Archimedes' method. The microstructures of composites were analyzed using SEM with EDS. The TEM analysis was also conducted to identify unreacted C and residual Si phases in the matrix. The mechanical properties of all composites were evaluated at room temperature using three point bending test. The dimensions of the test sample were $2 \times 4 \times 25$ mm^3. The dimensions of the test sample for Hi-Nicalon/SiC composites were $1 \times 4 \times 25$ mm^3. The span length and the crosshead speed were 18 mm and 0.5 mm/min, respectively.

RESULTS AND DISCUSSION

Density and microstructure

Figure 1 shows the density of MI-SiC$_f$/SiC composites depending on the

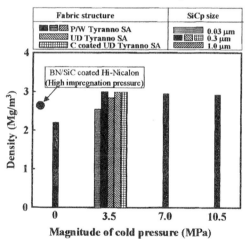

Figure 1 Density of MI-SiC$_f$/SiC composites depending on the preparation routes of fiber preform.

preparation routes of fiber preform associated with matrix slurry infiltration methods, staring SiC powder sizes, and fabric structures. The slurry infiltration process of low impregnation pressure and different magnitudes of cold pressure for the preparation of fiber perform provided a sufficient density and a sound morphology for MI-SiC$_f$/SiC. Especially, the P/W-Tyranno SA/SiC showed a good density (about 3.0 Mg/m^3), which was higher than that of Hi-

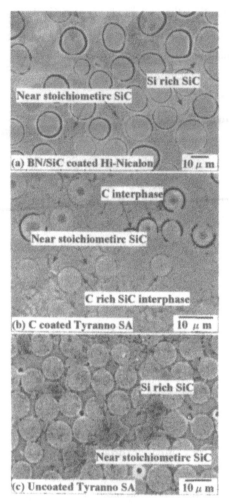

(a) BN/SiC coated Hi-Nicalon   10 μ m

Near stoichiometirc SiC

Si rich SiC

C interphase

Near stoichiometirc SiC

C rich SiC interphase

(b) C coated Tyranno SA   10 μ m

Si rich SiC

Near stoichiometirc SiC

(c) Uncoated Tyranno SA   10 μ m

Figure 2 Intra-fiber bundle microstructure of MI-SiC$_f$/SiC composites.

Nicalon/SiC fabricated by the high pressure impregnation (about 2.8 Mg/m^3), when the fiber preform containing starting SiC particle of 0.3 μm was prepared by the slurry infiltration process of an impregnation pressure of 0.9 MPa and a cold pressure of 3.5 MPa. However, the density of P/W-Tyranno SA/SiC had a similar level with the increase of cold pressure magnitudes. The induction of cold pressures higher than 3.5 MPa also led to some amount of large pores at the intersections of P/W fabrics.

Figure 2 shows the intra-fiber bundle microstructure of BN/SiC coated Hi-Nicalon/SiC, C coated UD-Tyranno SA/SiC, and uncoated P/W-Tyranno SA/SiC composites. The fiber performs of Hi-Nicalon/SiC and Tyranno SA/SiC were prepared by the high impregnation pressure of 6.0 MPa and the slurry infiltration process of an impregnation pressure of 0.9 MPa and a cold pressure of 3.5 MPa, respectively. The composition of each portion depicted in the figure was identified by the EDS analysis. MI-SiC$_f$/SiC showed a dense SiC matrix with some pores, due to the infiltration of molten Si into matrix pores or openings, even if there was severe degradation of fibers in Hi-Nicalon/SiC. However, a large amount of Si rich SiC phases with a Si/C ratio of about 1.7 were greatly created in intra-fiber bundles, compared to that of near stoichiometric SiC

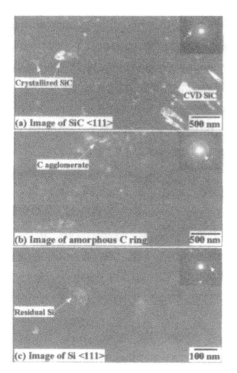

Figure 3 TEM micrographs and electron diffraction patterns for the matrix region of Hi-Nicalon/SiC composites.

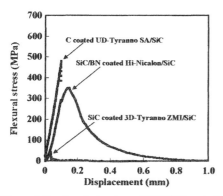

Figure 4 Fracture behaviors of BN/SiC coated Hi-Nicalon/SiC, C coated UD-Tyranno SA/SiC and SiC coated Tyranno ZMI/SiC composites.

phases with the ratio of Si/C around 0.9. In the case of C coated UD-Tyranno SA/SiC, majority of C interfacial layers were also transformed into C rich SiC phases, which resulted form the reaction of molten Si and C phases.

Figure 3 shows TEM micrographs for the matrix region of Hi-Nicalon/SiC composites. The fine β-SiC phases smaller than starting SiC particle (4.0 µm) used for the matrix slurry was created in the matrix. Moreover, there were clearly amounts of unreacted C phase and residual Si phase in the matrix. Such a presence of unreacted C and residual Si phases is considered to lead to the compositional fluctuation of the MI-SiC matrix. Therefore, it can be concluded that the matrix of MI-SiC$_f$/SiC is composed of at least three different phases, crystallized β-SiC, unreacted C and residual Si phases.

Mechanical properties

Figure 4 shows the fracture behaviors of MI-SiC$_f$/SiC composites. Hi-Nicalon/SiC and Tyranno ZMI/SiC showed noncatastrophical behavior. However, C coated UD-Tyranno

SA/SiC exhibited a typical brittle fracture behavior, in spite of the presence of C interfacial layer. The mechanical properties of MI-SiC$_f$/SiC composites depending on the preparation routes of fiber preform are summarized in Table 1. The slurry infiltration process by the combination of low impregnation pressure and cold pressing for the preparation of fiber perform improved the mechanical properties of Tyranno SA/SiC, even if their mechanical properties greatly decreased with the increase of cold pressure magnitudes, due to large scale of matrix pores at the intersection of P/W fabrics. Especially, in the case of fiber preform with an impregnation pressure of 0.9 MPa and the cold pressure of 3.5 MPa, the flexural strength of uncoated P/W-Tyranno SA/SiC and C coated UD-Tyranno SA/SiC represented about 500 MPa, which were superior to those reinforced by BN/SiC coated Hi-Nicalon fiber and SiC coated Tyarnno ZMI fiber. The drastic property degradation of Tyranno SA/SiC by the addition of starting SiC particles of 0.03 μm is caused by the formation of large amount of matrix cracks or debondings and non-homogeneous matrix.

Figure 5 shows the fracture profiles of MI-SiC$_f$/SiC composites. BN/SiC coated Hi-Nicalon/SiC and SiC coated Tyranno ZMI/SiC mainly displayed pull-out of fibers and interfacial delamination, which led to the noncatastrophical fracture behavior. However, the severe damage of Tyranno ZMI fibers by the molten Si was clearly observed. This is considered as a main factor to decrease the mechanical properties of Tyranno ZMI/SiC. Majority of Tyranno SA fibers, in

Table 1 Mechanical properties of MI-SiC$_f$/SiC composites.

Reinforcement	Matrix slurry		Interphase	Fiber volume fraction (%)	Slurry impregnation step (MPa)	Flexural Strength (MPa)	Elastic Modulus (GPa)
	SiC (μ m)	C (nm)					
Braided Hi-Nicalon	4.0		BN/SiC	30	HIP (6.0)	330.5	247.0
3D-Tyranno ZMI	0.3		SiC	48.3	LIP (0.9)	68.9	71.9
2D-Tyranno SA	0.3		-	15	LIP (0.9)	150.9	110.5
	0.3		-	15	LIP (0.9) + CP (3.5)	497.7	318.4
	0.3	85	-	15	LIP (0.9) + CP (7.0)	340.2	191.5
	0.3		-	15	LIP (0.9) + CP (10.5)	335.7	169.2
	1.0			15	LIP (0.9) + CP (3.5)	382.7	172.4
	0.03			15		178.5	134.6
UD-Tyranno SA	0.3		-	10	LIP (0.9) + CP (3.5)	398.2	247.0
	0.3		C	10		504.6	275.8

(HIP: High Impregnation Pressure, LIP: Low Impregnation Pressure, CP: Cold Pressure)

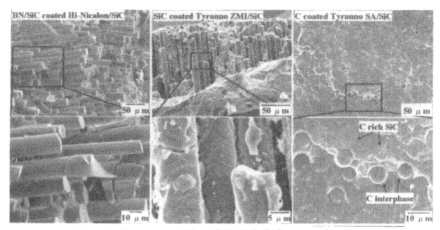

Figure 5 Fracture profiles of MI-SiC$_f$/SiC composites

which the C interfacial layer was transformed into C rich SiC phase during the MI process, mainly showed a fracture profile without pull-out of fibers and interfacial delamination. On the contrary, small amount of Tyranno SA fibers with pure C interfacial layer represented fiber pull-out phenomena.

CONCLUSIONS

1. The slurry infiltration process by the combination of low impregnation pressure and cold pressing for fiber preform preparation was effective for improving the densities and the mechanical properties of MI-SiC$_f$/SiC. The proper condition of slurry infiltration process can be selected as an impregnation pressure of 0.9 MPa and a cold pressure of 3.5 MPa.

2. The density and the fracture strength of uncoated P/W-Tyranno SA/SiC and C coated UD-Tyranno SA/SiC showed about 3.0 Mg/m^3 and about 500 MPa, respectively, which were higher than those of BN/SiC coated Hi-Nicalon/SiC fabricated by the high pressure slurry impregnation.

3. MI-SiC$_f$/SiC represented the chemical fluctuation in the intra-fiber bundle matrix, because of the coexistence of near stoichiometric SiC, Si rich SiC, residual Si and unreacted C phases.

4. C interfacial layer around SiC fiber can be selected as an alternative interphases

for MI-SiC$_f$/SiC. However, the reaction of molten Si and C interfacial layer must be controlled to avoid its transformation into C rich SiC phase.

ACKNOWLEDGEMENT

This study was supported by the CREST-ACE program sponsored by Japan Science and Technology Corporation (JST).

REFERENCES

[1] A. Hasegawa, A. Kohyama, R. H. Jones, L. L. Snead and P. Fenici, "Critical issues and current status of SiC/SiC composites for fusion", *Journal of Nuclear Materials*, **283-287** 128-137 (2000).

[2] D. Brewer, "HSR/EPM combustor material development program", *Materials Science & Engineering*, **A261** 284-291 (1999).

[3] Y. Katoh, M. Kotani, H. Kishimoto, W. Yang and A. Kohyama, "Properties and radiation effects in high-temperature pyrolyzed PIP-SiC/SiC", *Journal of Nuclear Materials*, **289** 42-47 (2000).

[4] H. Araki, W. Yang, S. Sato, T. Noda and A. Kohyama, "Bending properties of CVI composites at elevated temperatures, *Ceramic Engineering and Science Proceedings*. **20** [4] 371-377 (1999).

[5] T. Yano, Y. Yoshida and T. Seki, "Fabrication of silicon carbide fiber-reinforced silicon carbide composite by hot-pressing", *Fusion Engineering and Design*, **41** 157-163 (1998).

[6] T. Kameda, S. Suyama, Y. Itoh and Y. Goto, "Development of Continuous SiC Fiber-Reinforced Reaction Sintered SiC matrix Composites", *Journal of the Ceramic Society of Japan*, **107** [4] 327-334 (1999).

[7] K. Yoshida, M. Imai and T. Yano, "Improvement of the mechanical properties of hot-pressed silicon-carbide-fiber-reinforced silicon carbide composites by PCS impregnation", *Composite Science and Technology*, **61** 1323-1329 (2001).

[8] M. Kotani, A. Kohyama and Y. Katoh, "Development of SiC/SiC composites by PIP in combination with RS", *Journal of Nuclear Materials*, **289** 37-41 (2001).

Advanced SiC/SiC Ceramic Composites

Processing for SiC/SiC Composite Contituent

MECHANICAL, THERMOCHEMICAL AND MICROSTRUCTURAL CHARACTERIZATION OF AHPCS-DERIVED SiC

Leonard V. Interrante,
Kevin Moraes
Department of Chemistry,
Rensselaer Polytechnic Institute
Troy, NY 12180-3590, U.S.A.

Leo MacDonald, Walter Sherwood
Starfire Systems, Inc.,
877 25th St., Watervliet, NY 12189,
U.S.A.

ABSTRACT

Previous studies of the polycarbosilane, "HPCS", and its partially allyl-substituted derivative, AHPCS, have indicated a hyperbranched molecular structure consisting of a mixture of terminal $-CH_2SiR_3$, and internal $-CH_2SiR_n(CH_2-)_{3-n}$ (n = 0,1,2) units ($[R_3SiCH_2-]_x[-SiR_2(CH_2-)]_y[-SiR(CH_2-)_{1.5}]_z$ $[-Si(CH_2-)_2]_1$ (R = H or allyl ($-CH_2CH=CH_2$)). This hyperbranched, liquid, polycarbosilane undergoes crosslinking above ca. 250°C via loss of H_2 from the $-SiH_n$ (n = 2,3) groups and, in the case of AHPCS, through hydrosilation of the allyl ($-CH_2CH=CH_2$) groups to form a hard, glassy solid. Further heating to 1000°C produces an amorphous "SiC" which then undergoes conversion on further heating to 1600°C to produce a partially (nano)crystalline β-SiC ceramic. The conversion of the amorphous "SiC_xH_y" material obtained from AHPCS by partial pyrolysis to 800-1000°C has now been studied by means of DSC/TGA, solid state NMR, Raman, XRD, and TEM. Moreover, the polymer has been processed into monolithic notched-beam specimens suitable for hardness and toughness measurements. The results of some of these studies are summarized, and examples of composite ceramic objects obtained by using AHPCS as a SiC-matrix source are provided.

INTRODUCTION

In 1991, we reported the preparation of a hyperbranched carbosilane having the composition, SiCH$_4$, which was given the name "HPCS" (HydridoPolyCarboSilane) [1]. This carbosilane was obtained in good yield (60-70%) from chloromethytrichlorosilane (ClCH$_2$SiCl$_3$, ClMTS) by Grignard coupling with Mg, followed by reduction with LiAlH$_4$, as a mixture of oligomers and low molecular weight polymer, with an average degree of polymerization of ca. 15. The ClMTCS starting material can, in turn, be obtained from CH$_3$SiCl$_3$ (MTS) in high yield by chlorination, resulting in a low materials cost for this carbosilane. In addition to the identification of its hyperbranched structure by NMR spectroscopy, initial studies of this carbosilane demonstrated its potential utility as a precursor to essentially stoichiometric SiC, although due to its relatively low average molecular weight and tendency to partially vaporize before crosslinking, the char yield for this precursor was relatively low (50-60%). Subsequent studies revealed the potential for structural modification of the intermediate chloro('poly')carbosilane through reaction of the Si-Cl groups and led to the preparation of a partially allyl-substituted ('poly')carbosilane, "AHPCS" [2], as well as to alkoxy-substituted derivatives that could be converted, by sol-gel processing, to a [Si(O)CH$_2$]$_n$ gel and subsequently, through pyrolysis, to a high surface area silicon oxycarbide [3]. Various other types of derivatives were also obtained in later work that confirmed the hyperbranched ([R$_3$SiCH$_2$-]$_x$[-R$_2$SiCH$_2$-]$_y$ [-RSi(CH$_2$-)$_{1.5}$]$_z$[-Si(CH$_2$-)$_2$]$_1$) core structure for this carbosilane system [4]. Since 1995, through further improvements in the synthesis, which made it possible to control the molecular weight distribution, composition and degree of branching of these carbosilanes, AHPCS and subsequently HPCS, have become commercially available as, respectively, high molecular weight, high yield (ca. 80%) polymeric and volatile CVD precursors for SiC [5]. These precursors have been further studied by other groups [6] and are currently under evaluation worldwide as a source of hard, protective coatings for various substrates, CVI SiC sources (HPCS), binders for ceramic powders, and for Polymer Infiltration and Pyrolysis (AHPCS) [6,7]. Additionally, AHPCS has been used to bond monolithic and composite ceramics [6]. Examples of some of the various kinds of ceramic parts that have been obtained by using AHPCS are shown in Figure 1. The mechanical properties of the fiber-reinforced, SiC matrix composites that have been obtained by using AHPCS are comparable to, or better than, those obtained by using CVI, while the use of a relatively air-stable, controllable viscosity, liquid as a SiC precursor permits application of more conventional, lower cost, approaches to composite fabrication.

These liquid, relatively air-stable, carbosilanes yield a near-stoichiometric, amorphous "SiC" on pyrolysis to 1000°C, and crystallize to nanocrystalline β-SiC by 1600°C. The pyrolysis chemistry of HPCS has been elucidated by means of multinuclear Solid State NMR (SSNMR) spectroscopy, TGA, DSC, IR, and evolved gas analysis [2a] and found to behave in a quite similar manner on pyrolysis to its structurally related and·compositionally

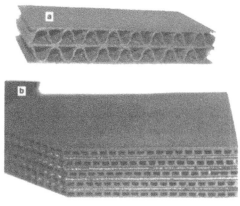

Figure 1: Examples of parts (heat exchangers) fabricated (by Starfire Systems, Inc.) by using AHPCS; (a) Counter flow heat exchanger (Sylramic/SiC); (b) Compact cross-flow heat exchanger (C/SiC).

equivalent linear polymer analog, polysilaethylene $[SiH_2CH_2]_n$ (PSE) [8]. Both polymers evolve H_2 above ca. 300°C by a 1,1-elimination process to form a crosslinked network, which then rearranges, while eliminating further H_2 from both SiH_x (x=1,2,3) and CH_x groups, between ca. 450 and 1000°C, to an amorphous "SiC" structure [8]. The addition of as little as 5-10 mol% of allyl groups (i.e., AHPCS) was found to increase the char yield considerably (up to ca. 80% for 10%-AHPCS), presumably due to thermally-induced crosslinking via hydrosilation of the terminal olefin groups. In the case of PSE, the "amorphous" SiC obtained on pyrolysis shows signs of crystalline order (by XRD and TEM) even at 1000°C, whereas for AHPCS, the small amount of excess carbon effectively inhibits crystallization to somewhat higher temperatures, and even after a 1600°C annealing, the crystallite sizes are limited to < 20 nm. We have now completed further microstructural and mechanical property measurements on the AHPCS-derived SiC and report here the results of preliminary findings from this study.

EXPERIMENTAL SECTION
Simultaneous DSC-TGA
    Samples for DSC/TGA were prepared as follows: A 5% AHPCS sample (nominally $[Si(CH_2CH=CH_2)_2CH_2]_{0.05}[SiH_2CH_2]_n)_{0.95}$) obtained from Starfire Systems, Inc. [5], was cured at 400°C for 0.5 h and partially pyrolyzed by heating at 3 °C/min to 600°C under nitrogen. This was then ground to a fine powder, using a boron carbide mortar and pestle. The fraction that passed through a 200 Mesh ASTM screen and was retained above a 230 Mesh screen (63-75 µm) was used as the starting material. This sample was heat treated by

heating at 800, 850, 900, 950 and 1000°C, as follows. Approximately 1 g of sample was heated under argon at 3 °C/min to the heat treatment temperature, held there for 4 h and cooled at 3 °C/min to room temperature.

These samples that were heat treated to between 800°C and 1000°C were used as the starting material for further study by DSC-TGA. Weight loss and phase transitions occurring in the AHPCS amorphous material were studied by using high temperature simultaneous DSC and TGA. Simultaneous measurements of both the temperatures and heat flow associated with transitions as well as weight changes in a sample as a function of temperature in an Argon atmosphere were measured on an TA instruments model SDT 2960, simultaneous DSC-TGA (TA Instruments, New Castle, DE).

Transmission Electron Microscopy

The same 600°C powder (63-75 μm) used above was heat treated to 1000, 1200, 1400 and 1600°C. Heat treatment was done under Argon at 3 °C/min to the heat treatment temperature, held there for 4 hr and cooled at 3 °C/min to room temperature. Transmission electron diffraction patterns were obtained using a Philips CM-12 operated at 120 kV.

Mechanical Property (Fracture Toughness and Vickers hardness) Studies of Monolithic Samples

For the mechanical property studies, the fraction of the 600°C partially pyrolyzed 5% AHPCS material that passed through an ASTM 200 mesh screen (<75 μm) was used in the subsequent steps. This material was heated to 1000°C under nitrogen to complete the conversion to amorphous SiC and then mixed with 5-7% (by wt.) of the same AHPCS precursor, cold pressed into pellets (dimensions 75 x 12 x 4 mm) and pyrolyzed to 1000°C under nitrogen. Further infiltration and pyrolysis cycles were used to fully densify these samples. The full details of the preparation and subsequent heat treatment of these samples, as well as the methods used to measure their Vickers hardness and toughness have been previously described [9].

RESULTS

DSC/TGA and TEM measurements

The DSC results from the simultaneous DSC-TGA runs for a series of powder samples that were preheated to various temperatures are shown in Figure 2. The simultaneously determined TGA curves obtained for these samples (not shown) indicate only a very slight mass loss (<1%) in this temperature range.

Electron diffraction patterns of the precursor derived material heat treated to various temperatures are shown in Figure 3. The diffraction patterns of samples heated to 1200°C appear quite amorphous with only a broad, amorphous, halo seen where a ring corresponding to the [111] β-SiC plane would be expected. However the sample heated to

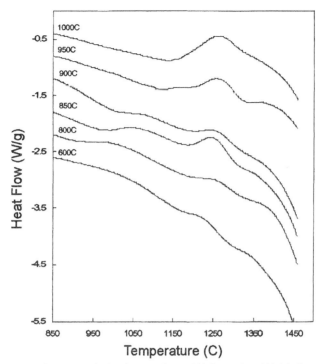

Figure 2. DSC traces of AHPCS derived amorphous SiC heated at 10°C/min under an Argon atmosphere. The lowest trace is from the sample heated to 600°C for 0.5 hr. The top five traces are from the AHPCS-SiC that has been heat treated at 1000, 950, 900, 850 and 800°C (from top to bottom) for 4 hrs under Argon. The origins of the curves have been shifted vertically for clarity.

1400°C and above appears crystalline with rings corresponding the [111], [220] and [311] planes of β-SiC clearly visible. Solid State NMR (SS NMR), XRD and TEM studies [13] also indicate that below 1200°C AHPCS derived SiC is amorphous and above this temperature it is partially crystalline.

The DSC curves for the different samples show a sharp exothermic transition in the material at ca. 1250-1260°C with no associated sharp change in the TGA curve. There is also an underling broad exotherm prior to this sharp exotherm. The onset temperature of this broad peak is dependent on the heat-treatment temperature and shifts to higher temperatures with higher heat-treatment temperatures. The DSC curves also show evidence for an endothermic process which starts at ca. 1400°C. As is discussed later, this probably corresponds to the decomposition of the oxycarbide phase in the AHPCS derived SiC and

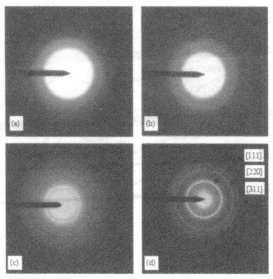

Figure 3. Electron diffraction patterns for AHPCS-SiC obtained on powder samples heated to various temperatures; (a) 1000°C, (b) 1200°C, (c) 1400°C (d) 1600°C.

the loss of SiO and CO gaseous species [10].

Elemental Analysis and Microstructure of Monolithic Samples

The composition of the 1000°C powder that was used for all sample preparations was found to be $SiC_{1.13}O_{0.28}$ (Wt.% C, Si, O = 27.7, 57.3, 9.1; Calc'd for SiC = 29.9, 70.1, 0). The C:Si atomic ratio is similar to those previously found for AHPCS-silicon carbide [7]; however the oxygen content in the sample is considerably higher. Similar values were also obtained from SS NMR studies of these samples (to be reported elsewhere [11]) through integration of the intensities of the various ^{29}Si regions contributing to the Si NMR spectrum. These higher values may be due to adsorbed water or to surface oxidation of the samples during the grinding, and subsequent handling, of the 600°C partially pyrolyzed material in air.

Monolithic samples for mechanical property measurements were prepared by the PIP (Polymer Infiltration and Pyrolysis) process, which was intended to simulate the formation of the SiC matrix material in a fiber-, or particulate-reinforced composite [7]. This involved cold pressing a mixture of the 1000°C pyrolyzed AHPCS-powder with AHPCS, followed by pyrolysis again to 1000°C and 10 subsequent PIP cycles [9]. This process produced specimens with low (<2%) open porosity in the case of the 1000, 1200 and 1400°C specimens. However, reinfiltration of the specimens with AHPCS after heat treatment was not effective in reducing the porosity of the sample prepared at 1600°C, in which the open porosity was found to have increased to as much as 9%. The Archimedes densities for these samples varied from 2.3 to 2.9 g/cm^3 for the 1000°C to 1600°C samples, depending on the time at the final temperature (1 vs. 4 h) as well as the final processing temperature. Optical microscope images of polished cross-sections for the 1200 and 1600°C samples after the final infiltration and pyrolysis indicate a fairly uniform microstructure for both samples, with occasional small pores and grains up to ca. 80 μm in size for the 1200°C sample and considerably larger, and more frequent, pores for the 1600°C sample. The grains in both of these samples appear to be of about the same size and shape as the powder particles that were initially used to prepare these samples and are distinguishable from the surrounding matrix material that presumably came from the added polymer. These particles appear, in these images and in the SEM images obtained of the fracture cross-sections, to be well-bonded to the surrounding matrix, forming an essentially continuous microstructure.

Fracture Toughness

For all of the specimens tested a smooth fracture surface was observed, with no obvious fragmentation of the sample, or particle pull-out, accompanying fracture. The observed fracture toughness increased from an average of 1.43±0.08 MPam$^{1/2}$ for the 1000°C samples to 1.65±0.09 MPam$^{1/2}$ for those heated to 1200°C and 1.67±0.07 MPam$^{1/2}$ for the 1400°C samples, and then dropped to 1.46±0.08 MPam$^{1/2}$ after the 1600°C treatment [12]. The fracture toughness of the samples heated to 1200°C was significantly higher ($p \leq 0.05$) than those heated to 1000°C. There is no statistical difference ($p > 0.05$) in the measured toughness of the 1200 and 1400°C samples. However, the porosity of these samples, while remaining fairly constant at < 2% for the first three samples, was observed to be considerably higher for the 1600°C sample (9 vol%).

DISCUSSION

DSC-TGA

It appears from the DSC curves that there are at least two separate exothermic processes that occur when AHPCS-SiC is heat-treated from 800 to 1450°C. A comparison of the TGA curves obtained simultaneously does not suggest any appreciable weight loss. However, this does not preclude the possibility of a small amount of volatile gas (e.g. H_2) produced in this temperature range. Indeed, a gradual H_2 loss between 600 and 1000°C was evidenced by mass spectrometry in the case of the closely related PSE ($[SiH_2CH_2]_n$) [8]. The second narrow peak at 1250 –1262°C is attributed to the crystallization of the amorphous silicon carbide, as was evidenced by the dramatic sharpening of the electron diffraction pattern (Figure 2) and the ^{29}Si NMR peaks between 1200 and 1350°C.

From electron diffraction (Figure 2) and ^{29}Si NMR spectroscopy [11] it is apparent that AHPCS-derived SiC is amorphous at temperatures below 1200°C. The broad peak that begins at lower temperatures is due to some transformation of the amorphous phase and thus is not directly linked to the crystallization of the amorphous matrix into SiC. It is also apparent that the onset temperature for this process shifts to higher temperatures with increasing heat treatment temperature. This initial exothermic release appears irreversible, i.e. if the thermal history of the sample already includes a heat treatment to 900°C, this thermal transition occurs only at a temperature above 900°C. All of this evidence leads to the conclusion that this initial exothermic release seen in the DSC corresponds to a one-time, irreversible process that occurs in the amorphous sample when it is first heated.

Several possible causes for this broad exotherm in addition to the SiC crystallization exotherm have been considered: (i) crystallization or phase transformation of phases other than SiC, (ii) structural relaxation, (iii) chemical condensation. Both crystallization and phase transformation are first order transformations. From ^{29}Si NMR spectroscopy [11] and electron diffraction results, early crystallization of the SiC can be ruled out as a reason for the low-temperature exotherm. The ^{29}Si NMR spectra also do not show any peaks that are characteristic of Si-Si bonded species, such as would likely be seen in the case of crystallization of Si phases. Crystalline SiO_2 is also not seen in the ^{29}Si NMR spectra nor expected under these conditions (below 1000°C). Raman spectroscopy (reported elsewhere [13]) does show evidence for 'crystalline' carbon. However its formation is by the process of transformation of previously formed sp^3 carbon to graphite-like sp^2 carbon. The relatively small amount of excess carbon and the negligibly small enthalpy change that is typically observed for this transformation (the standard heat of formation of diamond(sp^3)-like carbon being only slightly greater (1.9 kJ/mol [14]) than that of graphite(sp^2)-like carbon), make it quite unlikely that it would contribute appreciably to the observed exotherm. Similarly, although there is some oxygen present in these samples (as Si-O-), and it is known that Si-O/Si-C bond rearrangement processes occur in this temperature range [3], such rearrangement processes, while possibly facilitating structural relaxation and the eventual crystallization of SiC, should be essentially thermoneutral, and therefore should not contribute substantially to this exotherm.

Structural relaxation of the amorphous AHPCS derived SiC network could, in part, explain the irreversible exotherm prior to crystallization. These structural relaxations may involve the local breaking and remaking of bonds, a reduction in the average bond angle strain and bond angle fluctuations, and a possible evolution in the average structure through a succession of nonequilibrium states to one that approaches the metastable equilibrium state at that temperature. To the best of our knowledge, this relaxation behavior has not been reported in polymer precursor derived silicon containing ceramics. However such an exotherm prior to crystallization has been reported in other systems. It is most evident in amorphous materials that have been formed such that they are far from their equilibrium structure. For example it has been observed in melts of ionic [15] and metallic [16] glasses that have been rapidly cooled. It has also been reported for isomorphs of SiC, such as in amorphous Ge [17,18] and Si [19,20,21]. The structural relaxation in these cases is irreversible and occurs over a wide temperature range up to the crystallization temperature. Moreover, it was observed that heating the amorphous sample to a temperature below the crystallization temperature caused the structure to partially relax, with further relaxation only occurring at higher temperatures.

It is not hard to see that the amorphous material prepared from the AHPCS polymer precursor would have a structure that is very far from its equilibrium structure and would undergo structural relaxation to reduce its internal energy. In fact, we see evidence for this in the change in the ^{29}Si NMR spectra [11], where the peaks are seen to narrow with increasing heat treatment temperature from 1000 to 1600°C. However the shift observed in the ^{29}Si NMR peak position indicates that in addition to structural relaxation, the chemical composition is also changing in this temperature range.

Mass spectrometric analysis of the gas evolved during the pyrolysis of polycarbosilanes such as PSE and HPCS shows that although most hydrogen evolution occurs at temperatures below 800°C a significant amount of hydrogen evolution occurs above 800°C [8]. The loss of this hydrogen results in further condensation of the amorphous structure. Because of the relatively high H_2 bond energy, relative to that of other E-H bonds, such condensation reactions are typically exothermic [22].

Although this exothermic process of relaxation and condensation of the amorphous network prior to crystallization has not been reported in the case of polymer derived silicon carbide, it has been observed in hydrogenated polymorphs of SiC like a-C:H [23], a-Si:H [24,25] and a-Ge:H [23] and also in hydrogenated amorphous silicon carbide prepared by a sputter assisted plasma chemical vapor deposition [26,27], and in a-SiC deposited by glow discharge [28]; both from a mixture of silane and methane gas. In these cases the authors were able to detect two separate exotherms that correspond to structural relaxation and condensation, the latter occurring concomitantly with the loss of hydrogen. Two distinct exothermic events for the amorphous AHPCS-derived SiC could not be detected because they occur simultaneously. This is unlike the case of the CVD-deposited silicon carbide [26,28] where the hydrogenated SiC is formed at temperatures where the Si-H and C-H bonds are stable and structural relaxation can occur and be observed by DSC, before the onset of decomposition and condensation of the SiC:H.

For the case of only structural relaxation in a material, it is seen that the relaxation exotherm ends at the crystallization temperature. This is because when only structural relaxation is involved, at the crystallization temperature the structure can relax quickly enough to be completely relaxed during the course of the experiment. This has been observed in the case of amorphous Si [19] and Ge [18]. However, in the case of AHPCS derived SiC, the broad exotherm corresponding to relaxation and condensation continues beyond the crystallization exotherm. It is seen that the SiC crystallization exotherm is superimposed over the broad exotherm due to relaxation and condensation of the amorphous network. This is yet another reason to expect the existence of chemical condensation in addition to structural relaxation. Here, the two processes (relaxation-condensation and crystallization) occur simultaneously, with crystallization beginning in some regions even before the relaxation-condensation process is complete in the rest of the amorphous SiC.

At this point it is not clear what implications this might have for the mechanism of crystallization - but it does raise some important questions on the nature of nucleation and growth of crystalline SiC from the amorphous, precursor-derived, SiC. For example, does nucleation occur in the hydrogen depleted, relaxed, regions of the amorphous SiC? If this were the case, would the number of nuclei formed increase by relaxing the structure completely prior to crystal growth at higher temperatures? This understanding might allow for a process to produce fine-grained microstructures with better tensile strength required in fibers or coarse microstructures with better creep properties necessary for the matrix, as required.

Finally, there is growing evidence from studies of related polymer precursor derived ceramics that decomposition of the polymer precursor derived amorphous matrix and crystallization occurs at about the same time [29]. This is seen in the SiC-O, SiC-N and SiCBN systems where they remain amorphous to temperatures at which the amorphous matrix does not decompose. On heating to higher temperatures, crystallization is seen to occur with the decomposition of any unstable phase. At this point it is not clear if phase separation due to crystallization results in decomposition of the amorphous polymer precursor derived matrix, or decomposition of the amorphous matrix causes crystal growth.

Microstructure and Mechanical Properties of the Monolithic Samples

In the case of the monolithic specimens prepared for toughness measurements, heat treatment above 1000°C resulted in an increased shrinkage and increase in Archimedes density of the specimens. The strategy of interrupting the heat treatment with reinfiltration cycles resulted in low levels of open porosities in specimens heat-treated to 1200 and 1400°C. The specimens heat-treated at 1600°C for 1 h suffered a much larger weight loss and increase in porosity compared to the specimens heat treated at lower temperatures. After 4 h of heating at 1600°C (including three intermediate cycles of reinflitration with AHPCS) this porosity increased to 9%, while the density at both 1600 and 1400°C also increased slightly. This increase in porosity is probably due to the decomposition of the oxycarbide phase present in the specimens [10]. The oxycarbide phase is unstable above 1450°C and decomposes to SiO and CO gas. However the rate of decomposition is quite slow even at 1600°C. Considering that the samples prepared for mechanical testing were heat treated at 1600°C for 4 h, a portion of this oxycarbide phase would have decomposed, resulting in an increase in porosity compared to the samples heat-treated at lower temperatures. In addition, shrinkage accompanying the increase in density at higher temperatures may account for some of the increase in open porosity.

The most notable observation from the mechanical property measurements on these samples is the generally lower value of toughness in the precursor-derived material (1.5 - 1.7 MPam$^{1/2}$), compared to those reported in the literature for crystalline sintered silicon carbide samples (between 3 – 6 MPam$^{1/2}$ [30,31]). These K_{1c} values are also lower than the toughness of 2.5 MPam$^{1/2}$ reported by Kodama and Miyoshi on a sintered polycarbosilane derived SiC (<1% porosity) processed to 1800 °C with an average crystallite size of 220 nm [32], and the value of 2.1 MPam$^{1/2}$ reported by T. Nishimura et. al. [33] on a polymer precursor derived Si-C-N material processed to 1000°C (9% porosity). However, these latter measurements were performed by Vickers indentation and the authors suggested that the true toughness must be lower than the 2.1 MPam$^{1/2}$ reported.

The lower toughness measured for the 1600°C samples, relative to the ones processed at lower temperatures, can be at least partially attributed to the much higher porosity in these samples. It is well known that porosity reduces fracture toughness. To determine the effect of heat treatment temperature on fracture toughness the effect of this porosity must be accounted for. Rice and other [34] have shown that the intrinsic toughness can be related to the measured toughness as,

$$\frac{K_{1c}}{K_{1c}^{o}} = e^{-b \bullet P} \tag{1}$$

Where, K_{1c}, and K_{1c}^{o} are the measured and intrinsic (zero porosity) toughness, P is the porosity and b is a charateristic parameter that depends on pore stacking.

Rice [35] has summarized, and others [36,37] have reported, b values for fracture toughness ranging from 2 to 5.2 for a number of ceramics fabricated by different methods. The value of the parameter b has not been determined for AHPCS-SiC. However, for the purposes of identifying a trend, if any, even if we assume a conservative value of 2 for b, a significantly higher value of toughness for the AHPCS-SiC heat-treated to 1600°C is obtained, as is shown in Table 2. Thus, it appears that, as the density of the AHPCS-SiC increases, the intrinsic toughness increases. This is consistent with the observation by Kodama and Miyoshi that in sintered, polycarbosilane-derived, SiC, fracture toughness increases with increasing density [32].

CONCLUSIONS

Amorphous AHPCS-SiC produced by pyrolysis below 1000°C undergoes two distinct exothermic events in the DSC on heating: a broad one that corresponds to structural relaxation and chemical condensation and a second, much sharper, one that corresponds to crystallization. This relaxation-condensation process is irreversible and occurs before, during, and even after, the onset of crystallization. The onset of the relaxation-condensation process depends on the initial pyrolysis temperature (held for 4 h) and was typically 100 to 175°C higher than the pyrolysis temperature. The magnitude of the heat released in this process is largest for the sample that was only partially pyrolyzed to ca. 600°C; this may relate to the amount of hydrogen remaining in the sample, which is known to gradually decrease as the pryrolysis temperature increases [8]. In contrast, the peak corresponding to partial crystallization of the AHPCS-SiC occurs at ca. 1250-1260°C almost independent of the initial pyrolysis temperature. However, the magnitude of the exotherm associated with this crystallization process appears to increase as the initial pyrolysis temperature increases and, among the samples tested, was largest for the sample that was preheated to the highest temperature short of the crystallization temperature (the 1000°C pyrolyzed sample).

Cold pressing partially pyrolysed AHPCS-SiC polymer, followed by reinfiltration with the polymer to reduce porosity, was found to be an effective method to fabricate monolithic specimens of precursor-derived ceramics for conventional mechanical property testing. The fracture toughness (K_{1c}) of AHPCS-SiC for such samples heat treated for 4 hrs at 1000, 1200, 1400 and 1600°C (consisting of four cycles of heating to the indicated temperature for 1 h, reinfiltration with AHPCS and then reheating to this same final temperature) was found to be 1.40 ±0.08, 1.65 ±0.09, 1.67 ±0.07 and 1.46 ±0.08 MPam$^{1/2}$ respectively. Porosity in the samples was below 2% except for the 1600°C samples where it was 9%. The density of the samples increases with increasing heat treatment temperature. If corrected for porosity, it appears that the intrinsic fracture toughness may also increase with the heat treatment temperature.

ACKNOWLEDGEMENTS

The authors would like to thank Dr. Wei-Ping Pan of the Materials Characterization Laboratory, Western Kentucky University, Bowling Green, KY for the DSC/TGA data and the National Science Foundation for support under Grant No. CHE-9812191.

REFERENCES

[1]C. Whitmarsh and L.V. Interrante, "Synthesis and structure of a highly branched polycarbosilane derived from chloromethyltrichlorosilane", *Organometallics*, **10**, 1336-1344 (1991); C.W. Whitmarsh and L.V. Interrante, U.S. Patent No. No. 5,153,295, "Carbosilane polymer precursors to silicon carbide ceramics", 10/6/92.

[2]L.V. Interrante, C.W. Whitmarsh, W. Sherwood, H.-J. Wu, R. Lewis, and G. Maciel, "Hydridopolycarbosilane precursors to silicon carbide. Synthesis, pyrolysis and application as a SiC matrix source", NATO ASI Series E, Volume 297, J.F. Harrod and R.M. Laine, eds., Kluwer Academic Publishers, (1995) pp. 173-183; L.V. Interrante, J.M. Jacobs, W. Sherwood and C.W. Whitmarsh, "Fabrication and properties of fiber- and particulate-reinforced SiC matrix composites obtained with (A)HPCS as the matrix source", "Key Engineering Materials", Vols. 127-131 (1997), Transtec Publications, Switzerland, pp. 271-278.

[3]G.D. Soraru, Q. Liu, L.V. Interrante and T. Apple, "The role of precursor molecular structure on the microstructure and high temperature stability of silicon oxycarbide ceramics", *Chem. Mater.*, **10**, 4047 (1998).

[4]I. Rushkin, Q. Shen, S.E. Lehman, and L.V. Interrante, "Modification of a highly branched hydridopolycarbosilane as a route to new polycarbosilanes", *Macromolecules*, **30**, 3141-46 (1997).

[5]Starfire Systems, Inc., 877 25th St, Watervliet, NY, 12189; nominal composition "[Si(C$_3$H$_5$)$_x$H$_{1-x}$CH$_2$]", x = ca. 0.05.

[6]J. Zheng and M. Akinc, *J. Am. Ceram. Soc.*, **84**(11), 2479 (2001);); M.Z. Berbon, D.R. Dietrich, and D.B. Marshall, "Transverse thermal conductivity of thin SiC/SiC composites fabricated by slurry infiltration and pyrolysis", , *J. Am. Ceram. Soc.* **84**(10), 2229-2234 (2001); J. Zheng, S.P. Beckman, J.N. Gray, and M. Akinc, *J. Am. Ceram. Soc.* **84**(9), 1961 (2001); F.I. Hurwitz, *Ceram. Eng. Sci. Proc.*, **21**(4), 45 (2000); M.Z. Berbon and D.B. Marshall, *Ceram. Trans.*, **103**, 319 (2000); G.A. Danko, R. Silberglitt, P. Colombo, E. Pippel, J. Woltersdorf, *J. Am. Ceram. Soc.*, **83**(7), 1617 (2000); R.L. Bruce, S.K. Guharay, F. Mako, W. Sherwood, and E. Lara-Curzio, "Polymer-derived SiC$_f$/SiC$_m$ Composite Fabrication and Microwave Joining for Fusion Energy Applications", Proceedings of the 19th IEEE/NPSS Symposium on Fusion Engineering, Atlantic City, NJ, Jan. 22-25, 2002.

[7]L.V. Interrante, C.W. Whitmarsh, C-Y. Yang, and W. Sherwood, "Processing of Si-based ceramic using hydridopolycarbosilane (HPCS)", *Ceram. Trans.*, **42**, 57-69 (1994); L.V. Interrante, C.W. Whitmarsh, and W. Sherwood, "Fabrication of SiC matrix composites by liquid phase infiltration with a polymeric precursor", *Mater. Res. Soc. Sympos. Proc.*, **346**, 593-603 (1994); L.V. Interrante, C.W. Whitmarsh, and W. Sherwood,

"Fabrication of SiC matrix composites using a liquid polycarbosilane as the matrix source", *Ceram. Trans.*, **58**, 111-18 (1995); J.M. Jacobs, W.J. Sherwood, and L.V. Interrante, *Ceram. Eng. Sci. Proc.*, **21**(3), 289 (2000); L.V. Interrante, K. Moraes, W. Sherwood, J. Jacobs and C.W. Whitmarsh, "Low-cost, near-net shape ceramic composites by polymer infiltration and pyrolysis", Proceedings of the 8th Japan-U.S. Conference on Composite Materials, G.M. Newaz and R.F. Gibson, eds., Technomic Publishing Co., Inc., Lancaster, PA, (1998), pp. 506-515; K. Moraes, J.M. Jacobs, W.J. Sherwood, and L.V. Interrante, "The effect of particulate fillers on the processing and properties of a fiber-reinforced polymer precursor derived SiC-matrix composite", *Ceramic Engrg. and Sci. Proceedings*, **21**(3), 289-295 (2000).

[8]Q. Liu, H.-J. Wu, R. Lewis, G.E. Maciel, and L.V. Interrante, "Investigation of the pyrolytic conversion of poly(silylenemethylene) to silicon carbide", *Chem. Mater.*, **11**, 2038-2048 (1999).

[9]K. Moraes and L.V. Interrante, "Processing, fracture toughness and Vickers hardness of allylhydridopolycarbosilane-derived silicon carbide", *J. Amer. Ceram. Soc.*, in press (2002).

[10]R. H. Baney, K. Eguchi, G. A Zank, Soc. Chem. Japan, 8, 23, 1996; Soraru, G. D.; Suttor, D.; "High temperature stability of sol-gel-derived SiO_xC_y glasses", *J. Sol-Gel Sci. & Technol.*, **14**(1), 69-74 (1999).

[11] Christopher J. Wiegand, Ph.D. thesis, "NMR Studies of Novel Substituted Poly(silylenemethylene)s, Motion in Polysilaethylene, Allyl-hydridopolycarbosilane-derived Silicon Carbide and Inorganic PMMA Composites", Department of Chemistry, RPI, 2002.

[12]Each value represents the average of measurements carried out on 6-10 different samples; the value is followed by the standard deviation for each set of measurements.

[13]Kevin Moraes, Ph.D. thesis, "The Densification, Crystallization and Mechanical Properties of Allylhydridopolycarbosilane Derived Silicon Carbide", Department of Materials Engineering, RPI, 2000

[14]R. Chang, "Essentials of Chemistry", pg. 668, McGraw-Hill, New York (1996)

[15]C. T. Moynihan, A. J. Bruce, D. L. Gavin, S. R. Loerh, S. M. Opalka and M. G. Drexhage, "Physical aging of heavy metal fluoride glasses-Sub-T_g enthalpy relaxation in a ZrF_4-BaF_2-LaF_3-AlF_3 glass", *J. Poly. Eng. Sci.*, **24**, 1117-1122 (1984).

[16] J. Zhu, M. T. Clavaguera-Mora and N. Clavaguera, "Relaxation process of Fe(CuNb)SiB amorphous alloys investigated by dynamic calorimetry", *Appl. Phys. Lett.*, **70**, 1709-1711 (1997).

[17]H. S. Chen and D. Turnbull, "Specific heat and heat of crystallization of amorphous germanium", *J. Appl. Phys.*, **40**, 4214-4215 (1969).

[18]E. P. Donovan, F. Spaepen, D. Turnbull, J. M. Poate and D. C. Jacobson, "Calorimetric studies of crystallization and relaxation of amorphous Si and Ge prepared by ion implantation", *J. App. Phys.*, **57**, 1795-1804 (1985).

[19]S. Roorda, S. Doorn, W. C. Sinke, P. M. L. O. S. Scholte and E. van Loenen, "Calorimetric evidence for structural relaxation in amorphous silicon", *Phys. Rev. Lett.*, **62**, 1880-1883 (1989).

[20] S. Roorda, "Low temperature relaxation in amorphous silicon made by ion implantation", *Nucl. Instrum. Methods Phys. Res. B*, **148**, 366-369 (1999).

[21]K. H. Tsang, H. W. Kui and K. P. Chik, "Calorimetric studies of the heat capacity and relaxation of amorphous Si prepared by electron beam evaporation", *J. Appl. Phys.*, **74**, 4932-4935 (1993).

[22]CRC Handbook of Chemistry and Physics, 60th Edition, R.C. Weast, editor, CRC Press, Inc. (1979).

[23]L. Battezzati, F. Demichelis, C. F. Pirri, A. Tagliaferro and E. Tresso, "Investigation on structural changes in amorphous tetrahedral alloys by means of differential scanning calorimetry", *J. Non-Cryst. Sol.*, **137-138**, 87-90 (1991).

[24]B. G. Budaguan, A. A. Aivazov and A. Yu. Sazonov, "Calorimetric investigation of relaxation processes in disordered semiconductors", *Mat. Res. Soc. Symp. Proc.*, V **420**, 635-640 (1996); B. G. Budaguan, A. A. Aivazov, M. N. Meytin, A. Yu. Sazonov and J. W. Metsellar, "Relaxation processes and metastability in amorphous hydrogenated silicon investigated with differential scanning calorimetry", *Physica B*, **252**, 198-206 (1998).

[25] P. Delli Veneri, C. Privato and E. Terzini, "Changes of hydrogen evolution thermodynamics induced by He and H_2 dilution in PECVD a-Si:H films: influence on thermal crystallization", *J. Non-Cryst. Sol.*, **266-269**, 635-639 (2000).

[26]L. Battezzati, F. Demichelis, C. F. Pirri and E. Tresso, "Differential scanning calorimetry (DSC) studies of hydrogenated amorphous semiconductor alloys", *Physica B*, 176, 73-77 (1992).

[27]G. DellaMea, F. Demichelis, C. F. Pirri, P. Rava, V. Rigato, T. Stapinski and E. Tresso, "Influence of hydrogen on the evolution of structural properties of amorphous silicon carbide", *J. Non-Cryst. Sol.*, **137-138**, 95-98 (1991).

[28]F. Demichelis, C. F. Pirri, E. Tresso, V. Rigato and G. DellaMea, "Hydrogen diffusion and related defects in hydrogenated amorphous silicon carbide", *J. Non-Cryst. Sol.*, **128**, 133-138 (1991).

[29] Y. Iwamoto, W. Voelger, E. Kroke, R. Riedel, T. Saitou, and K. Matsunaga, Crystallization Behavior of Amorphous Silicon Carbonitride Ceramics Derived from Organometallic Precursors", *J. Am. Ceram. Soc.*, **84**, 2170-78 (2001).

[30] J. P. Schaffer, A. Saxena, S. D. Antolovich, T. H. Sanders, Jr. and S. B. Warner, "The Science and Design of Engineering Materials", McGraw-Hill, Boston, pp. 788 (1999).

[31] Y.-W.Kim, M. Mitomo and H. Hirotsuru, "Grain growth and fracture toughness of fine-grained silicon carbide", *J. Am. Ceram. Soc.*, **78**, 3145-48 (1995).

[32] H. Kodama and T. Miyoshi, "Study of fracture behavior of very fine grained silicon carbide ceramics", *J. Am. Ceram. Soc.*, **73**, 3081-3086 (1990).

[33] T. Nishimura, R. Haug, J. Bill, G. Thurn, F. Aldinger, "Mechanical and thermal properties of Si-C-N material from polyvinylsilazane", *J. Mater. Sci.*, **33**, 5237-41 (1998).

[34] R. W. Rice, "Evaluation and extension of physical property-porosity models based on minimum solid area", *J. Mater. Sci.*, **31**, 102-118 (1996).

[35] R. W. Rice, "Grain size and porosity dependence of ceramic fracture energy and toughness at 22 °C", *J. Mater. Sci.*, **31**, 1969-1983 (1996).

[36] C. Reynaud and F. Thevenot, "Porosity dependence of mechanical properties of porous sintered SiC. Verification of the minimum solid area model", *J. Mater. Sci. Lett.*, **19**, 871-874 (2000).

[37] D-M. Liu, B-W. Lin and C-T. Fu, "Porosity dependence of mechanical strength and fracture toughness in SiC-Al₂O₃-Y₂O₃ ceramics", *J. Ceram. Soc. Jap.*, **103**, 878-881 (1995).

POLYSILANE-BASED PRECURSORS FOR SiC/SiC COMPOSITES

Masaki Narisawa, Kiyohito Okamura, and Takashi Iseki
Graduate School of Engineering, Osaka Prefecture University
1-1, Gakuen-Cho
Sakai 599-8531, Japan

Kunio Oka, and Takaaki Dohmaru
Research Institute of Advanced Science and Technology
Osaka Prefecture University
1-2, Gakuen-Cho
Sakai 589-8570, Japan

ABSTRACT

Various Poly(methylsilane-dimethylsilane) (P(MS-DMS)) co-polymers were synthesized under the controlled methylsilane (MS:-SiH(CH$_3$)-) unit contents. Introduction of MS units in the co-polymers was effective to promote the cross-linking in the case of reflux treatments. In particular, the ceramic yields of the co-polymers subjected to the reflux treatments at 623K were substantially improved. By adjusting the MS units in the co-polymers, the C/Si ratios of the resulting ceramic products were also controllable. P(MS-DMS) co-polymers are considered to be progressive precursors as matrix sources in Ceramic Matrix Composites (CMC).

INTRODUCTION

Since the Scilling's early work on polymethylsilane (PMS) synthesis,[1] numerous studies have been performed on PMS for silicon carbide (SiC) precursor uses. In PMS, Si-H bonds often play an important role in improving the final ceramic yields by promoting cross-linking. The cross-linking mechanism in PMS and their pyrolysis process have been widely investigated.[2-4]

During thermal treatments, PMS shows a large mass loss and a large amount of gaseous silicon compound evolution at 500-600K. These phenomena likely reveal occurrence of fragmentation of PMS chains at this temperature range, which must be accompanied by the cleavage of Si-Si backbone. The spectroscopic analyses, however, suggest the disappearance of Si-H bonds and the formation of Si-Si crosslinks ($\underline{Si}$CSi$_3$: -75ppm) at the same temperature range, 500-600K.[5,6] In order to explain this conflict, "Redistribution Reaction" based on radical or silylene intermediate behavior has been proposed.[4,7] By the reflux treatment at 500-600K, the ceramic yields of PMS are remarkably improved, which indicates the trapping effect of active intermediates of organosilicon fragments formed during thermolysis.

A few research groups made good use of the reactivity of PMS for the cross-linking of other compounds. Boury et al. reported that PMS consisted of methylsilane (MS: -SiH(CH$_3$)-) units reacted with polyvinlysilane via hydrosilylation to form cross-linked polymers, giving high ceramic yields.[8] Gozzi also obtained SiC in high yields by reacting PMS with tetra-allylsilane.[9] The drawback of this method is that blending two compounds sometimes brings about phase separation if their compatibility is low.

When another monomer units are chemically bonded to MS units, the resultant polymers are expected to become homogeneous. The co-polymers (poly(methylsilane-dimethylsilane)s: {[-Si(Me)H-]$_x$[-SiMe$_2$-]$_{1-x}$}$_n$), in which MS units are chemically bonded with dimethylsilane units (DMS: -Si(CH$_3$)$_2$-), were prepared as precursors for the SiC ceramics. Polydimethylsilane (PDMS) comprised of only DMS units gives a negligible ceramic yield because thermal Si-Si cleavages take place and the resultant fragments are vaporized during the pyrolysis. Therefore, if MS units are incorporated into PDMS structures, the ceramic yields are expected to improve.

Such PMS-based precursors, containing a considerable amount of MS units in their molecular structures, should be promising matrix sources in SiC/SiC composite fabrication, which make possible the control of the final ceramic yield, C/Si molar ratio, pore distribution, and the interface properties.[10,11] The efficiency of MS units as cross-linking agents, however, will strongly depend on the thermal histories and the heat treatment environments on the tailored precursors, because the trapping of active intermediates by Si-H is intrinsically a condensation reaction proceeding in gaseous, liquid, and solid polymer (oligomer) mixtures.

In this paper, we will discuss the pyrolysis process of the P(MS-DMS) co-polymers synthesized in our laboratory from the viewpoint of heat treatment environments, spectroscopic analyses, and the resulting chemical compositions.

EXPERIMENTAL

Figure 1 shows the synthesis process for the starting poly(methylsilane-dimethylsilane) co-polymers, which are composed of methylsilane (MS: -SiH(Me)-) and dimethylsilane (DMS: -Si(Me)$_2$-) units. The co-polymers prepared by Wurtz-type coupling had nominal MS/DMS molar ratios (3/7, 5/5, and 7/3) after feeding the molar ratios of methyldichlorosilane /dimethyldichlorosilane (3/7, 5/5, and 7/3), and are designated as P(MS-DMS)-3, -5 and -7, respectively. For instance, P(MS-DMS)-3 possesses 30 % of the MS units and 70 % of the DMS units. The feature of the as-prepared co-polymers is summarized in Table 1. All the co-polymers obtained were oily liquids. The polymers exhibited relatively low molecular weights indicative of oligomeric character.

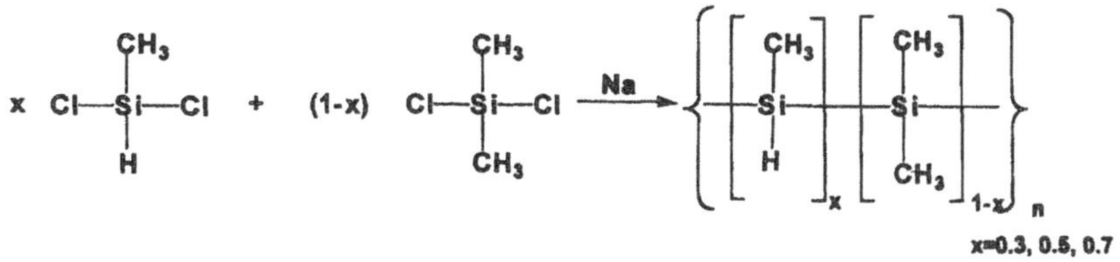

Fig. 1. Preparation of P(MS-DMS) co-polymers.

Table 1. Properties of the starting co-polymers.

Co-Polymers	Appearance	M_n	M_w/M_n
P(MS-DMS)-7	White Liquid	930	2.8
P(MS-DMS)-5	White Liquid	1170	2.9
P(MS-DMS)-3	Yellow Liquid	750	2.1

For the reflux treatment of the co-polymers, 1.0 g of the starting co-polymer was placed in a Pyrex tube with a reflux condenser.   After the evacuation, Ar gas was filled in the tube and was flowed out continuously (300 ml/min). The tube was then heated from room temperature to the defined temperatures at a rate of 10 K/min and held for 2h.   The co-polymers with the reflux treatment was pyrolyzed at 1273K at a heating rate of 5 K/min and held for 2h in an Ar gas flow. The pyrolysis products were subjected to the isothermal pyrolysis at 1773K for 15min in an Ar gas flow.   The detail of the heat treatment and pyrolysis process was described in the previous study.[6]   Infrared spectra and ^{29}Si NMR spectra measurements were performed on the co-polymers after the reflux treatments. Thermogravimetric analyses were carried out at a heating rate of 10 K/min in an Ar gas flow.   The samples pyrolyzed at 1773K were analyzed by the X-ray powder diffractometer and the electron probe micro analyzer.[6]

RESULTS AND DISCUSSION

Thermogravimetric Behavior of Precursors

Figure 2 shows the TG curves for the co-polymers as well as PMS and PDMS (Nippon Soda Co.).   PDMS consisted of only DMS units gives a negligible ceramic yield (1 %) at 1273K because gas evolution due to Si-Si cleavage takes place at 700K.   On the other hand, the ceramic yields of the co-polymers with MS units are improved (14-27 %) and approach to that of PMS when MS units increase.   This suggests that the MS units work as an inhibitor of the gas

Advanced SiC/SiC Ceramic Composites

evolution by forming cross-linking bridges. However, the ceramic yields of the co-polymers are still not high enough.

Reflux Heat Treatment of Precursors

To enhance a ceramic yield, the cross-linking of the polymers is essential. We have described the efficiency of the reflux treatment for providing cross-linked PMS, which offers SiC in high yields.[5,6] Such reflux treatments were also applied to the synthesized co-polymers. After the reflux treatments, the appearance of the refluxed co-polymers changed from liquid to rigid solids as the temperature increased and the color of the co-polymers turned yellow at 673K. These results suggest the cross-link formation. The polymer recoveries after the reflux treatment were 66-80 mass% at the reflux temperature of 623K. In general, the polymer recoveries are decreased as the reflux temperature increases. Also, the recovery is enhanced as the molar ratio of MS units increases in the polymer structure at the same reflux temperatures.

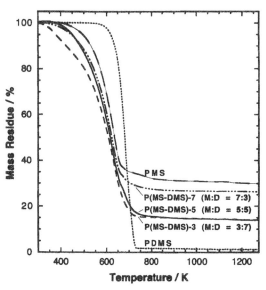

Fig. 2. TG-curves of the co-polymers.

Thermogravimetric Behavior of Refluxed Polymers

Figure 3 shows the TG curves of the refluxed co-polymers: P(MS-DMS)-3, -5, and -7 at 623K; PMS at 623K; PDMS at 673K. The starting masses before the temperature rising correspond to the polymer recoveries after the individual reflux treatments on the co-polymers. Thus, the finishing points of this graph reveal the overall ceramic yields at 1273K from the original co-polymers. The refluxed PDMS shows a large mass loss during the pyrolysis and gives no ceramic yield (~0 %), whereas it has a high polymer recovery (~80 %) after the refluxing.

Such a large mass loss is probably caused by its oligomeric carbosilane structure with a less amount of Si-H groups. In contrast, it is noted that the co-polymers with MS

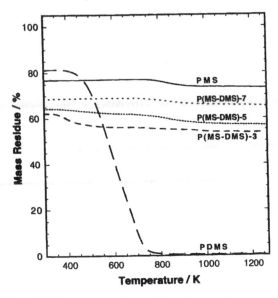

Fig. 3. TG curves of the refluxed co-polymers.

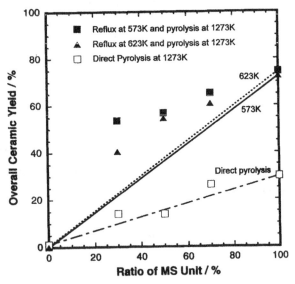

Fig.4. Relationship between the MS unit content and the overall ceramic yield.

units offer considerably higher ceramic yields (623K reflux treatment, 1273K pyrolysis: 54-65 %).

Figure 4 summarizes the relationship between the molar ratios of the MS units in the co-polymers and the overall ceramic yields. Open squares correspond to the direct pyrolysis of the co-polymers up to 1273K without passing the reflux treatments. In this case, the resulting ceramic yields are low, which show good fits on the ceramic yields expected linearly from those of PMS and PDMS without the reflux treatments. Black squares or triangles in Fig. 4 correspond to the pyrolysis of co-polymers up to 1273K with passing the reflux treatment at 573 or 623K. In these cases, the resulting ceramic yields are usually high, which exceed the values expected linearly from the ceramic yields of PMS and PDMS. In particular, misfits in the overall ceramic yields are remarkable in the P(MS-DMS)-3, which contains the smallest amount of MS units. These results suggest that the interaction between MS units and DMS units occurs enhancing the ceramic yields only in the case of the refluxing.

Structural Transformation of P(MS-DMS)-3 by Refluxing

In order to examine the enhancement of the overall ceramic yields for the refluxed co-polymers, the structural transformation of P(MS-DMS)-3 according to the reflux temperature was examined. The IR spectra of the refluxed polymers (Fig. 5) show no substantial change up to 523K. However, the new peaks, probably -CH$_2$- deformation absorption, appear at 1350 and 1030 cm^{-1} at 573K. This indicates the occurrence of the Kumada rearrangement converting methyl groups (Si-CH$_3$) into methylene bridging (Si-CH$_2$-Si).[12] An additional structural change is also observed in the ^{29}Si NMR spectra as seen in Fig. 6. The resonance of (CH$_3$)SiSi$_3$ at -78 ppm increases as the reflux temperature increases.[13,14] In the refluxed PMS, the (CH$_3$)SiSi$_3$ units are almost the sole units at 573K. Therefore, the Si-Si cross-linking is also expected to proceed prior to the conversion of methyl into methylene (Kumada rearrangement) in the PMS pyrolysis.

In the case of the P(MS-DMS) co-polymers, however, a part of DMS units near MS units is probably incorporated into the congested polymer frameworks up to 573K to increase the overall ceramic yield (40 %).

Above 623K, the NMR spectra show the appearance of C_3Si H (-15 ppm) and C_4Si (0 ppm) units,[15-17] and disappearance of the $(CH_3)SiSi_3$ units at -78 ppm, suggesting that the Kumada rearrangement becomes predominant. Since the Kumada rearrangement forms Si-H and Si-CH$_2$-Si groups in the polymer structures, Si-C cross-linking will be additionally provided by the inter-molecular reaction owing to these Si-H groups formed beyond 573K. Chemical conversion on the DMS unit moiety proceeds only above 573K, while structural change in the MS unit moiety proceeds even below 573K. The DMS units would be incorporated into the cross-link effectively above 573K. This is probably one reason why the overall ceramic yields are high in the case of the 623K reflux treatments as compared with the case of the 573K treatments as shown in Fig. 4.

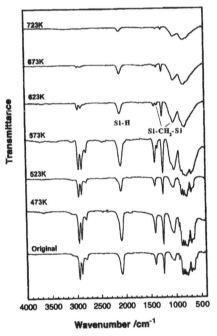

Fig. 5. IR spectra of the refluxed P(MS-DMS)-3 co-polymer.

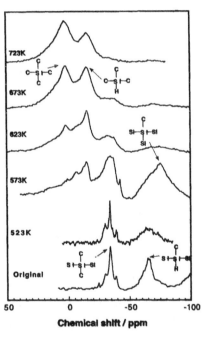

Fig. 6. ^{29}Si NMR spectra of the refluxed P(MS-DMS)-3 co-polymer.

Advanced SiC/SiC Ceramic Composites

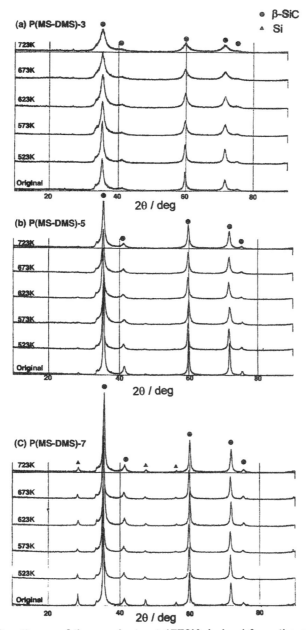

Fig. 7. XRD patterns of the products at 1773K derived from the copolymers with with various reflux temperatures.

Ceramic Products

XRD patterns of the ceramics derived from the co-polymers are shown in Fig. 7. The peaks due to β-SiC are primarily observed in all ceramic products. Small peaks attributable to metallic silicon are also seen in the ceramic products from P(MS-DMS)-5 and -7 as observed in the ceramics from PMS. The gases, which reduce carbon content (Me_2SiH_2, Me_3SiH, and CH_4), are evolved during the pyrolysis of PMS. Likewise, the co-polymers would release similar gaseous fragments during the pyrolysis and thus results in the formation of silicon in the ceramic products. The elemental analyses of the ceramics from the co-polymers (Table 2) support that the increase of the C/Si ratio in the original polymers reduces the amount of metallic silicon.

Table 2. Chemical composition of the pyrolysis products at 1773K.

Precursors	Pre-treatment	Chemical Composition
P(MS-DMS)-7	None	$SiC_{0.89}$
	Reflux at 623K	$SiC_{0.86}$
P(MS-DMS)-5	None	$SiC_{0.94}$
	Reflux at 623K	$SiC_{0.95}$
P(MS-DMS)-3	None	$SiC_{1.02}$
	Reflux at 623K	$SiC_{1.05}$

SUMMARY

Poly(methylsilane-dimethylsilane) co-polymers (P(MS-DMS)-7, -5, and -3) were cross-linked by refluxing to afford the higher overall ceramic yields (40-65 %) than the same co-polymers without the reflux treatments (14-27 %). The methylsilylene (MS) units play an important role in the formation of the Si-Si cross-link structures by refluxing at <573K. The MS units probably serve for the effective incorporation of a part of DMS units by the Si-Si cross-linking via silylenes. The first Si-Si cross-linking formation leads to further Si-C cross-linking above 623K because the DMS units forms reactive Si-H bonds by the Kumada rearrangement, which contribute to cross-linking. Therefore, the ceramic yields of the refluxed samples above 623K were improved, compared with those below 573K.

The crystallization of β-SiC in the ceramic products was suppressed by increasing the amounts of DMS units and the reflux temperature of the precursors. This is probably because the more cross-linked polymers give the smaller amounts of silicon.

REFFERENCES

[1] C. L. Schilling, J. P. Wesson, and T. C. Williams, "Polycarbosilane Precursors for Silicon Carbide," *Am. Ceram. Soc. Bull.*, **62**(8), 912-15 (1983).

[2] Z. Zhang, F. Babonneau, R. M. Laine, Y. Mu, J. F. Harrod, and J. A. Rahn, "Poly(methylsilane)- A High Ceramic Yield Precursor to Silicon Carbide," *J. Am. Ceram. Soc.*, **74**(3), 670-73 (1991).

[3] D. Seyferth, T. G. Wood, H. J. Tracy, and J. L. Robisoe, "Near-Stoichiometric Silicon Carbide from Economical Polysilane Precursor," *J. Am. Ceram. Soc.*, **75**(5), 1300-302 (1991).

[4] R. M. Laine, and F. Babonneau, "Preceramic Polymer Routes to Silicon Carbide," *Chem. Mater.*, **5**(3), 260-79 (1993).

[5] T. Iseki, M. Narisawa, K. Okamura, K. Oka, and T. Dohmaru, "Reflux Heat-Treated Polymethylsilane as a Precursor to Silicon Carbide," *J. Mater. Sci. Lett.*, **18**(3), 185-87 (1999).

[6] T. Iseki, M. Narisawa, Y. Katase, K. Okamura, K. Oka, and T. Dohmaru, "An Efficient Cross-Linking Process of Polymethylsilane for SiC Ceramics," *Chem. Mater.*, **13**(11), 4163-69 (2001).

[7] B. Boury, N. Bryson, and G. Soula, "Borate-Catalyzed Thermolysis of Polymethylsilane," *Chem. Mater.*, **10**(1), 297-303 (1998).

[8] B. Boury, N. Bryson, and G. Soula, "Stoichiometric Silicon Carbide from Borate-Catalyzed Polymethylsilane-Polyvinylsilane Formulations," *Appl. Organomet. Chem.*, **13**, 419-30 (1999).

[9] M. F. Gozzi, M. D. Goncalves, and I. V. P. Yoshida, "Near-Stoichiometric Silicon Carbide from a Poly(methylsilylene)/Tetra-Allylsilane Mixture," *J. Mater. Sci.*, **34**(1), 155-59 (1999).

[10] R. Jones, A. Szeda, and D. Petrak, "Polymer Derived Ceramic Matrix Composites," *Composites*, **Part A 30**, 569-75 (1999).

[11] M. Kotani, A. Kohyama, and Y. Katoh, "Development of SiC/SiC Composites by PIP in Combination with RS," *J. Nuclear Mater.*, **289**, 37-41 (2001).

[12] K. Shiina and K. Kumada, "Thermal Rearrangement of Hexamethyldisilane to Trimethyl(dimethylsilylmethyl)silane," *J. Org. Chem.*, **23**, 139 (1958).

[13] M. F. Gozzi, and I. V. P. Yoshida, "Thermal and Photochemical Conversion of Poly(methylsilane) to Polycarbosilane," *Macromolecules*, **28**(21), 7235-40 (1995).

[14] P. Czubarow, T. Sugimoto, and D. Seyferth, "Sonochemical Synthesis of a Poly(methylsilane), a Precursor for Near-Stoichiometric SiC, *Macromolecules*, **31**(2), 229-38 (1998).

[15] G. D. Soraru, F. Babonneau, and J. D. Mackenzie, "Structural Evolutions from Polycarbosilane to SiC Ceramics," *J. Mater. Sci.*, **25**(9), 3886-93 (1990).

[16] A. T. Hemida, M. Birot, J. P. Pillot, J. Dunogues and R. Pailler, "Synthesis and Characterization of New Precursors to nearly Stoichiometric SiC Ceramics," *J. Mater. Sci.*, **32**(13), 3475-83 (1997).

[17] Q. Liu, H.-J. Wu, R. Lewis, G. E. Maciel, and L. V. Interrante, "Investigation of the Pyrolytic Conversion of Poly(silylenemethylene) to Silicon Carbide," *Chem. Mater.*, **11**, 2038-48 (1999).

PRESENT STATUS AND FUTURE TREND ON DEVELOPMENT AND APPLICATION OF CONTINUOUS SIC FIBERS

Hiroshi Ichikawa
Nippon Carbon Co, Ltd.
2-6-1, Hatchobori, Chuo-ku, Tokyo
104-0032, Japan

ABSTRACT

Polymer-derived SiC fiber Nicalon has been produced commercially and widely used as a reinforcement of Polymer Matrix Composites (PMCs) and Ceramic Matrix Composites (CMCs). Main applications of Nicalon fibers are PMC for structural materials of aerospace and CMC for high temperature components of aerospace engine.

In recent years, it has been increasing in demand for high performance CMC for high temperature application. Ceramic matrix composites are most promising materials for high temperature structural materials of gas turbines for aerospace and power generation. Mechanical performances of CMCs are highly dependent on properties of the reinforcement. We have been improving fiber properties for CMCs by reducing oxygen content and excess carbon in chemical composition of the SiC fibers.

The oxygen free SiC fiber (Hi-Nicalon) was developed and has been commercially produced using an electron beam curing process. By reducing oxygen, this fiber has a higher elastic modulus and creep resistance, and thermal stability up to 1,600°C than those of Nicalon fiber. High temperature mechanical properties of CMCs using Hi-Nicalon have significant improved. However, creep resistance of Hi-Nicalon CMC is not satisfactory as structural materials at high temperature, because Hi-Nicalon mainly consists of SiC micro-crystals and excess carbon. Recently, the SiC fiber (Hi-Nicalon type S) having stoichiometric SiC composition and high crystallinity have been developed by Nippon Carbon. By reducing excess carbon and putting C/Si ratio close to stoichiometric composition, Type S fiber has high elastic modulus of 420 GPa, high thermal conductivity, and excellent creep resistance. In this paper, the recent development of these oxygen free SiC fibers and results of physical and mechanical properties, thermal stability, and an environmental resistance are reported. Hi-Nicalon Type S fibers should be the best candidates for the reinforcement of ceramic matrix composites.

Application of SiC fibers is introduced. Now, Nicalon fiber/ceramic matrix composites are used for the exhaust flaps and seals of jet engine. Hi-Nicalon Type S fiber/BN/SiC composites are being developed as the components of gas turbine for aerospace and power generation such as shrouds and combustor.

INTRODUCTION

In recent years, it has been increasing in demand for high performance ceramic matrix composites (CMCs) on high temperature application. CMCs are required with heat resistance above 1,500 °C as structural materials for space plane and high-temperature gas turbine applications. CMCs are highly dependent on the properties of the reinforcement. A reinforcing fiber should have high environmental stability and sufficient mechanical properties even at high temperature. SiC fibers that have high tensile strength, high elastic modulus, and good thermal stability are one of the best candidates for reinforcement. Especially polymer derived SiC fibers offer advantage of flexible, fine diameter form over those from CVD or sintering process.[1]

The Si-C-O fiber Nicalon synthesized from polycarbosilane(PCS) has been produced industrially and applied widely as reinforcements for polymer, metal, and ceramic matrix composites.[2)-3)] Nicalon performs well as a reinforcement for CMCs.[4] However, the conventional Nicalon has a temperature limit of 1,200 °C in heat resistance.[5]

Recently, oxygen-free Si-C fibers (Hi-Nicalon) have been developed. The fibers were produced from polycarbosilane (PCS) by electron beam curing and pyrolysis. The Hi-Nicalon fiber had a high elastic modulus, and creep resistance, and much improved thermal stability over the Nicalon fiber. [6).7)] Therefore, Hi-Nicalon has realized wide variety of fabrication processing and desirable mechanical properties of CMCs. However, the Hi-Nicalon fibers mainly consist of SiC micro crystals and amorphous carbon. These micro crystals would cause creep deformation more easily at high temperature, and excess carbon can degrade oxidation resistance.

We have synthesized the oxygen-free S-C fibers with various C/Si atomic ratios by electron beam curing and pyrolysis from PCS fibers. Based on the study of thermo-mechanical properties with various C/Si fibers, the fiber with the C/Si atomic ratio of 1.05 was selected as a new grade of Hi-Nicalon. This fiber is pyrolyzed at high temperature under special conditions, and SiC crystals grew without decreasing the fiber strength. The stoichiometric, highly crystalline SiC fiber (Hi-Nicalon type S) has been developed .[8]

In this study, recent results of these fibers' physical and mechanical properties, thermal stability and environmental resistance, and industrial applications are reported.

PREPARATION OF SIC FIBERS

Fig.1 shows the fabrication process of SiC based fibers. A preceramic polymer, polycarbosilane (PCS), was synthesized from dimethyldichlorosilane as a starting material. The SiC fibers were prepared by the melt spinning of polycarbosilane, followed by curing and pyrolysis[2), 3)]. While conventional Nicalon fiber was prepared by oxidation curing of polycarbosilane fiber, Hi-Nicalon fiber was made by radiation curing using an electron beam irradiation from 10 to 15 MGy in a helium gas flow.

CHARACTERISTICS OF NICALON FIBER

Nicalon fiber consists of a ternary compound of silicon, carbon and oxygen, and has an atomic ratio of approximately Si_3C_4O. The existence of excess carbon relative to silicon, comes from the polycarbosilane containing a larger ratio of carbon to silicon. Oxygen is introduced from the curing process.

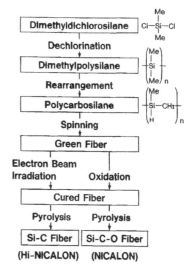

Figure 1. Fabrication process of
the SiC Based fibers
(Nicalon and Hi-Nicalon)

Table I. Typical properties and applications of Nicalon fiber

		Ceramic Grade NL–200	HVR Grade NL–400	LVR Grade NL–500	Carbon Coated NL–607
Diameter	(μm)	14/12	14	14	14
Numbers of filaments	(fil./yarn)	250/500	250/500	500	500
Tex	(g/1000m)	105/210	110/220	210	210
Tensile strength	(MPa)	3000	2800	3000	3000
Tensile modulus	(GPa)	220	180	220	220
Elongation	(%)	1.4	1.6	1.4	1.4
Density	(kg/m³)	2550	2300	2500	2550
Specific resistivity	(ohm–cm)	10^3–10^4	10^6–10^7	0.5–5.0	0.8
Co–efficient of thermal expansion					
	(10⁻⁶/K)	3.1	—	—	3.1
Specific heat	(J/(kg·K))	1140	—	—	1140
Thermal conductivity	(W/(m·K))	12	—	—	12
Dielectric constant	(at 10 GHz)	9	6.5	20–30	—
Applications		Heat resistant materials PMC, MMC, CMC	PMC	PMC	CMC

Table I shows the typical properties of commercial Nicalon fibers. Nicalon fibers have high tensile strength and high modulus. Four grades of Nicalon fibers are shown. Ceramic grade has been used for reinforcement of PMC and CMC. HVR grade has a high electrical resistivity and LVR grade has a low electrical resistivity. Both grades have been mainly used as reinforcement for PMC. Nicalon has the characteristics of excellent heat resistance to oxidation at high temperatures in air. Figure 2 shows the change in tensile strength after exposure at high temperatures in air. Nicalon fibers have maintained a 2 GPa tensile strength after exposure for 1000 h at 1000°C or 10 h at 1200°C. However, Nicalon has a lower tensile strength after exposure for a short time at 1400°C in air.

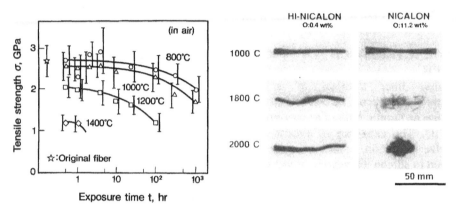

Figure 2. Tensile strength of Nicalon fibers after thermal exposure

Figure 3. Macroscopic view of SiC fibers after thermal exposure

LOW OXYGEN CONTENT SiC FIBERS HI-NICALON

Nicalon fiber degraded on thermal exposure at over 1300°C with the evolution of carbon monoxide. The degradation of the fiber resulted from the oxygen presence in the Nicalon fiber. Thus, this degradation could be prevented by minimizing the oxygen content in the Nicalon fiber. A new curing process was developed that the polycarbosilane fibers are cured using electron beam irradiation in an oxygen free atmosphere as shown in Figure 1. This development was carried out jointly with the Japan Atomic Energy Research Institute.

Figure 3 shows the macroscopic view of the fibers after thermal exposure at high temperature. The oxygen content of the fibers was 0.4 wt% for the irradiation curing and 11.2 wt% for the oxidation curing, which corresponds to ordinary Nicalon fiber, both fibers being pyrolyzed at 1000°C. The fiber having 11.2 wt% oxygen content lost its fibrous form and changed to a powder-like material with extreme crystal growth after thermal exposure at 1800 and 2000°C. On the other hand, it should be noted that the fiber having 0.4 wt% oxygen content maintained the same appearance and flexibility even after exposure at 1800 and 2000°C.

Table II shows the low oxygen content "Hi-Nicalon" SiC fiber. Hi-Nicalon fiber has remarkably low 0.5wt % oxygen content. This continuous Ni-Nicalon fiber has the same value of filament diameter and number of filaments per yarn as Nicalon fiber. However, it has a slightly increased density and a sharply increased tensile modulus compared to Nicalon.

STOICHIOMETRIC SiC FIBERS HI-NICALON TYPE S

The radiation cured polycarbosilane fibers were pyrolyzed in hydrogen gas flow with various conditions, and the fibers with various C/Si atomic ratios were prepared. As a function of C/Si atomic ratio,the density of the Si-C fibers undergoes a maximum when the C/Si ratio is close to the stoichiometric C/Si atomic ratio of 1.0. The stoichiometric composition yields the highest density, 3.02 g/cm^3. When the Si-C fiber with near stoichiometric C/Si ratio (1.05) was pyrolyzed at 1,500°C, the density of the fiber reached 3.1 g/cm^3, which is close to theoretical value of SiC, 3.2 g/cm^3.

The strength of each fiber is over 2.5 GPa. The tensile strength of the fiber seems to independent of its chemical composition. On the other hand, the elastic modulus varies dramatically as a function of C/Si. It exhibits a maximum for a C/Si ratio close to 1. The near stoichiometric fiber exhibited the highest modulus 400 GPa. The modulus appears to be related to density.

Based on the study of the various C/Si fibers, we have selected the fiber with C/Si atomic ratios of 1.05 as a new grade of Hi-Nicalon fiber.

Table II. Typical properties of PCS-derived SiC based fibers

Properties		Nicalon NL-200	Hi-Nicalon	Hi-Nicalon type S
Fiber diameter	(μm)	14	14	12
Number of filaments	(fil./yarn)	500	500	500
Tex	(g/1000m)	210	200	180
Tensile strength	(GPa)	3.0	2.8	2.6
Tensile modulus	(GPa)	220	270	420
Elongation	(%)	1.4	1.0	0.6
Density	(g/cm^3)	2.55	2.74	3.10
Specific resistivity	(ohm-cm)	10^3-10^4	1.4	0.1
Specific heat 25 °C	(J/g·K)	0.71	0.67	0.70
500 °C		1.17	1.17	1.15
Heat conductivity 25 °C	(W/m·K)	2.97	7.77	18.4
500 °C		2.20	10.1	16.3
Coefficient of thermal expansion 25~500 °C	(10^{-6}/K)	3.2	3.5	-
	Si (wt.%)	56.6	62.4	68.9
Chemical composition	C	31.7	37.1	30.9
	O	11.7	0.5	0.2
	C/Si (atomic)	1.31	1.39	1.05

The fiber was pyrolyzed at high temperature by special conditions, and the crystallite size of the fiber increased without decreasing the strength. Then, the stoichiometric SiC fiber (Hi-Nicalon type S) as a new grade of Hi-Nicalon was developed.

CHARACTERISTICS OF Si-C FIBERS
Physical and mechanical properties

Table II shows the typical properties of Hi-Nicalon type S as compared to other PCS-derived SiC fibers. The properties of type S are preliminary ones. The fiber diameter and tex of type S are rather smaller than the other two types of fibers. Tensile strength is 2.6 GPa, which is sufficiently high for the reinforcement of CMCs. Type S shows the highest modulus of 420 GPa and the highest density, 3.1 g/cm^3.

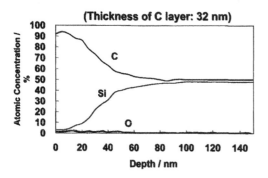

Figure 4.   AES depth profile of Hi-Nicalon type S fiber

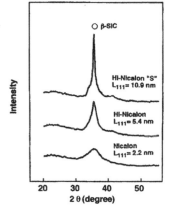

Figure 5. XRD patterns of three types of the SiC based fibers

The specific resistivity of type S is much smaller, which can be attributed to the existence of a carbon layer at the fiber surface.

Fig. 4 shows AES depth profiles of Hi-Nicalon type S fiber. The fiber was observed with the stoichiometric SiC composition, and covered with a carbon layer whose thickness was approximately 32 nm on the surface. Fig. 5 shows the XRD patterns of the fibers. In all three types of the fibers, only β-SiC diffraction peaks were observed. Excess carbon in the Nicalon and Hi-Nicalon was not detected. The SiC crystallite size of these fibers was 2.2, 5.4, and 10.9 nm, respectively.

The TEM micrographs of SiC-based fibers are shown in Fig. 6. The grain size of three types of the fibers were obviously different from each other, and increased in the order of Nicalon, Hi-Nicalon, and type S fiber. Each fiber is

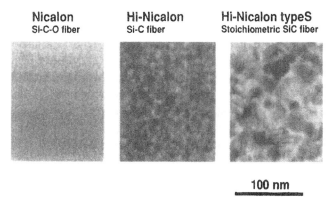

| Nicalon | Hi-Nicalon | Hi-Nicalon typeS |
| Sl-C-O fiber | Si-C fiber | Stoichiometric SiC fiber |

100 nm

Figure 6. TEM micrographs of the SiC based fibers

approximately 5, 10, and 100 nm, respectively. The sizes were larger than those calculated by XRD line broadening.

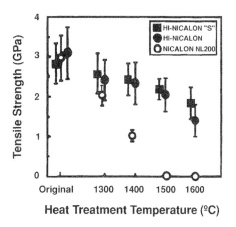

Figure 7. Tensile strength of the SiC-based fibers after 10 hours exposure in an argon gas

Thermal stability

Fig.7 shows the tensile strength of the fibers after the thermal exposure test. Nicalon fiber exhibited low strength after exposure at 1,400 °C and no strength after 1,500 °C exposure. On the other hand, Hi-Nicalon and type S fibers retained good strength even after 1,600°C exposure. Hi-Nicalon type S fibers showed the highest strength, 1.8 GPa, after 10 hours exposure in argon gas at 1,600 ℃.

Further tests at higher temperature were performed on Hi-Nicalon and type S. After exposure for one hour at 2,000℃ in argon gas, Hi-Nicalon fiber maintained the same appearance and flexibility as that before exposure and had moderate strength of 1.3 GPa. Hi-Nicalon type S after one hour exposure in an argon gas at 1,800℃ was quite stable chemically, since no structural decomposition occurred and it exhibited a good strength of 1.9 GPa. It is evident that SiC-based fibers with reduced oxygen content exhibits superior thermal stability.

Oxidation resistance

The tensile strength of the SiC-based fibers after exposure at 1,400℃ for 10 hours in dry air is shown in Fig. 8. All tested fibers showed a decrease in their strength due to oxidation. The Hi-Nicalon type S fiber showed highest residual strength, 1.8 GPa. After oxidation, the oxide layer thickness on all of the fibers was almost the same value, $0.3 \sim 0.4$ μ m. The strength was not in good agreement with the oxide layer thickness. Nicalon might be thermally degraded, in addition to being oxidized. In any case, the Hi-Nicalon type S showed the best oxidation resistance on the strength basis.

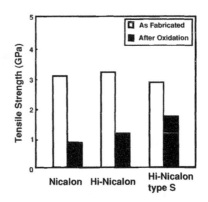

Figure 8. Tensile strength of the SiC-based fibers
after 10 hours exposure at 1,400℃ in dry air

Creep resistance

Fig. 9 shows the creep properties of some representative PCS-derived SiC fibers as compared to those of various ceramic fibers reported by DiCarlo[9), 10)]. Hi-Nicalon and type S fibers show much better creep resistance than conventional Nicalon fibers. Excellent creep resistance was obtained in the case of the annealed

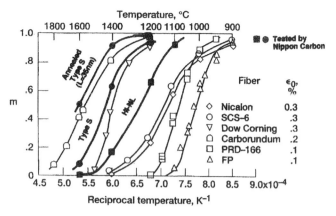

Figure 9. One-hour stress relaxation ratio of the PCS-derived SiC fibers
and other polycrystalline SiC and Alumina fibers[10]

type S fiber, which is a stoichiometric composition with 35 nm crystallite sizes. TEM observation shows that this annealed fiber has a SiC grain size of approximately 200 nm, which is about 10 times larger than that of the as-fabricated fiber.

SUMMERY OF THE SIC FIBER

Table III represents the properties of PCS-derived fibers, Nicalon, Hi-Nicalon, and Hi-Nicalon type S. Oxygen-free Si-C fibers (Hi-Nicalon) has been developed. The fibers were produced from polycarbosilane (PCS) by electron beam curing and pyrolysis. Hi-Nicalon had a high tensile strength and a high elastic modulus.

The fiber retained a strength and modulus after thermal exposure at 1,600 °C

Table III. Comparison of properties on PCS-derived SiC fibers

⊙ excellent ○ good ▲ moderate	Nicalon ➡ Si-C-O Fiber	Hi-Nicalon ➡ Si-C Fiber	Hi-Nicalon type S Stoichiometric SiC Fiber
Chemical C/Si composition O	1.31 12 wt%	1.39 0.5 wt%	1.05 0.2 wt%
Tensile strength	⊙ 3.0 GPa	○ 2.8 GPa	○ 2.6 GPa
Elastic modulus	○ 200 GPa	○ 270 GPa	⊙ 420 GPa
Thermal stability	○ 1473 K	⊙ 2073 K	⊙ 2073 K
Oxidation resistance	▲	▲	○
Creep resistance	▲	○	⊙

in argon atmosphere. Furthermore, the stoichiometric crystalline SiC fiber (Hi-Nicalon Type-S) has been developed. The fiber has been prepared from E-beam cured PCS fibers using special process conditions. The Hi-Nicalon S fiber has a excellent thermal stability up to 1,600 °C, as does the Hi-Nicalon fiber. The fiber also has higher elastic modulus of 420 GPa, excellent creep resistance, and better oxidation resistance than other Si-C fibers. Hi-Nicalon and Hi-Nicalon Type S fibers should be the best candidates for the reinforcement of ceramic matrix composites. Hi-Nicalon fibers have been produced by Nippon Carbon Co., Ltd. commercially, and Hi-Nicalon Type S fibers also have been provided to customers.

CHARCTERISTICS OF SIC FIBER/ CERAMIC MATRIX COMPOSITES

The SiC fiber/SiC matrix composites were prepared. Table IV shows the structure of the composites. Reinforcements are 8-harness satin cloth of Hi-Nicalon and Type S. The matrix is the silicon carbide converted from polycarbosilane PIP (Polymer Impregnation Pyrolysis) method. Interphase between fiber and matrix is CVD boron nitride.

The bending properties of SiC/BN/SiC composites were tested at room temperature and 1400. Fig. 10 shows stress-displacement curves of SiC/BN/SiC composites on bending test.

Table IV. The structure of SiC/BN/SiC composites

Symbol	Reinforcement	Interphase	Matrix	Vf (%)
S	Hi-Nicalon Type S 8 Harness Satin Weave	CVD-BN	PIP-SiC	32.6
H	Hi-Nicalon 8 Harness Satin Weave	CVD-BN	PIP-SiC	35.5

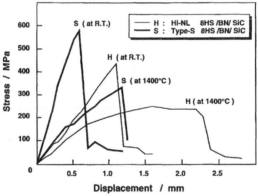

Fig.10. Stress-displacement curves of SiC/BN/SiC composites on bending test

In Type S fiber composites, both of proportional and maximum stress are higher than that of Hi-Nicalon composites at room temperature and 1400℃. It depends on higher modulus of type S fiber than Hi-Nicalon fiber.

APPLICATION OF SIC FIBERS AND CERAMIC MATRIX COMPOSITES
Ceramic Matrix Composites for Gas Turbine Components
SiC fiber reinforced Ceramic Matrix Composites (CMCs) have high strength and toughness in both ambient and high temperature, and also excellent thermal shock resistance. Nicalon fiber/ ceramic matrix composites are used for the exhaust fraps and seals of jet engine (Fig. 13). Hi-Nicalon fiber/SiC composites are also being developed as the components of gas turbine for aerospace and power generation. Examples of components are as follows; transition ducts, nozzles, combustor, shrouds, turbine rotor, flame holder, and exhaust liner (Fig.11, Fig.12).

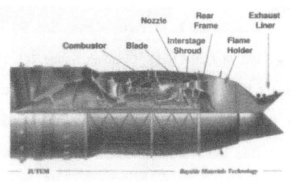

Figure 11. CMC for gas turbine application

Figure 12   Combustor liners made of SiC/SiC Composites

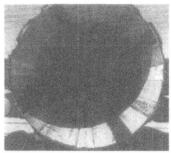

Figure 13 Exhaust flaps and seals made of SiC/C composites

CMC for Industrial use

Industrial application of the ceramic fiber CMCs are being developed and demonstrated as follows; hot gas filter, the immersion tube for metal casting, a hot gas air circulating fan in high temperature furnace, and the porous radiant burner (Fig. 14).

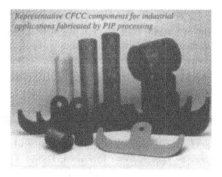

Figure 14. Various SiC/SiC components for industrial applications

RE FERENCES

1) J. R. Strife, J. J. Brennan and K. M. Prewo, Ceram. Eng. Sci. Proc., **11**[7-8], 871, (1990)

2) S. Yajima, J. Hayashi, M. Omori and K. Okamura, Nature, **261**, 683 ,(1976)

3) T. Ishikawa, Composites Science and Technology, **51**, 135 ,(1994)

4) K.M.. Prewo, J. J. Brennan and G. K. Layden, Am. Ceram. Soc. Bull., **65** [2], 305 ,(1986)

5) T. Ishikawa, H. Ichikawa and H. Teranishi, Proc. Electrochem. Soc. 88-5; Symp. High Temp. Mater. Chem., **4**, 205 ,(1987)

6) K. Okamura, M. Sato, and T. Seguchi, Proc. 1st Jap. Int. SAMPE Symp., 929, (1989)

7) M. Takeda, Y. Imai, H. Ichikawa, T. Ishikawa, N. Kasai, T. Seguchi and K. Okamura, Ceram. Eng. Sci. Proc., **14** [9-10], 540, (1993)

8) M. Takeda, Y. Imai, H. Ichikawa and T. Ishikawa, Ceram. Eng. Sci. Proc.,**15** [4], 133 ,(1994)

9) G. N. Morsher, J. A. DiCarlo and T. Wagner, Ceram. Eng. Sci. Proc., **12** [7-8], 1032 ,(1991)

10) J. A. DiCarlo, Composites Science and Technology, **51**, 213 ,(1994)

PROPERTIES OF BN COATING ON SiC FIBER BY CONTINUOUS CVD PROCESS

Michiyuki Suzuki, Yoshizumi Tanaka, Yoshiyuki Inoue, Norihumi Miyamoto, Mitsuhiko Sato
Ube Industries, Ltd.
1978-5 Kogushi, Ube, Yamaguchi 755-8633, Japan

Koichi Goda
Dept. of Mechanical Engineering, Yamaguchi University
2-16-1 Tokiwadai, Ube, Yamaguchi 755-8611, Japan

ABSTRACT

BN coating on sintered SiC fibers (Tyranno SA) by continuous CVD process was developed for the interphase of SiC/SiC composites. BN coating by the continuous CVD process was conducted at 1580°C and pressure of around 1 Torr. The reactant gases were NH_3 and BCl_3 and diluted by Ar. And yarn speed was 4.5 mm/s. Conventional experiments showed that the BN coating thickness across the fiber tow varied significantly. Hence, the CVD process was modified by not only adjustments of the amount of reactant gases, but also loosing the yarn bundle by reduction of the yarn tension and setting a new furnace for removing the sizing agent of the fiber tow prior to the CVD furnace. The BN coating thickness by the modified process was much more uniform than that by the conventional process. The coated BN was ordered structure, nearly stoichiometric and pure with little O and C. The strength of the BN coated SA fiber retained 90% of the strength of as-received SA fiber.

INTRODUCTION

The interphase of SiC/SiC composites has significant effects on the properties of SiC/SiC composites. The current choice for interphase is mainly BN [1]. So far, two coating techniques of BN coating on SiC fiber have been developed. One is processed at low (~ 1000°C) temperatures and the other is processed at high (>1400°C) temperatures [2]. The low temperature processed BN can make a uniform coating on the fibers, but may contain large amounts of O or C causing degradation. Furthermore, the BN does not possess well ordered microstructures. The high temperature processed BN is more pure with little O or C and turbostratic with the c-axis well aligned normally to the fiber axis. A significant problem for the high temperature processed BN is a coating nonuniformity across the fiber tow [3,4]. While the problem of the coating nonuniformity still remains, recently we confirm that a vender of the high temperature processed BN stopped the fiber-coating business.

Under these situations described above, we have started to make researches for developing the high temperature processed BN coating technology, focusing on the coating uniformity of BN thickness across the tow. In this paper, properties of BN coating on SiC fiber by the newly developed continuous CVD process were evaluated.

EXPERIMENTAL PROCEDURES

SiC fiber used in this study was a sintered SiC fiber (Tyranno SA, Ube Industries, Ltd.; diameter 7.5 μm, 1600 filaments per tow, sized by 1wt% polyethylene oxide (PEO)). Hereinafter, the sintered SiC fiber is denoted as the "SA fiber".

Conventional coating equipment is shown in Fig. 1. The length of the carbon heater in the CVD furnace was 360 mm and the hot zone was about 100 mm. SA fiber yarn was continuously coated by fed through the furnace from feed spool to collecting spool. The whole system was evacuated using a vacuum booster pump backed by a rotary piston vacuum pump. A coating condition was as follows. The reactant gases were NH_3 and BCl_3 and diluted by Ar. Flow rates were 150 ml/min. for NH_3, 50 ml/min. for BCl_3 and 400 ml/min. for Ar. The coating temperature was 1580°C, and pressure is around 1 Torr. Yarn speed was 4.5 mm/min. and yarn tension was about 100 g, controlled by adjustment between the feed roller and torque limiter of the collecting spool.

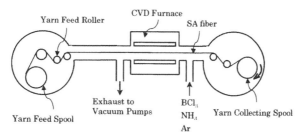

Fig. 1. Schematic illustration of conventional CVD coating equipment.

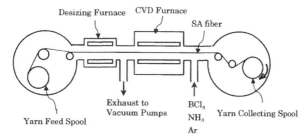

Fig. 2. Schematic illustration of modified CVD coating equipment.

Modification of coating equipment and coating condition was conducted to improve a nonuniformity of BN coating thickness. Modified coating equipment is shown in Fig. 2. New small furnace for removing the sizing agent (1wt% PEO) of the fiber tow was set prior to the CVD furnace, and the feed roller was removed to decrease the yarn tension. As a result, the yarn tension was reduced up to one-tenth of the initial value. The modified coating condition was as follows. Flow rates were increased to 300 ml/min. for NH_3, 100 ml/min. for BCl_3 and 1000 ml/min. for Ar. The yarn tension was about 10 g, and the desizing furnace was set at 800°C. The coating temperature, the pressure and the yarn speed were the same as the conventional condition.

Microstructures of BN coated SA fiber were observed by SEM and TEM. Auger electron spectroscopy was used to record depth profile of composition of the coated BN on SA fiber. X-ray diffraction (XRD) was used to confirm the crystal structure and measure the interlayer spacing ($c_0/2$) of the coated BN. Tensile strength of the BN coated and uncoated SA fibers was measured by single filament method. The filament gage length was 25 mm and cross head speed was 2 mm/min. 15 filaments were measured for each fiber. For the strength calculation of the BN coated SA fiber, thickness of BN layer was excluded from the cross section area of the fiber.

RESULTS AND DISCUSSION

Properties of BN coated SA fiber by conventional coating technique

Figure 3 shows SEM images of BN coated SA fibers by the conventional coating technique. Some of outermost filaments were well coated with thickness of about 1 μm as shown in Fig. 3(a). However, wall-like debonded BN layer around some outermost filaments was also observed as shown in Fig. 3(b). Furthermore, most of the innermost filaments were rarely coated as shown in Fig. 3(c). This means, the conventional coating technique generates a significant nonuniformity of the BN thickness across the fiber tow. From those observations, it is considered that the reactant gases (NH_3 and BCl_3) did not diffuse into the fiber tow enough in depositing BN on innermost filaments. Under the conventional coating technique, the yarn tension was 100 g so that the filaments bundled very tightly. The observed very thin coating on innermost filaments due to the insufficient diffusion of the reactant gases may be attributed to the tightly bundled filaments. To loose the yarn bundle, the yarn tension reduced to 10 g and a small furnace to remove the PEO sizing agent was set prior to the CVD furnace, as described in the previous section.

Properties of BN coated SA fiber by modified coating technique

SEM images of outer BN coated SA fibers and interior BN coated SA fibers

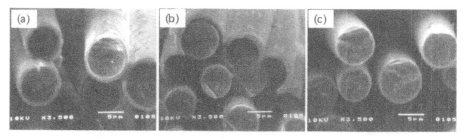

Fig. 3. SEM images of BN coated SA fibers by the conventional coating technique; (a), (b) outermost filaments and (c) innermost filaments.

of the fiber tow by the modified coating technique are shown in Figs. 4(a) and (b), respectively. From Fig. 4(a), outermost filaments were well coated with thickness of about 1 μm and no wall-like BN layer shown in Fig. 3(b) was found around the tow. In addition, innermost filaments were also coated with thickness of around 0.5 μm as shown in Fig. 4(b). Approximate ratio of exterior coating thickness to interior coating thickness was less than 5, which was smaller than that (6 to 6.6) shown by the vender of the high temperature processed BN [3,4]. Comparing Figs. 3 and 4, it is obvious that the uniformity of BN thickness across the tow was drastically improved by the modified coating technique. From this observation, loosing the yarn bundle by reducing the yarn tension and removing the sizing agent prior to the CVD furnace were considered to be quite effective in promoting the diffusion of reactant gases and Ar into the tow to deposit BN on innermost filaments.

Figures 5(a) and (b) show higher magnification SEM images of the outer BN coated SA fiber and Figs. 5(c) and (d) show higher magnification SEM images of the inner BN coated SA fiber, respectively. From Figs. 5(a) and (b), the surface of the coated BN was relatively smooth, and a layered-structure parallel to the fiber surface was observed in the coated BN on the outer fiber, which implied that the c-axis of BN was aligned normal to the fiber axis. However, the surface of the coated BN on the inner fiber was rougher than that on the outer fiber, and layered-structure was not well developed compared with that on the outer fiber. Those differences of structures of the coated BN between the outer fiber and the

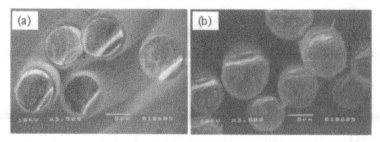

Fig. 4. SEM images of BN coated SA fibers by the modified coating technique for (a) outermost filaments and (b) innermost filaments, respectively.

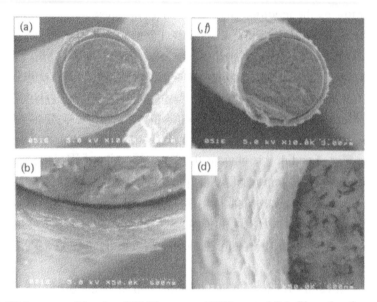

Fig. 5. Higher magnification SEM images of BN coated SA fibers by the modified coating technique for (a), (b) outermost filaments and (c), (d) innermost filaments, respectively.

inner fiber implied that the BN deposition condition of the inner region in the fiber tow was not same as that of the outer region. The possible reason would be that the gas flow affecting the nucleation of BN on fiber surface in the inside of the tow may be different from the outside of the tow because of its narrow space among many fibers (The fiber tow was composed of 1600 filaments). Enough gas flow into the inside must be considered as a future issue to be solved.

Auger depth profile of the BN coated SA fiber revealed the coated BN was nearly stoichiometric and contents of C and O were below 5at.% as shown in Fig. 6. Figure 7 shows XRD pattern of the BN coated SA fiber. It was confirmed that peaks were corresponding to h-BN and β-SiC. Hence, the coated BN was hexagonal structure and β-SiC peaks must be due to the SA fiber. From the peak at 2θ=26.38°, interlayer spacing ($c_0/2$) of BN was 3.38Å which was closer to the ideal value of 3.33Å than that (3.44Å) of the vender of the high temperature processed BN [2]. Table I shows tensile strength of the uncoated and the BN coated SA fiber. The strength of BN coated SA fiber retained 90% of the strength of uncoated SA fiber. In spite of high temperature (1580°C) coating

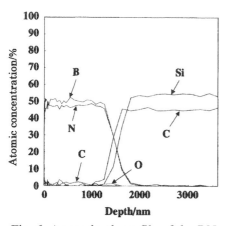

Fig. 6. Auger depth profile of the BN coated SA fiber.

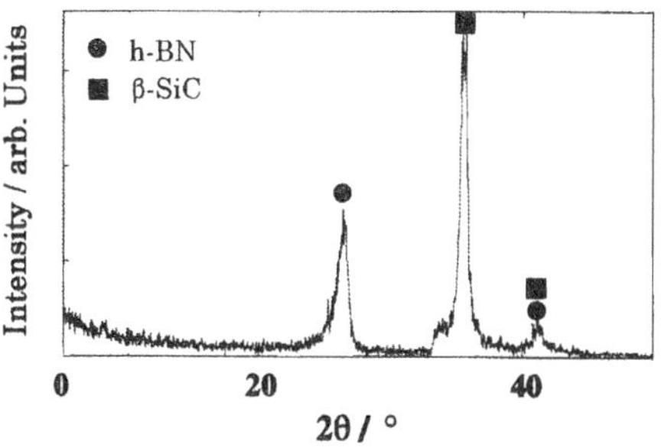

Fig. 7. XRD pattern of the BN coated SA fiber.

Table I. Tensile strength of the uncoated and the BN coated SA fiber.

	Tensile strength / GPa	Retention ratio (%)
Uncoated SA fiber	2.75	
BN coated SA fiber	2.47	90

process, only a small degradation of the fiber strength was observed. This is probably due to a short exposure time during CVD coating and high heat resistance of the SA fiber [5].

SUMMARY

BN coating on SiC fibers by continuous CVD process was developed for an interphase of SiC/SiC composites. Loosing the yarn bundle by reducing the yarn tension and removing the sizing agent prior to the CVD furnace improved a nonuniformity of BN thickness across the fiber tow drastically. The coated BN was ordered structure, pure with little O and C (<5at.%), nearly stoichiometric, and close to the ideal interlayer spacing. Furthermore, the strength of the coated SA fiber retained 90% of the strength of as-received SA fiber. The above mentions indicate that the BN coating obtained in this study could be applicable for an interphase material for SiC/SiC composites.

REFERENCES

[1]R. Naslain, O. Dugue and A. Guette, "Boron Nitride Interphases in Ceramic-Matrix Composites," *J. Am. Ceram. Soc.*, **74**, 2482-2488 (1991).

[2]N. P. Bansal and Y. L. Chen, "Chemical, Mechanical and Microstructural Characterization of Low-Oxygen Containing Silicon Carbide Fibers with Ceramic Coatings," *J. Mater. Sci.*, **33**, 5277-5289 (1998).

[3]G. N. Morscher, D. R. Bryant and R.E. Tressler, "Environmental Durability of BN-Based Interphases (For SiC$_f$/SiC$_m$ Composites) in H$_2$O Containing Atmospheres at Intermediate Temperatures," *Ceram. Sci. Eng. Proc.*, **18**, 525-534 (1997).

[4]G. N. Morscher, "Tensile Stress Rupture of SiC$_f$/SiC$_m$ Minicomposites with Carbon and Boron Nitride Interphases at Elevated Temperatures in Air," *J. Am. Ceram. Soc.*, **80**, 2029-2042 (1991).

[5]T. Ishikawa, Y. Kohtoku, K. Kumagawa, T. Yamamura and T. Nagasawa, "High-Strength Alkali-Resistant Sintered SiC Fibre Stable to 2,200℃," *Nature*, **391**, 773-775 (1998).

SiC CERAMIC FIBERS SYNTHESIZED FROM POLYCARBOSILANE-POLYMETHYLSILANE POLYMER BLENDS

Masaki Narisawa, Masaki Nishioka, and Kiyohito Okamura
Graduate School of Engineering
Osaka Prefecture University
1-1, Gakuen- Cho
Sakai 599-8531, Japan

Kunio Oka, and Takaaki Dohmaru
Institute of Advanced Science and Technology
Osaka Prefecture University
1-2, Gakuen-Cho
Sakai 589-8570, Japan

ABSTRACT

Polymethylsilane (PMS) synthesized by Wurtz type polycondensation reaction was blended to polycarbosilane (PCS) as an additive in melt-spinning process. The prepared blend precursors can be spun into fiber form in the range of 0-1 mass % PMS contents. At 1273K for pyrolysis temperature, the resulting fiber strength decreased as the PMS content increased. At 1573K, however, the strength of the fiber derived from the PMS blend precursor was higher than that of the fiber derived from pure PCS. In the fiber derived from PMS blend precursors, mechanical degradation and crystallite growth at 1673-1773K were reduced.

INTRODUCTION

Since the original work of SiC fiber synthesis from polycarbosilane (PCS),[1] many efforts have been aimed at the design of new SiC precursors. Various innovations in the synthesis technique have contributed new SiC fiber production with controlled chemical compositions, microstructures, and improved mechanical properties.[2,3] There are now a great number of reviews on the SiC fiber production from the viewpoints of synthetic chemistry, ceramic science and engineering applications.[4,5]

Today in industry, we can find that various polymer-based materials have been produced from rather limited types of fundamental low-price polymers. In these cases, blending, cross-linking, and selection of filler are key technologies. In the field of the SiC fiber production, the use of blend techniques for the fiber spinning process has been studied in recent years. By adjusting polymeric additives, as spinning aids or cross-linking agents, the fibers with decreased oxygen content can be synthesized by skipping the oxidation curing step.[6-8] The use of such additives, however, has been examined mainly in the dry spinning process. In the dry spinning process, the viscosity control of the prepared solution and the environment control of the solvent removal process are sometimes difficult, particularly in mass scale production. Utilization of blend techniques in the case of melt-spinning process is more favorable. The use of polyvinylsilane (PVS) in melting state has contributed to thin SiC fiber production.[9]

The addition of cross-linking agent in melting state, which is sensitive to temperature, will be particularly interesting, because the degree of cross-linking will be controllable by adjusting thermal history of the precursor melts. We selected polymethylsilane (PMS) to be blended for PCS in the melt-spinning process. PMS is known to form Si-Si cross-linking in the temperature range of 473-573K, which often contributes the resulting ceramic yields.[10,11] In this article, we reported the pyrolysis process and the resulting properties of the Si-C-O fibers derived from the blend precursor containing small amounts (0-1 mass%) of PMS with major PCS.

EXPERIMENTALS

PCS, in transparent solid form, was supplied from NCK. PMS was synthesized by a Wurtz type coupling reaction of methyldichlorosilane with sodium. Outline of the SiC fiber synthesis procedure is shown in Fig. 1. PCS-PMS mixed solutions in tetrahydrofuran (0-5 mass% PMS to PCS) were prepared, and dried under vacuum sufficiently to obtain a blended solid precursor. The solid precursors (10-15g) were put into a small batch melt spinning unit built in house. After the oxidation curing (453K in air), the cured fibers were pyrolyzed at 1273K. In order to investigate thermal degradation process, the fiber pyrolyzed at 1273K were heat-treated at 1573-1773K using the furnace equipped with molybdenum silicide heating elements.

RESULTS AND DISCUSSION

The obtained precursor can be spun into fiber form at 600K with the 0-1 mass% PMS contents in an Ar atmosphere. Precursor with 5mass% PMS, however, can not be spun into fiber form because of the melt hardening during the temperature rising. Precursor with 0.5mass% PMS indicates the best spinnability with the decreased line scissions (maximum drum speed: 8 m/sec).

After the oxidation curing, IR spectra of the obtained fibers showed the broad absorption peaks at 3600-3400 cm⁻¹ assigned to Si-OH groups and shoulders at 1100 cm⁻¹ assigned to Si-O-Si bridges. The absorption at 2100 cm⁻¹ assigned to Si-H groups

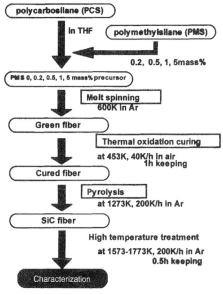

Fig. 1. Experimental Procedure.

decreased by the oxidation curing. The addition of PMS, however, had no influence on these changes in spectra. Perhaps, the amount of PMS addition was too small to influence thermal oxidation process of the blend fibers.

TG analyses of the cured fiber, however, indicate that the PMS addition increase the residual mass after pyrolysis (Fig. 2). In particular, the mass loss at 500-700K, which has been assigned to oligomer evolution from the precursor fiber, is reduced, while the mass loss at 850-1000K assigned to CH_4 evolution is not influenced by the PMS addition.

In the range of 0-1 mass% PMS, all the fibers maintain smooth appearance and the fractured surfaces show glassy appearances. Table 1 shows the diameters and the strength of the fibers pyrolyzed at 1273K. In spite of well spinnability of the 0.5 mass% PMS precursor, there are no differences in the fiber diameters. On the other hand, the strength of the fiber decreases as the PMS content increases.

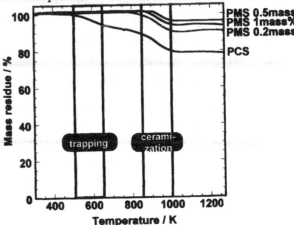

Fig. 2. TG-curves of the precursor fibers (10 K/min in Ar).

Table1 Diamaters and strength of the fibers after pyrolysis at 1273K.

	PCS	PMS0.2	PMS0.5	PMS1
Diameters after melt spinning	20.2 µm	20.1 µm	19.9 µm	20.4 µm
Diameters after curing	19.4 µm	20.3 µm	21.3 µm	20.2 µm
Diameters after pyrolysis	13.1 µm	13.4 µm	13.7 µm	14.3 µm
Strength of the pyrolyzed fibers	2.32 GPa	2.14 GPa	2.05 GPa	1.97 GPa

Table2 Tensile strength of the fibers after heat treatments.

	1573K	1673K	1773K
The fiber derived from PCS	2.20 GPa	0.86 GPa	0.20 GPa
The fiber derived from PMS0.5	2.42 GPa	1.12 GPa	0.49 GPa

These fibers pre-pyrolyzed at 1273K were heat-treated at 1573-1773K to evaluate the heat resistance. Table 2 shows the resulting tensile strength after high temperature treatments. The strength of the fiber derived from 0.5 mass% PMS precursor is increased with heat treatment at 1573K, while that of the fiber derived from pure PCS is slightly decreased with heat treatment at 1573K. At higher temperatures, the fiber strength decreases. The fiber derived from the 0.5 mass% PMS blends, however, always maintains higher strength than the fiber derived from pure PCS.

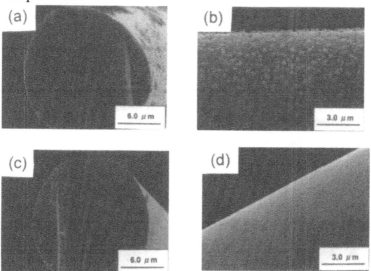

Fig. 3. SEM micrographs after heat treatment at 1673K: (a)-(b) PCS, (c)-(d) 0.5 mass% PMS.

Up to 1573K, all the fibers exhibited a smooth appearance. At 1673K, the fiber derived from pure PCS indicates the precipitation of SiC large crystallites on its surface,[12] while the fiber derived from 0.5 mass% PMS precursor reveals a smooth surface (Fig. 3). Apparent crystallite sizes in the heat-treated fibers estimated from the XRD patterns, however, indicate that the difference in the crystallite size becomes substantial at 1773K (Fig. 4). The SiC crystallite precipitation at 1673K (0.5 mass% PMS precursor) is probably limited at the surface area. At 1773K, inner area of the fiber also becomes crystalline.

TEM micrographs of the inner area of the fibers after heat treatment at 1773K are shown in Fig. 5. The observed crystallite sizes in the TEM micrographs are somewhat heterogeneous in the case of pure PCS precursor as compared with that observed in the case of 0.5 mass% PMS precursor. Such reduced crystallite growth not only at the surface area but also at the inner area probably contributes the strength maintenance even after high temperature degradation.

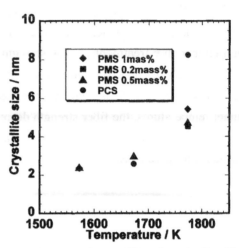

Fig. 4. Apparent crystallite sizes in the fibers after heat treatments.

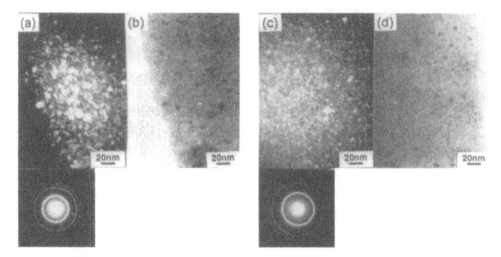

Fig. 5. TEM micrographs of the inner area of the fibers after heat treatment at 1773K: (a)-(b) PCS, (c)-(d) 0.5 mass% PMS.

Advanced SiC/SiC Ceramic Composites

Since the amount of PMS is fairly small, the influence of PMS addition to the resulting chemical compositions (Si, C and O) is hardly expected. Cross-linkings introduced in the precursor fibers by the PMS addition may act as inclusions, which reduce the sudden crystallite growth at high temperature region. The quantitative investigation of cross-linking degree in the blend precursors and the resulting SiC crystallite growth will shed light on the detailed mechanism.

CONCLUSIONS

The blend of polymethylsilane (PMS) to polycarbosilane (PCS) contributes the cross-linking formation in the melting state. A small amount of the PMS addition appears to improve the spinnability, while a large amount of the addition causes the melt hardening. The fiber derived from the precursor with 0.5 mass% PMS shows the highest strength at 1573K, while the fiber derived from pure PCS shows the highest strength at 1273K. SEM micrographs shows the large SiC crystallite precipitation on the fiber surface is reduced in the fiber derived from 0.5 mass% PMS precursor. Inner area crystallization proceeding at 1773K is also reduced. TEM micrographs of the fibers indicate that the distribution in crystallite sizes is homogeneous in the case of PMS blending as comared with the case of pure PCS.

REFERENCES
[1]S. Yajima, J. Hayashi, and M. Omori, "Continuous Silicon Carbide Fiber of High Tensile Strength," *Chem. Lett.*, **1975**, 931-34 (1975).

[2]G. Chollon, R. Pailer, R. Naslain, F. Laanani, M. Monthioux, and P. Oley, "Thermal Stability of a PCS-Derived SiC Fibre with a Low Oxygen Content (Hi-Nicalon)," *J. Mater. Sci.*, **32**, 327-47 (1997).

[3]T. Ishikawa, Y. Koutoku, K. Kumagawa, T. Yamamura, and T. Nagasawa, "High-Strength Alkali-Resistant Sintered SiC Fibre Stable to 2,200 °C," *Nature* **381**, 773-75 (1998).

[4]K. Okamura, "Ceramic Fibres from Polymer Precursors," *Composites*, **18**, 107-20 (1987).

[5]J. Bill, and F. Aldinger, "Precursor-Derived Covalent Ceramics," *Adv. Mater.*, **7**, 775-87 (1995).

[6]W. Toreki, G. J. Choi, C. D. Batich, M. D. Sacks, and M. Saleem, "Polymer-Derived Silicon Carbide Fibers with Low Oxygen Content," *Ceram. Eng. Sci. Proc.*, **13**, 198-208 (1992).

[7]M. D. Sacks, G. W. Scheiffele, L. Zhang, Y. Yang, and J. J. Brennan, "Polymer-Derived SiC –Based Fibers with High Tensile Strength and Improved Creep Resistance," *Ceram. Eng. Sci. Proc.*, **19**, 73-86 (1998).

[8]F. Cao, D.-P. Kim, and X.-D. Li, "Preparation of Hybrid Polymer as a Near-Stoichiometric SiC Precursor by Blending of Polycarbosilane and Polymethylsilane," *J. Mater. Chem.*, **12**, 1-6 (2002).

[9]A. Idesaki, M. Narisawa, K. Okamura, M. Sugimoto, S. Tanaka, Y. Morita, T. Seguchi, and M. Itoh, "Fine SiC Fiber Synthesized from Organosilicon Polymers: Relationship between Spinning Temperature and Melt Viscosity of Precursor Polymers," *J. Mater. Sci.*, **36**, 5565-5569 (2001).

[10]B. Boury, N. Bryson, and G. Soula, "Borate-Catalyzed Thermolysis of Polymethylsilane," *Chem. Mater.*, **10**, 297-303 (1998).

[11]T. Iseki, M. Narisawa, Y. Katase, K. Oka, T. Dohmaru, and K. Okamura, "An Efficient Process of Cross-Linking Poly(methylsilane) for SiC Ceramics," *Chem. Mater.*, **13**, 4163-69 (2001).

[12]R. Bodet, X. Bourrat, J. Lamon, and R. Naslain, "Tensile Creep Behaviour of a Silicon Carbide-Based Fibre with a Low Oxygen Content," *J. Mater. Sci.*, **30**, 661-77 (1995).

EFFECT OF RESIDUAL SILICON PHASE ON REACTION-SINTERED SILICON CARBIDE

Shoko Suyama and Yoshiyasu Itoh
Power & Industrial Systems R & D Center, Toshiba Corporation
2-4 Suehiro-cho, Tsurumi-ku
Yokohama 230-0045, Japan

Akira Kohyama and Yutai Katoh
Institute of Advanced Energy, Kyoto University, CREST
Gokasho
Uji, Kyoto 611-0011, Japan

ABSTRACT

SiC/SiC composite has been regarded as a candidate for the construction material for future fusion reactors because of its high temperature stability and low activation. Typical fabrication processes for the SiC/SiC composites are chemical vapor infiltration (CVI), precursor impregnation and pyrolysis (PIP), hot pressing (HP), reaction sintering (RS) and their hybrid processes. The reaction sintering process realizes full density and near net shape capabilities of the large and complex shape parts. However, the residual silicon phase is retained in the RS-SiC matrix during the reaction-sintering process, and could affect the radiation response and creep behavior. The processing approach to reduce residual silicon phase and the effect of raw material composition (C/SiC) as well as the size on the residual silicon content in final material is presented.

INTRODUCTION

Ceramic matrix composite is one of the most promising candidates for high temperature structural material. SiC matrix composite is expected to be used as a material for future fusion reactors because of its low activation [1]. With regard to high thermal conductivity and high Young's modulus, a dense matrix is favorable. Reaction sintering process is one of the suitable processes to form dense matrix without firing shrinkage, and near net shape capabilities for the large and complex shape parts [2-4].

RS-SiC is typically produced by infiltrating a porous compact of SiC and carbon with liquid silicon. Upon infiltration silicon reacts with carbon to form new SiC phase, which bonds the original SiC together, and the remaining porosity in the body is filled by residual silicon from the melt [5-9]. The major disadvantage of RS-SiC materials results from the presence of the residual silicon phase, which is thought to be detrimental to the radiation response and creep resistance properties [1,6].

RS-SiC matrix composite was fabricated by pressure impregnation and casting of the slurry that was a mixture of SiC and carbon powder with water, and subsequent reaction sintering with molten silicon [3,10]. In order to decrease the residual silicon content in RS-SiC matrix, the processing conditions of pressure casting of the slurry casting were studied. In this work, the effect of the raw material composition ratio (C/SiC) and original SiC particle size on residual silicon phase was examined.

EXPERIMENTAL

The process flowchart of RS-SiC was shown in Fig. 1. The slurry was a mixture of SiC powder (Showa denko co.), carbon powder (Cancarb Ltd.) and water with some dispersant. The green body of $45 \times 45 \times 5$mm plate was prepared by pressure casting of the slurry. The green body was dried and then reaction-sintered with Si melt at 1693K in vacuum for 1 hour.

The density of the sintered body was measured by the Archimedes principle using water, after surface machining. The polished surface was observed with an optical microscope. The volume fraction of residual silicon phase in the sample was calculated from the measured density of the sintered body and the theoretical densities of SiC and Si. The mean size of residual silicon phase was measured by the mercury intrusion method, after heat treatment. The sintered body was heated at 1873K in vacuum to vaporize the residual silicon phase [11], and the mean pore size in the sample was measured [10].

RESULTS AND DISCUSSION

The effect of the raw material composition ratio (C/SiC) and the original SiC particle size on the green and sintered density are shown in Fig. 2 (a) and (b). It was found that the green density was increased by increasing the C content ratio and the SiC particle size. It was found that the sintered density was increased by increasing the C content ratio and SiC particle size. In this work, the sintered density was increased up to 3.1 Mg/m^3 at the processing condition which was 0.5 of C/SiC for 0.1 μ m of SiC particle size.

Fig. 3 shows the material composition of RS-SiC before and after the reaction sintering. In the reaction sintering, macroscopic volume change was known to be less than 1%. Then, the material composition after the reaction sintering was calculated on the assumption that the volume is equal. It was made clear that the increasing of C content ratio could be contributed to decrease the residual Si by volume expansion of new SiC formed.

The effect of raw material composition (C/SiC) and SiC particle size on volume fraction of residual silicon phase are shown in Fig. 4 (a). The volume fraction of residual silicon phase was decreased by increasing the C content ratio and SiC particle size. The volume fraction of residual silicon phase was reduced to 12.1% at the C/SiC ratio of 0.5 for 1 μ m SiC powder. The effect of raw material composition (C/SiC) and SiC particle size on size of residual silicon phase are shown in Fig. 4 (b). The size of a residual silicon phase showed the tendency to be decreased by increasing ratio of carbon content and decreasing the SiC particle size. The size of residual silicon phase was reduced to 0.14 μ m at the C/SiC ratio of 0.55 for 1 μ m SiC powder. The size of residual silicon phase seemed to be decreased by increasing of C content ratio and decreasing of the SiC particle size.

Fig. 5 (a) shows the microphotographs of RS-SiC samples at 0.1, 0.3 and 0.4 of C/SiC for 4 μ m SiC powder. Fig. 5 (b) shows the microphotograph of RS-SiC samples at 0.4, 0.45, 0.5 and 0.55 of C/SiC for 1 μ m SiC powder. Fig. 6 shows the microphotograph of RS-SiC samples at 4 μ m, 2 μ m and 1 μ m of SiC powders for C/SiC=0.4. The gray grain was SiC particle and the white range around it was residual Si phase. All RS-SiC samples were homogeneous and isotropic microstructure on the whole and the pore and un-reacted carbon was not observed.

CONCLUSIONS

In order to develop lower Si content of RS-SiC matrix that was fabricated by pressure casting of a slurry and a subsequent reaction sintering process, the processing conditions were examined;

At the C/SiC ratio of 0.5 for 1 μ m SiC powder, residual silicon phase was reduced to 12.1%, which mean size of residual silicon phase was 0.17 μ m.

REFERENCES

[1] R.W. Conn, E.E. Bloom, J.W. Davis, R.E. Gold, R. Little, K.R. Schultz, D.L. Smith and F.W. Wiffen, "Lower Activation Materials and Magnetic Fusion Reactors", *Nuclear Technology / Fusion*, **5**, 291-310 (1984).

[2] K.L. Luthra, R.N. Singh and M.K. Brun, American Ceramic Society Bulletin, "Toughened Silicomp Composites – Process and Preliminary Properties", *Am. Ceram. Soc. Bull.*, **72** [7] 79-85 (1993).

[3] T. Kameda, Y. Itoh, T. Hijikata and T. Okamura, "Development of Continuous Fiber Reinforced Reaction Sintered Silicon Carbide Matrix Composite for Gas Turbine Hot Parts Application", *Proceeding of ASME TURBOEXPO 2000*, 2000-GT-67.

[4] G.S. Corman, A.J. Dean, S. Brabetz, M.K. Brun, K.L. Luthra, L. Tognarelli, M. Pecchioli, "Rig and Engine Testing of Melt Infiltrated Ceramic Composites for Combustor and Shroud Application", *Proceeding of ASME TURBOEXPO 2000*, 2000-GT-638.

[5] P. Popper, "The Preparation of Dense Self-bonded Silicon Carbide", pp.209-19 in *Special Ceramics*, Heywood & Co., London, (1960).

[6] B.A. Fields and S.M. Wiederhorn, "Creep Cavitation in a Siliconized Silicon Carbide Tested in Tension and Flexure", *J. Am. Ceram. Soc.*, **79**, 977-86 (1996).

[7] E. Scafe, G. Giunta, L. Fabbri, L.Di Rese, G.De Portu and S. Guicciardi, "Mechanical Behavior of Silicon-Silicon Carbide Composites", *J. European Ceram. Soc.*, **16**, 703-13 (1996).

[8] R.F. Davis, C.H. Cater Jr., S.R. Nutt, K.L. More and S. Chevacharoenkul, "Structure and Chemistry of Interfaces in Silicon Carbide-Containing Materials", *Mater. Sci. Res.*, **21**, 897-910 (1987).

[9] S. Ramanathan, R. Prasad and C.K. Gupta, "Reaction Sintering Studies of Silicon Carbide", *Transaction of the Indian Ceramic Society*, **46**, 107-14 (1989).

[10] S. Suyama, Y. Itoh, S. Nakagawa, N. Tachikawa, A. Kohyama and Y. Katoh,

"Effect of Residual Silicon Phase on Reaction-Sintered Silicon Carbide", The 3rd International Energy Agency Workshop on SiC/SiC ceramic composites for fusion structural application, 108-112 (1999).

[11]K. Shobu, E. Tani, M. Akiyama, and T. Watanabe, "High-Temperature Strength of Melt-Infiltrated SiC-Mo(Al,Si)$_2$ Composites", *J. Am. Ceram. Soc.*, **79**, 544-46 (1996).

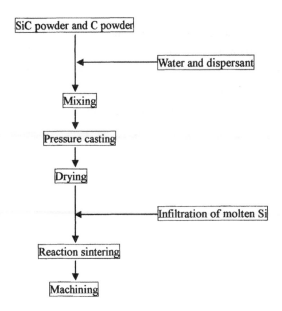

Fig. 1. Flow chart of fabrication process for reaction-sintered silicon carbide.

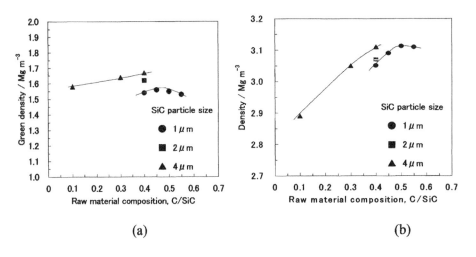

Fig. 2. Effect of raw material composition (C/SiC) and SiC particle size on green and sintered density; (a) green density, (b) sintered density.

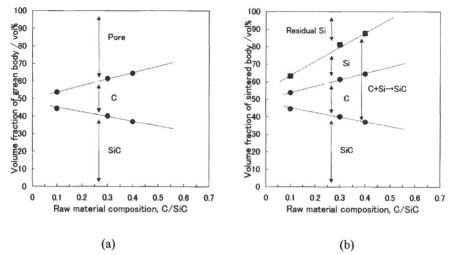

(a)   (b)

Fig. 3. Material composition of RS-SiC before and after reaction sintering; (a) green body, (b) sintered body.

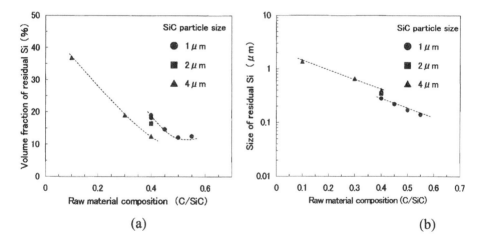

(a)   (b)

Fig. 4. Effect of raw material composition (C/SiC) and SiC particle size on volume fraction and size of residual silicon phase; (a) volume fraction of residual silicon phase, (b) size of residual silicon phase.

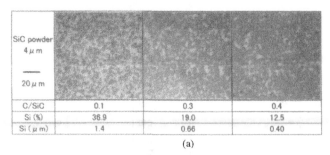

	C/SiC	0.1	0.3	0.4
SiC powder 4 μ m	Si (%)	36.9	19.0	12.5
20 μ m	Si (μ m)	1.4	0.66	0.40

(a)

SiC powder 1 μ m	C/SiC	0.4	0.45	0.5	0.55
	Si (%)	19.0	14.6	12.1	12.5
20 μ m	Si (μ m)	0.28	0.22	0.17	0.14

(b)

Fig. 5. Microphotographs of RS-SiC samples; (a) C/SiC=0.1, 0.3 and 0.4 for 4 μ m SiC powder, (b) C/SiC=0.4, 0.45, 0.5 and 0.55 for 1 μ m SiC powder.

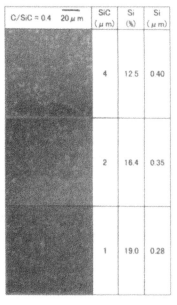

C/SiC = 0.4 20 μ m	SiC (μ m)	Si (%)	Si (μ m)
	4	12.5	0.40
	2	16.4	0.35
	1	19.0	0.28

Fig. 6. Microphotographs of RS-SiC samples; 4 μ m, 2 μ m and 1 μ m of SiC powders for C/SiC=0.4

Characterization of Thermomechanical Performance

DEVELOPMENT OF TEST STANDARDS FOR CONTINUOUS FIBER CERAMIC COMPOSITES IN THE UNITED STATES

Edgar Lara-Curzio
Oak Ridge National Laboratory
Oak Ridge, TN 37831-6069

ABSTRACT

The development of standardized test methods for the mechanical evaluation of ceramic matrix composites in the United States is reviewed. In particular, the activities of sub-committee C28.07 on Ceramic Matrix Composites of the American Society for Testing and Materials (ASTM) are reported. Efforts for international harmonization of test methods for these materials and their constituents are also discussed.

INTRODUCTION

Over the last twenty years continuous fiber-reinforced ceramic composites (CFCCs) have been the focus of intensive developmental efforts in the U. S., Europe and Japan. These efforts have been driven to a large extent by the promise of substantial economic and environmental benefits if CFCCs were used in military, energy-related and other industrial applications [1]. For example, SiC/SiC and other CFCCs are leading candidate materials for the fabrication of components in advanced energy systems such as the primary wall of nuclear fusion reactors and combustor liners for gas turbine engines.

CFCCs are attractive materials for these applications because they are notch insensitive, they exhibit a non-catastrophic mode of failure and have the potential of retaining their properties at elevated temperatures. However, their unique mechanical behavior requires that new and special methodologies be used for designing and fabricating components for the aforementioned applications. Traditionally designers and engineers perform preliminary designs of components using computer-aided design tools. Then, making a material selection from a materials database they evaluate structural requirements using finite-element stress analyses, and assess the long-term resistance of the design to environmental degradation and service loads. Although the stochastic nature of the matrix and fiber strength in CFCCs and other factors that differentiate CFCCs from conventional engineering materials will necessitate the use of probabilistic design methods and specialized design tools, experimental measurements of material properties, according to standardized test methods, are and will continue being the data most needed by designers (Figure 1). In the United States, the American Society for Testing and Materials (ASTM) has spearheaded the widespread introduction of standard test methods for advanced ceramics and ceramic composites. Concurrent with these efforts, committees within the Military Handbook 17 and the ASME Pressure Vessel Code [2] are actively developing databases and codes for designing engineering structures with CFCCs. Similar efforts have been initiated in Europe and Japan [3].

STANDARD TEST METHODS

Since its inception in 1988, ASTM committee C-28 on Advanced Ceramics has been responsible for the creation of 42 standard test methods. These standards range from clones of prior ASTM standards (with some new provisions) to complex, innovative documents tailored specifically to advanced ceramics. Committee C28 of ASTM is structured into ten subcommittees as indicated in Figure 2. Since its establishment in 1991, subcommittee C28.07 has been responsible for formalizing 10 full consensus standard test methods for CFCCs and for drafting several other documents that are currently undergoing through ASTM's internal balloting process. Table 1 lists the standards for CFCCs that have been formalized by subcommittee C28.07, along with drafts that are in the process of becoming standard test methods. In addition to its main mission of developing standards, subcommittee C28.07 has been involved in disseminating information in the US CFCCs community about progress in the standardization process through the organizations of workshops and symposia[1]. The work of subcommittee C28.07 is organized around task groups, which are responsible for the preparation of drafts of test methods, and for the progress of these documents as they evolve through the balloting and approval process at the subcommittee, committee and society levels. Table 2 lists the task groups active in C28.07.

ASTM thrives for well-written test methods that specify control over such factors as the test equipment, the test environment, the qualifications of the operator, the preparation of test specimens, and the operating procedure for using the equipment in the test environment to measure some property of the test specimens [4]. The test method should also specify the number of test specimens required and how measurements on them are to be combined to provide a test result. During their development, many standard test methods are subjected to a screening procedure and ruggedness tests in order to establish the proper degree of control over factors that may affect the test results. One of the most important features of ASTM test standards is the requirement to include precision and bias statements. A statement of precision allows potential users of a test method to assess in general terms the test method's usefulness with respect to variability in proposed applications [4]. A statement on precision provides guidelines as to the kind of variability that can be expected between test results when the method is used in one or more reasonably competent laboratories. A statement on bias furnishes guidelines on the relationship between a set of typical test results produced by the test method under specific test conditions and a related set of accepted reference values [4]. ASTM has established procedures for determining precision and bias statements through round robin testing programs. The U.S. Department of Energy through the Continuous Fiber-Reinforced Ceramic Matrix Composites Program and the U.S. Air Force have sponsored two round robin program for determining precision and bias statements for several ASTM standard test methods for CFCCs. In the following sections, the more relevant features of existent ASTM standard test methods for CFCCs are reviewed.

In-Plane Tensile Testing

The first standard test method developed for CFCCs by ASTM sub-committee C28.07 was test method C1275 for *Monotonic Tensile Strength of Continuous Fiber-Reinforced Advanced Ceramics with Solid Rectangular Cross-Section Specimens at Ambient Temperatures*. Later on, this standard test method became a template for other documents involving tensile testing, such as C1337 *Standard Test Method for Creep and Creep Rupture of Continuous Fiber-Reinforced Advanced Ceramics under Tensile Loading*, C1359 *Standard Test Method for Monotonic Tensile Strength Testing of Continuous Fiber-Reinforced Advanced Ceramics with Solid Rectangular Cross Sections at Elevated Temperatures*, and C1360 *Standard Practice for Constant-Amplitude, Axial, Tension-Tension Cyclic Fatigue of Continuous Fiber-Reinforced Advanced Ceramics at Ambient Temperatures*.

---

[1] "Workshop on Thermal and Mechanical Test Methods and Behavior of CFCCs," Montreal, Canada June 1994; "Workshop on Thermomechanical Tests for CFCCs," Bozeman, Montana, October 18, 1994; "Symposium on Thermal and Mechanical Tests Methods and Behavior of CFCCs," Cocoa Beach, FL. January 8-9, 1996, . "Symposium on Mechanical, Thermal and Environmental Characterization of CFCCs," Seattle, WA June 2000.

The most important issues addressed by these test methods are: gripping devices, test system alignment, strain measurements, specimen geometry, specimen preparation, mode and rate of testing, and data acquisition. Although C1275 allows the use of any specimen geometry if it meets the gripping, fracture location, and fracture requirements prescribed in the document, in the case of two-dimensionally reinforced (2-D) CFCCs it recommends the use of specimens with a contoured gauge section (Figure 3). This document also addresses the need to test specimens having dimensions (e.g., volume) that are consistent with the ultimate use of the tensile data. However, design codes will ultimately provide guidance on how to interpret test results obtained with standardized test specimens to design larger components. This is important because those mechanical properties of CFCCs that are dominated by the reinforcing fibers (e.g.- in-plane tensile strength) are known to depend on the volume of material being stressed.

During tensile testing the applied force is transferred to the test specimens through gripping devices. Gripping devices can have either active (e.g., hydraulically-actuated grips) or passive (e.g., edge or pin-loaded arrangements) interfaces, and the availability of a given type of testing arrangement will dictate the type of specimen geometry and vice versa. Gripping devices are typically attached to the test system through couplers, which can be classified either as fixed or non-fixed. However, regardless of the type of coupler used, C1275, C1337, C1359 and C1360 mandate the verification of the alignment of the test system either prior to each test or before and after a series of tests. Although recent studies have revealed that the ultimate tensile strength of some CFCCs is relatively insensitive to percent bending[2], other properties, such as the proportional limit stress do depend on percent bending [5]. Therefore, the maximum allowable percent bending as defined in ASTM E1012 should not exceed five.

Test standards addressing the tensile evaluation of CFCCs also cover different techniques for strain measurements, such as optical methods using lasers and flags, or contact methods such as adhesively bonded strain gauges and extensometers. However, regardless of the type of extensometer used, these test methods require that extensometers satisfy Class B-1 requirements as outlined in Practice E83 [6]. Additional requirements for contact-type extensometers are that the extensometer should not damage the specimen, and to be externally supported so that its weight doesn't introduce bending strains greater than those allowed.

Standard test methods that address tensile testing at elevated temperatures (e.g.- C1337, and C1359) have special provisions in addition to the requirements outlined in C1275. These include, for example, issues such as heating methods and temperature measurements. These test methods also address the importance of mechanical testing rates, since CFCCs may exhibit time-dependent deformation at elevated temperatures. All of the existent tensile-based ASTM standard test methods for CFCCs address tests conducted in ambient air, but given the potential of environmental effects on the mechanical behavior of CFCCs, ongoing work has been focused on developing test methods to conduct tests in controlled environments other than air (e.g.- steam, inert gases, combustion gases).

The tensile behavior of a CFCC is not deterministic, but varies from one test specimen to another. Sources of variability include inherent variations in the properties of ceramic fiber reinforcements, ceramic matrices and fiber coatings, fiber architecture and volume fraction. Such variations can occur spatially within a given test specimen, as well as between different test specimens. A multiple laboratory round-robin test program sponsored by the U.S. Department of Energy and the U.S. Air Force, was conducted in 1998 to determine the precision of tensile properties measurements in accordance with Test Method C1275 for a commercially-available

---

[2] This apparently results from loading large aspect ratio-fibers when these bridge matrix cracks in the composite.

CFCC [7]. Repeatability and reproducibility were assessed only for modulus of elasticity, proportional limit stress (extension under load method at 0.001 mm/mm), ultimate tensile strength and strain at fracture. Ninety randomly divided test specimens were tested in sets of ten by nine different laboratories. Bias was not evaluated because there is no commonly recognized standard reference material for CFCCs.

The tensile test specimens (150 mm long X 10 mm wide with reduced gage sections of 35 mm long and 8 mm wide) were diamond-grit cut from three panels (nominally 3 mm thick) of a commercial Sylramic® S200 CFCC[3]. The panels were fabricated with eight plies of ceramic grade (CG)-Nicalon™ fabric (8–Harness Satin) in a SiNCO matrix. The Nicalon™ fiber tows had a proprietary boron nitride coating. The as-fabricated tensile test specimens had a nominal density of 2200 kg/m^3, a nominal fiber volume fraction of 45 % and average open porosity of 2.7 %. A statistical analysis of the tensile test results was performed using the procedures and criteria of Practice E 691. All the results for elastic modulus, proportional limit stress, ultimate tensile strength, and strain at fracture were determined to be valid and applicable and the repeatability and reproducibility results are listed in Table 3. Figure 4 shows the summary of the tensile strength results reported by each laboratory along with the value of the great mean.

At the time of this publication a round-robin test program was in progress to determine precision statements for standard test methods C1337 and C1359. To eliminate potential interference of environmental effects on the the high temperature tensile results, the material used in this study consists of an all-oxide CFCC.

Transthickness Tensile Testing

The in-plane tensile properties of 1-D and 2-D CFCCs are determined primarily by the ability of the fibers to debond and slide in the wake of advancing cracks in the matrix. Over the last 15 years significant progress has been made in the synthesis of fibers and fiber coatings, and in the engineering and design of fiber-matrix interfaces with low fracture toughness to optimize the in-plane tensile properties of CFCCs. However, the optimization of the in-plane tensile properties of CFCCs has been achieved at the expense of their transthickness and interlaminar properties. Consequently, when 1-D or 2-D CFCCs are subjected to tensile stresses in the transthickness direction the fiber-matrix interfacial region becomes the weakest link. Because the magnitude of the transthickness tensile strength of most 1-D and 2-D CFCCs is only a small fraction of their in-plane tensile strength the former will dictate the design of components that are subjected to multiaxial states of stress.

Standard test method C1468-00 *Test Method for Transthickness Tensile Strength of Continuous Fiber-Reinforced Advanced Ceramics at Ambient Temperatures* was approved in 2000 to determine the transthickness tensile strength of CFCCs under monotonic loading. Because most CFCCs are prepared as "thin" plates it is difficult to grip directly a test specimen to conduct a transthickness tensile test. Therefore C1468 recommends the use of adhesives to bond the specimen to fixtures that are pulled apart using a universal testing machine (Figure 5). Studies have shown that Nicalon™/MAS-5 and SCS-6/Si$_3$N$_4$ CFCCs square or circular test specimens with a cross-sectional area of about 282 mm^2 exhibit no geometry effect on the measured transthickness tensile strength [8]. However, work is still needed to determine the effect of actual volume or surface area on the transthickness tensile strength. A mini-round robin testing program involving 4 laboratories to determine the transthickness tensile strength of a 2-D Nicalon™/SiNCO CFCC revealed that the variability and scatter of the results was directly associated with the degree of microstructural uniformity in the material. Figure 6 shows the results of this study using test specimens with a square cross section (10 mm X 10 mm).

---

[3] COI Ceramics Inc. San Diego, CA 99999

Because this test method relies on the use of an adhesive to transfer forces from the load train to the test specimen, this test would be difficult to implement at elevated temperatures. This limitation has prompted efforts to develop alternative test methods to determine the transthickness tensile strength of CFCCs at elevated temperatures. For example, researchers at NASA Glenn Research Center have proposed the use of C-coupon specimens loaded in tension [9]. However, this test method requires the preparation of special test specimens, which may or may not have the same properties of the actual material. Furthermore, the state of stress in this specimen configuration is not uniform and the interpretation of the experimental results is not straightforward. Recent work has shown that the Brazilian test (diametral compression) is a good candidate for the determination of the transthickness tensile strength of these materials [10]. When a disk-shaped specimen is subjected to diametral compression, tensile stresses are induced normal to a planes parallel to the loading direction. Figure 7 shows a schematic of the loading configuration along with CFCC test specimens before and after being subjected to diametral compression. Current work within C28.07 has been focused on drafting a test method based on the Brazilian test, which has the advantage of being applicable for testing at elevated temperatures.

Shear Testing

Except for the torsion testing of thin-walled tubes, there is no single "good" test to measure the shear strength of CFCCs [11]. Because it would be prohibitively expensive to prepare and test tubular specimens to measure the shear strength of CFCCs, standard test method C1292 for *Shear Strength of Continuous Fiber-Reinforced Advanced Ceramics at Ambient Temperatures* addresses two popular, but less than perfect, test methods for determining the shear strengths of uni-directionally and two-directionally reinforced CFCCs: the compression of a double-notched specimen to determine interlaminar shear strength, and the Iosipescu test method to determine both in-plane and interlaminar shear strength. Schematics of the specimen geometries and test configurations for these two tests are shown in Figure 8. Since both of these test methods have been widely used for the evaluation of polymer matrix composites, much of the groundwork for the drafting of C1292 had already being laid down.

Although both the compression of double-notched specimens and Iosipescu's test have the advantage of requiring relatively small specimens and being simple to conduct, their main disadvantage is that both rely on the stress-concentration at the root of notches to initiate shear failure. As a consequence, in the case of the compression of double-notched specimens for example, the shear stress in the gauge section (the region between notches) is not uniform. Furthermore, it has been found that the apparent interlaminar shear strength depends on the distance between the notches [12].

A multiple laboratory round-robin test program sponsored by the U.S. Department of Energy and the U.S. Air Force, was conducted in 1998 to determine the precision of shear strength measurements in accordance with Test Method C1292 for a commercially-available CFCC [7]. Both the in-plane and interlaminar shear strength of the material were reported. Eighty randomly divided Iosipescu test specimens were tested in sets of ten by eight different laboratories to determine in-plane shear strength, whereas seventy randomly divided double-notched test specimens were tested in sets of ten by seven different laboratories to determine interlaminar shear strength. Bias was not evaluated, because there is no commonly recognized standard reference material for CFCCs.

The Iosipescu test specimens (78 mm long X 19 mm wide with V-notches) were diamond-grit cut from three panels (nominally 3 mm thick) of a commercial Sylramic® S200 CFCC[4]. The

---

[4] COI Ceramics Inc. San Diego, CA 99999

panels were fabricated with eight plies of ceramic grade (CG)-Nicalon™ fabric (8–Harness Satin) in a SiNCO matrix. The Nicalon™ fiber tows had a proprietary boron nitride coating. The as-fabricated shear test specimens had a nominal density of 2200 kg/m^3, a nominal fiber volume fraction of 45 % and an average open porosity of 2.7 %. The double-notched specimens (30 mm long X 15 mm wide with notches machined half-way on opposite sides of the specimen and separated 6 mm apart) were also diamond-grit cut from the same three panels from which the Iosipescu specimens were obtained.

A statistical analysis of the test results was performed using the procedures and criteria of Practice E 691. All the results for interlaminar and in-plane shear strength were determined to be valid and applicable and the repeatability and reproducibility results are listed in Table 4. Figures 9 and 11 show the summary of the shear strength results reported by each laboratory along with the value of the great mean. While the degree of scatter for the inter- and intralaboratory in-plane shear strength results is very small, there was a high degree of scatter for the interlaminar shear strength results. It was found that the lack of uniformity in the microstructure of the material and the location of the notches with respect to the microstructure (e.g.- interlaminar region) where responsible for the variability in the results. Figure 10 shows the edge of a double-notched test specimen after a test and illustrates the location where the interlaminar crack initiates and along which planes it propagates.

In 1999 *Standard Test Method C1425 for Interlaminar Shear Strength of 1–D and 2–D Continuous Fiber-Reinforced Advanced Ceramics at Elevated Temperatures* was formalized to determine the interlaminar shear strength of CFCCs at elevated temperatures based on the compression of double-notched test specimens.

Flexure Testing

One of the motivations for developing a standard test method to evaluate CFCCs in tension was attributed to the discrepancies between, and the wide range of, "strength" values reported when CFCCs were evaluated in flexure. The problem is that when testing CFCCs in flexure it is difficult to relate the forces and displacements measured during the test to key material properties (e.g. tensile strength, work of fracture) through a simple analysis. As a result, beam theory equations that are derived for linear elastic, isotropic, homogeneous materials that exhibit symmetric behavior in tension and compression, and that have been customarily used to calculate the so-called "flexural strength" of CFCCs are not applicable for these materials. In spite of this, flexural testing has been and remains a popular test method in industrial laboratories because of its simplicity and because it requires relatively small samples. As a result of the interest expressed by industry to continue using this method, standard test method C1341 *Standard Test Method for Flexural Properties of Continuous Fiber Reinforced Advanced Ceramics* was developed and approved in 1996. The scope of C1341 is limited to use "flexural data" for quality control and material development, and in contrast to flexural data for monolithic ceramics, the use of CFCC flexure data for design purposes is discouraged.

A multiple laboratory round-robin test program sponsored by the U.S. Department of Energy and the U.S. Air Force, was conducted in 1998 to determine the precision of flexural "strength" measurements in accordance with Test Method C1341 for a commercially-available CFCC [7]. Both the in-plane and interlaminar shear strength of the material were reported. One hundred randomly divided flexural test specimens were tested in sets of ten by ten different laboratories to determine their "flexural" strength. Bias was not evaluated, because there is no commonly recognized standard reference material for CFCCs.

The flexural test specimens (110 mm long X 9 mm wide) were diamond-grit cut from three panels (nominally 3 mm thick) of a commercial Sylramic® S200 CFCC[5]. The panels were fabricated with eight plies of ceramic grade (CG)-Nicalon™ fabric (8–Harness Satin) in a SiNCO matrix. The Nicalon™ fiber tows had a proprietary boron nitride coating. The as-fabricated flexural test specimens had a nominal density of 2200 kg/m^3, a nominal fiber volume fraction of 45 % and an average open porosity of 2.7 %. The tests were carried out in four-point bending. A statistical analysis of the test results was performed using the procedures and criteria of Practice E 691. All the results for "flexural" strength were determined to be valid and applicable and the repeatability and reproducibility results are listed in Table 5. Figure 12 shows the summary of the "flexural" strength results reported by each laboratory along with the value of the great mean.

Fiber Testing

In 1998 the jurisdiction of test method D3379 (*Test Method for Tensile Strength and Young's Modulus for High-Modulus Single-Filament Materials*) was transferred from ASTM committee D30 to subcommittee C28.07. This move coincided with on-going efforts within subcommittee C28.07 to develop a standard test method for the tensile evaluation of ceramic fibers. The development of a new standard test method was motivated by the shortcomings of D3379. Specifically, it had been found that the determination of fiber tensile strength according to D3379 was inappropriate, in particular when the diameter of fibers varied significantly among fibers in a bundle [13-15]. The new standard test method for fiber testing is currently at the last stages of ASTM's internal balloting process and addresses issues associated with sample preparation, determination of fiber dimensions, determination of strength and Young's modulus.

Other

Other documents dealing with the thermomechanical behavior of CFCCs, the determination of CFCC fiber-matrix interfacial properties and the hoop strength of CFCC tubular components are currently in draft form or undergoing the ASTM balloting process. In 2000 standard test method Standard Test Method C1469 for *Shear Strength of Joints of Advanced Ceramics at Ambient Temperature* was adopted for determining the shear strength of joints by asymmetric four-point bending.

INTERNATIONAL HARMONIZATION

The U.S. is represented at the technical committees (TC) of the International Organization for Standardization (ISO) through technical advisory groups (TAG) of the American National Standards Institute (ANSI). In 1994, ANSI named ASTM administrator of TAG-206 to represent the U.S. at ISO's TC206 on Fine (Advanced, Technical) Ceramics. In February of 1997, a working group (WG9) on tensile behavior of CFCCs was officially established as part of TC206 with the U.S. serving as the convenor. In 1999 this working group produced the first ISO standard for CFCCs. Working groups (WG20 and WG21) have also been established to develop ISO standards for the in-plane and interlaminar shear strength of CFCCs and documents are at the late stages of standardization. The significance of ISO standards is tremendous when one considers that disputes between countries that are members of the General Agreement for Tariffs and Trade (GATT) will be resolved using ISO standards.

ASTM Subcommittee C28.07 has established, and maintains, strong liaisons with other international standardization organizations such as the Comité Européen de Normalisation (CEN) Technical Committee 184. As part of a memorandum of understanding established in September of 1996 between ASTM C28.07 and subcommittee SC1 on Ceramic Matrix Composites of CEN TC184, mechanisms were put in place for exchanging documents, organizing joint round robin

---

[5] COI Ceramics Inc. San Diego, CA 99999

testing programs, technical meetings, and for synchronizing efforts towards the development of ISO standards[6].

SUMMARY

Since its establishment in 1991, ASTM Subcommittee C28.07 on ceramic matrix composites has actively promoted the development of standardized test methods for CFCCs. C28.07 has been responsible for formalizing ten consensus standard test methods for CFCCs and for drafting several other documents that are currently going through the internal ASTM balloting process. Through ANSI and ASTM, the U.S. participates in Technical Committee 206 of the International Standards Organization where the first international standard for the tensile evaluation of CFCCs was developed, and where there are standards under development for the determination of the shear strength of CFCCs. Although significant advances have been made in the area of standardization, work is still needed. For example to develop standard test methods for multiaxial evaluation, and for the mechanical evaluation of materials in special environments.

ACKNOWLEDGMENTS

This work was sponsored by the US Department of Energy, Assistant Secretary for Energy Efficiency and Renewable Energy, Office of Industrial Technologies, Continuous Fiber-reinforced Ceramic Composites Program under contract, DE-AC05-96OR22464 with UT-Battelle, LLC.

REFERENCES

1. M. A. Karnitz, D. F. Craig and S. L. Richlen, "Continuous Fiber Ceramic Composite Program," *Ceramic Bulletin*, **70**, 2 (1991) pp. 430-5
2. E. Lara-Curzio and M. G. Jenkins," Development of Test Standards for CFCCs in the US," *Composites Part A,* **30**, 4 (1999) pp. 561-567
3. Proceedings of *the 9th International Symposium on Ultra-high Temperature Materials*, Japan's Ultra-high Temperature Materials Research Institute, Tajimi City, Japan, September 9-10, 1999
4. E 177 – 96, "Standard Practice for Use of the Terms Precision and Bias in ASTM Test Methods," American Society for Testing and Materials, West Conshohocken, PA 19428
5. M. G. Jenkins, J. P. Piccola and E. Lara-Curzio, "Influence of Bending, Test Mode, Test Rate, Specimen Geometry and Grip System on the Tensile Mechanical Behavior of CFCCs," in *Thermal and Mechanical Test Methods and Behavior of Continuous Fiber Ceramic Composites*, ASTM STP 1309M. G. Jenkins, S. T. Gonczy, E. Lara-Curzio, N. E. Ashgaugh and L. P. Zawada Eds., American Society for Testing and Materials, Philadelphia, PA 1996
6. E 83 – 00 "Standard Practice for Verification and Classification of Extensometer System" American Society for Testing and Materials, West Conshohocken, PA 19428
7. M. G. Jenkins, E. Lara-Curzio, S. T. Gonczy, and L. P. Zawada, "Multiple-Laboratory Round-Robin Study of the Flexural, Shear and tensile Behavior of a Two-Dimensionally Woven Nicalon™/Sylramic™ Ceramic Matrix Composite," in *Environmental, Mechanical and Thermal Properties and Performance of CFCC Materials and Components, ASTM STP 1392*, M. G. Jenkins, E. Lara-Curzio and S. T. Gonczy, Eds., American Society for Testing and Materials, West Conshohocken, PA, 2000
8. L. P. Zawada and K. E. Goecke, "Testing Methodology for Measuring Transthickness Tensile Strength for Ceramic Matrix Composites," ibid 7.
9. A. Abdul-Aziz, A. M. Calomino and F. I. Hurwitz, "Analysis of CMC C-Coupon Specimens for Structural Evaluation," *Ceram. Eng. Sci. Proc.*, **22**, No. 3 (2001) pp. 569-576
10. E. Lara-Curzio, unpublished results. (2001)

---

[6] The European Community has no representation at the International Standards Organization

11. E. Lara-Curzio, M. K. Ferber and M. G. Jenkins," Methodologies for the Thermomechanical Characterization of Continuous Fiber-Reinforced Ceramic Matrix Composites: A Review of Test Methods," Proceedings *39th. International SAMPE Symposium, Society for the Advancement of Material Process and Engineering*, Anaheim CA, April 11-14 (1994) pp. 1780-9

12. E. Lara-Curzio and M. K. Ferber, "Shear Strength of Continuous Fiber Reinforced Ceramic Composites," ibid 5.

13. E. Lara-Curzio and C. M. Russ, "On the Relationship Between the Distributions of Fiber Diameters, Breaking Loads and Fiber Strengths," *J. Mater. Sci. Letters*, **18**, 24 (1999) pp. 2041-2044.

14. E. Lara-Curzio and C. M. Russ, "Why it is Necessary to Determine Each Fiber Diameter When Estimating the Parameters of the Distribution of Fiber Strengths," *Ceram. Eng. and Sci.*, **20**, (1999) pp. 681-688

15. E. Lara-Curzio and D. Garcia Jr., "Strength Statistics of Fiber Bundles: The Effect of Diameter Variation Along and Among Fibers, on the Determination of the Parameters of the Distribution of Fiber Strengths," *Ceram. Eng. Sci. Proc.*, **22**, 3 (2001) pp. 363-370.

Table I. Existent standards and drafts under the jurisdiction of ASTM subcommittee C28.07 on Ceramic Matrix Composites.

C1275-95	Standard Test Method for Monotonic Tensile Strength Testing of Continuous Fiber-Reinforced Advanced Ceramics with Solid Rectangular Cross Sections at Ambient Temperatures.
C1292-95	Standard Test Method for Shear Strength of Continuous Fiber- Reinforced Advanced Ceramics at Ambient Temperatures.
C1337-96	Standard Test Method for Creep and Creep Rupture of Continuous Fiber-Reinforced Advanced Ceramics under Tensile Loading
C1341-96	Standard Test Method for Flexural Properties of Continuous Fiber Reinforced Advanced Ceramics.
C1358-97	Standard Test Method for Monotonic Compressive Strength Testing of Continuous Fiber-Reinforced Advanced Ceramics with Solid Rectangular Cross Sections at Ambient Temperatures.
C1359-97	Standard Test Method for Monotonic Tensile Strength Testing of Continuous Fiber-Reinforced Advanced Ceramics with Solid Rectangular Cross Sections at Elevated Temperatures.
C1360-97	Standard Practice for Constant-Amplitude, Axial, Tension-Tension Cyclic Fatigue of Continuous Fiber-Reinforced Advanced Ceramics at Ambient Temperatures.
C1425-99	Test Method for Interlaminar Shear Strength of 1-D and 2-D CFCCs at Elevated Temperatures.
C1468-00	Test Method for Transthickness Tensile Strength of CFCCs at Ambient Temperature.
C1469-00	Test Method for Shear Strength of Joints of Advanced Ceramics at Ambient Temperatures.
CXXXX-XX	Test Method for Transthickness Tensile Strength of CFCCs by Diametral Compression at Ambient and Elevated Temperatures.
CXXXX-XX	Test Method for Hoop Strength of Tubular CFCC components by Internal Pressurization.
CXXXX-XX	Test Method for Tensile Strength and Young's Modulus of Ceramic Fibers.

Table II. Task groups within ASTM C28.07

Task Group	Activity
C28.07.01	Tension
C28.07.02	Compression
C28.07.03	Creep
C28.07.04	Flexure
C28.07.05	Shear
C28.07.06	Cyclic Fatigue
C28.07.07	Fibers
C28.07.08	Interfacial
C28.07.09	Thermal
C28.07.10	Environmental
C28.07.11	Thermomechanical Fatigue
C28.07.12	Structural/Components

Table IIIa. Interlaboratory tensile test results.

Tensile Property	Grand Mean ± 1 SD	Coefficient of Variation
Modulus of Elasticity (GPa)	92 ± 6	6.7 %
Poisson's Ratio	-	-
Proportional Limit Stress (MPa)	85 ± 3	4.0%
Tensile Strength (MPa)	251 ± 18	7.2%

Table IIIb. Coefficients of variation for tensile properties of Sylramic®200 obtained from round robin testing program. r: repeatability for 9 laboratories and 10 test each laboratory; R: Reproducibility.

	Elastic Modulus	Proportional Limit Stress	Tensile Strength	Strain at Fracture
CV% r	4.7	3.4	7.2	9.3
CV% R	5.0	4.1	7.2	9.2

Table IVa. Interlaboratory shear test results.

Property	Grand Mean ± 1 SD	Coefficient of Variation
Interlaminar Shear Strength (MPa)	33 ± 8	17.3%
In-plane Shear Strength (MPa)	110 ± 6	5.3%

Table IVb. Coefficients of variation for shear strength properties of Sylramic®200 obtained from round robin testing program. r: repeatability for 8 laboratories for in-plane shear strength, 7 laboratories for interlaminar shear strength, 10 test each laboratory; R: Reproducibility

	In-plane Shear Strength	Interlaminar Shear Strength
CV% r	2.0	7.6
CV% R	4.9	17.7

Advanced SiC/SiC Ceramic Composites

Table Va. Interlaboratory flexural test results.

Flexural Property	Grand Mean ± 1 SD	Coefficient of Variation
Modulus of Elasticity (GPa)	93 ± 6	6.9 %
"Flexural" Strength (MPa)	339 ± 37	10.9%

Table Vb. Coefficients of variation for flexural properties of Sylramic®200 obtained from round robin testing program. r: repeatability for 10 laboratories, 10 tests each laboratory; R: Reproducibility.

	Modulus of Elasticity	"Flexural" Strength
CV% r	4.4	9.9
CV% R	7.1	11.1

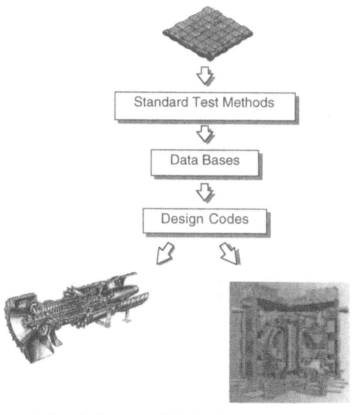

Figure 1. Methodology for the design and fabrication of components for advanced energy systems utilizing CFCCs. These methodologies rely on the use of standardized test methods for establishing data bases.

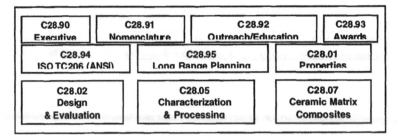

Figure 2. Structure of ASTM Committee C-28 on Advanced Ceramics.

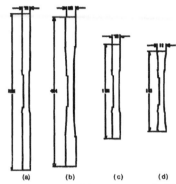

Figure 3. Examples of tensile specimen geometries recommended in ASTM C1275. (a-c) face-loaded specimens, (d) shoulder-loaded specimen. Typical dimensions in millimeters.

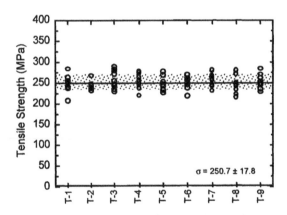

Figure 4. Round-robin tensile test results. Also shown are the grand mean value and 1 standard deviation about this value.

Advanced SiC/SiC Ceramic Composites

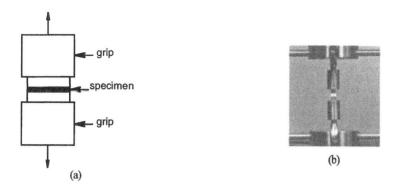

Figure 5 (a) Schematic of transthickness tensile test. (b) Photograph of specimen and grips.

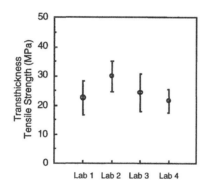

Figure 6. Results of mini-round robin program to evaluate the transthickness tensile strength of Sylramic® 200.

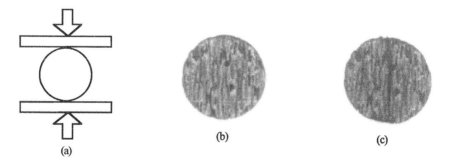

Figure 7. (a) Schematic of Brazilian test (diametral compression). SiC/SiC specimen (b) before and (c) after test. Specimen diameter is 3.0 mm.

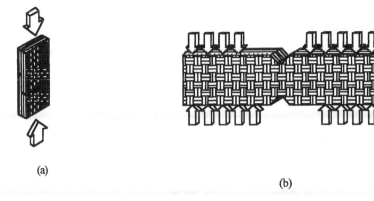

Figure 8. (a) Schematic of compression of double-notched specimen for the determination of interlaminar shear strength of CFCC. (b) Schematic of Iosipescu specimen for the determination of in-plane shear properties of CFCCs.

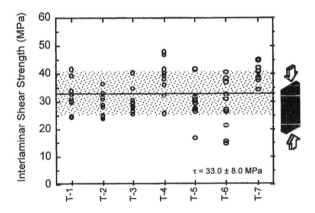

Figure 9. Round-robin interlaminar shear strength results.

Figure 10. Micrograph of double-notched Nicalon™/SiNCO specimen after test.

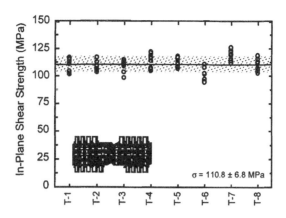

Figure 11. Round-robin in-plane shear strength results.

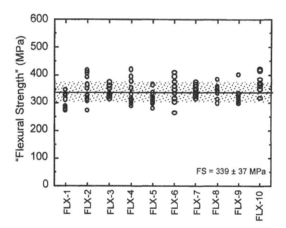

Figure 12. Round-robin "flexural" strength results.

EVALUATION OF LIFETIME PERFORMANCE OF HI-NICALON™ FIBER-REINFORCED MELT-INFILTRATED SiC CERAMIC COMPOSITES

Hua-Tay Lin
Metals and Ceramics Division
Oak Ridge National Laboratory
Oak Ridge, TN 37831-6068

Mrityunjay Singh
QSS Group, Inc
NASA Glenn Research Center
Cleveland, OH 44135

ABSTRACT

Evaluation of lifetime performance of Hi-Nicalon™ fiber-reinforced SiC ceramic composites was carried out in four-point bending over a temperature range of 700° to 1300°C in air. The effects of number of harness and type of coating on the lifetime response were also evaluated in the present study. The composites consisted of ~40 vol. % Hi-Nicalon™ fiber (8 or 5 harness satin weave) with a 0.5 μm BN (or C) fiber coating and a melt-infiltration SiC ceramic matrix. Lifetime results indicated that the Hi-Nicalon/MI SiC composites exhibited a stress-dependent lifetime at stress levels above an apparent fatigue limit, similar to the trend observed in CG-Nicalon™ fiber-reinforced CVI SiC matrix composites. The observed stress-dependent lifetime behavior was independent upon the number of harness and type of coating employed. However, results indicated that the lifetime performance of Hi-Nicalon/MI SiC composites could be greatly enhanced via the use of a higher number of harness satin weave and/or BN fiber coating. For SiC composites with BN fiber coating at temperatures ≤ 950°C, the lifetimes of Hi-Nicalon/MI SiC composites decreased with increases in applied stress level and test temperature. However, the lifetimes were extended as test temperature increased from 950° to 1150°C as a result of surface crack sealing due to a rapid SiO_2 formation by the oxidation of MI SiC matrix. At temperatures > 1150°C the lifetime performance of SiC composites was limited by the onset of creep of Hi-Nicalon™ fibers. The lifetime governing processes in an oxidizing environment were generally attributed to the progressive oxidation of fiber coatings and subsequent formation of glassy phase(s), which formed a strong bond between fiber and matrix, resulting in embrittlement of the composite with time.

I. INTRODUCTION

Continuous Nicalon™ fiber-reinforced silicon carbide (SiC) matrix composites (designated as Nicalon/SiC composites) are considered potential candidates for many high-temperature structural components in the energy and aerospace industries due to its promising mechanical performance and thermal properties at elevated temperatures [1-4]. These structural applications include various aerospace components, radiant burners, hot gas filters, high-pressure heat exchanger tubes, and combustor liners in industrial gas turbines. [5,6].

During the past decades extensive studies have been carried out to provide understanding of the effects of stress, temperature, and environments on the long-term mechanical reliability and lifetimes of CG-Nicalon/SiC composites [7-12]. It was previously shown that a CG-Nicalon/SiC composite with a 0.3 μm carbon fiber coating and without an outer seal coating exhibited stress-dependent lifetimes when the applied stresses were above a fatigue limit for a temperature range of 425° to 1150°C in air [13]. The time to failure in air decreased with increasing applied stress level and test temperature. Results also showed that at temperatures ≥ 950°C in air the lifetimes of composites were not influenced by the presence of an external SiC seal coat. The fatigue studies in four-point bending conducted by Lin *et al.* [9,10] indicated that the lifetimes of the CG-Nicalon/SiC composite above the threshold stress were controlled by the removal of the carbon fiber coating and subsequent oxidation of fibers and matrix. A tensile study in air conducted by Lara-Curzio *et al.* [14] on a CG-Nicalon/SiC composite with a ~ 0.3 μm carbon fiber coating further confirmed that the time-dependent degradation occurred at a relatively low temperature of 425°C.

On the other hand, studied also indicated that the lifetimes of CG-Nicalon/SiC composites can be significantly extended by employing the boron-containing fiber coatings [11,12, 15-17]. For instance, results showed that the lifetimes of CG-Nicalon/SiC composites with a 0.3 μm boron-doped fiber coating exhibited lifetimes that were, at least, 10 times longer than those fabricated with carbon fiber coating. Furthermore, the composites containing a 0.4 μm boron nitride (BN) fiber coating exhibited 100-1000 times longer lifetimes than those with carbon fiber coating. The lifetime limiting processes of CG-Nicalon/SiC composites with boron-containing fiber coatings were, in general, attributed to the oxidation of B-doped C and BN coatings and, thus, formation of a boron-containing glassy phase [11,18-20] that led to embrittlement of the composites. However, a recent study carried out by More *et al.* has shown that the lifetime of CG-Nicalon/SiC composites with a BN fiber coating can be significantly improved by reducing the oxygen content in the BN [21]. The substantial improvement in lifetime performance of composites with low oxygen-containing BN was attributed the increased resistance to oxidation and water vapor environments.

The present study was carried out to evaluate the effect of applied stress and temperature on the long-term mechanical performance and lifetimes for a Hi-Nicalon™ fiber-reinforced melt-infiltrated (MI) SiC matrix composites in air. The effects of number of harness and type of coating on the lifetime response were also evaluated in the present study. The damage evolution and lifetime limiting processes in terms of test temperature and stress level will also be examined in details via both optical and scanning electron microscopy analysis.

II. EXPERIMENTAL PROCEDURES

Time-to-failure studies as functions of the applied stress and test temperature were conducted in air on Hi-Nicalon™ fiber-reinforced MI SiC matrix composites with a dual fiber coating of BN (or C) and SiC. The composite consists of ~ 40 vol. % of Hi-Nicalon™ fibers (both 5 and 8 harness satin weave) and melt-infiltrated SiC matrix. The fiber preform was coated with a dual layer of fiber coating, consisting of ~ 0.5 μm BN (and also C) and ~ 2.5 μm SiC overlay by chemical vapor infiltration (CVI), prior to the final infiltration of SiC matrix via a melt infiltration process. The purpose overlay CVI SiC layer was to protect the BN- and C-coated fibrous preform during the final Si melt infiltration process. The melt-infiltration process, a low cost, robust processing technique, was developed by Singh *et al.* to fabricate near-net and complex shaped ceramic matrix composite components [22]. The details of composite fabrication can be found in Ref. 22.

The dimensions of test bend bars are 3 mm x 4 mm x 50 mm. All the bend bars were tested with machined surfaces. The 0° fibers in the test bend bars were in an orientation parallel to the tensile stress direction. The static fatigue in four-point bending was conducted at temperatures ranging from 700° to 1300°C in air. The test fixtures were fabricated from sintered α-SiC with inner and outer spans of 20 and 40 mm, respectively. A lever arm loading system was used to apply the load to the test bend bars through a sintered α-SiC pushrod. The test bars were held in the fixture with a small load (< 15 MPa outer fiber tensile stress) and heated to the desired test temperature and allowed to equilibrate for at least 30 min before increasing applied stress to the selected level. The applied stress was held constant until the test bar failed; at that point, sensors interrupted the furnace power supply circuit to allow the bend bars to cool quickly to minimize damage and oxidation of the fracture surface. The tests were terminated if the test bars did not fracture after 1000 h test. Both the optical and scanning electron microscopy (SEM) were used to characterize the microstructure of the as-received composites and the high temperature fracture surfaces as a function of stress level and test temperature. Selected bend bars were also polished and examined via SEM to understand the effect of temperature and stress level on the damage evolution and lifetime of the composites.

III. RESULTS AND DISCUSSION

Lifetime Performance Results

Figure 1 shows the stress-lifetime response of Hi-Nicalon/MI SiC composites with BN fiber coating tested at temperatures ranging from 700° to 1300°C in air. The lifetime performance of Hi-Nicalon/MI SiC composite was compared with a commercially available CG-Nicalon™ fiber-reinforced enhanced CVI SiC matrix composite (fabricated by Honeywell Ceramic Composites, and designated as enhanced SiC/SiC), as shown in Fig. 2. Enhanced SiC/SiC composite contains ~43 vol. % CG-Nicalon™ fibers and ~15 vol. % open porosity with a 0.1 μm BN fiber coating. Also note that enhanced SiC/SiC was tested with an external layer of CVD SiC seal coat (~50 μm thick). Due to the differences in the grade of fiber used and matrix processing methods, the results of enhanced SiC/SiC composite will be solely used as a benchmark in the present study.

Results showed that the Hi-Nicalon/MI SiC composite exhibited a stress-dependent lifetime behavior at applied stresses above a threshold level (here defined as an apparent fatigue limit). The observed trend of the stress-lifetime curves (lifetime decreased with increasing stress level) was similar to observed for Composite A and other previously investigated CG-Nicalon fiber-reinforced CVI SiC composites with a carbon-coated fibers. Results also showed that at temperatures ≤ 950°C the lifetimes of Hi-Nicalon/MI SiC composite decreased with an increase in test temperature. However, the lifetimes of Hi-Nicalon/MI SiC composite were extended as temperature increased from 950° to 1150°C. In addition, the Hi-Nicalon/MI SiC composites exhibited lifetimes that were at least 10 times longer than enhanced SiC/SiC composite under the same applied stress levels at the temperature range investigated. Also, the fatigue limit of Hi-Nicalon/MI SiC composites (200 MPa) was also higher than that obtained for enhanced SiC/SiC composite (150 MPa). The longer lifetimes plus higher fatigue limit obtained for the Hi-Nicalon/MI SiC composite were, in part, due to the higher temperature capability of Hi-Nicalon™ fibers as compared to CG-Nicalon™ fibers employed in the Composite A [23,24]. On the other hand, at temperature > 1150°C the lifetime of composites decreased with increasing test temperature. Specimens tested at 1300°C exhibited a permanent curvature after fracture, indicative of the occurrence of creep deformation. Previous study has shown that the Hi-Nicalon™ fiber began to exhibit substantial creep deformation and strength degradation (70 % decrease) at temperatures > 1200°C [24].

Figure 3 compares the lifetime performance of Hi-Nicalon/MI SiC composites fabricated with 8 harness and 5 harness satin weave. Results showed that both materials exhibited similar stress-dependent lifetime behavior under the test conditions employed. Also, results showed that the lifetimes of composite with 8 harness satin weave exhibited lifetimes were, at least, 10-times longer than those fabricated with 5 harness satin weave. In addition, the composite with 8 harness satin weave exhibited a relative higher fatigue limit (250 MPa) as compared with the value (200 MPa) obtained for the composite with 5 harness satin weave under the same test temperature. Note that there was no difference in theoretical density between these composites. Therefore, the increase in lifetime and apparent fatigue limit with increasing the number of harness could be attributed to the increased load carrying capacity of the reinforcing fiber per unit volume of material.

Results of the effect of fiber coating (BN vs. C) on the lifetime performance of the Hi-Nicalon/MI SiC composites are shown in Figure 4. As expected the lifetime of the Hi-Nicalon/MI SiC composites is substantially improved via the use of BN fiber coating. For instance, the composite with BN fiber coating exhibited lifetimes, which were about two orders of magnitude longer that those with C fiber coating under the same test condition, consistent with the results obtained previously for the CG-Nicalon/CVI SiC composites [11]. Note that the lifetime of composite with C fiber coating revealed a substantial decrease in loading carry capacity and lifetime when tested at 950°C. Also, the apparent fatigue limit of the composite with C fiber coating is ~ 100 MPa for both test temperature, which is lower than the values obtained for the composites with BN fiber coating. Note that the 100 MPa fatigue limit obtained is similar to the value reported for the composite reinforced with CG-Nicalon and also with C fiber coating [9,13].

Lifetime Limiting Processes

Figure 5 shows the SEM micrographs of Hi-Nicalon/MI SiC composite with a dual coating layer of BN and SiC tested at temperatures of 700° and 950°C in air under an applied stress of 250 MPa. Note that the specimen tested at 700°C/250 MPa did not fracture after 1000 h of testing, while the one tested at 950°C/250 MPa exhibited a lifetime of ~22 h. The SEM observations on tensile surface regions for specimen tested at 700°C revealed very limited fiber pullout, with the pullout length < one diameter length of fiber. On the other hand, the specimen tested at 950°C revealed a brittle fracture with no fiber pullout. This brittle fracture zone was limited to a depth ~ 30% of specimen thickness. Note that extensive fiber pullout was observed in the compressive surface regions of both specimens. Observation of specimens tested at 700°C revealed a recession of the BN fiber coating, leaving an open gap between fiber and matrix. The localized formation of droplet-like glassy phase containing B at the interface due to the oxidation of BN coating was also observed. For specimen tested at 950°C there was substantial formation of glassy phase that completely filled the gap where BN was present before the testing. Most of the matrix regions were also covered with a layer of glassy phase. Similar features of recession and glassy phase formation due to oxidation of BN coating have also been reported for other CG-Nicalon/SiC composites [11,25].

As shown in the previous studies, oxidation of BN and volatilization of its oxidation products will readily occur at temperatures < 1000°C in air [26-28]. It has been reported by More *et al.* [21] that BN oxidation and borosilicate-type glass formation at the interface occur simply due to the introduction of oxygen within the BN fiber coating during processing. The oxygen from an oxidative environment, which could ingress along cracks generated by the applied stress and/or interfaces, would further enhance the oxidation reaction. The progressive oxidation of the BN fiber coating and the formation of glassy droplets and/or layers at the interfaces will eventually develop a strong interfacial bond between fiber and matrix and thus increase the debond and interfacial shear stresses, resulting in embrittlement of the composites with time. In addition, interactions with the boron-containing silicate glass could also degrade the fiber strength by localized surface attack (pitting effect), and, thus, weaken the composite.

Results showed that specimens tested at 1150°C exhibited longer lifetimes than those tested at 950°C under the same stress level. This observed lifetime/temperature behavior was also reported in a CG-Nicalon/SiC matrix composite with a BN fiber coating [11]. The extension of lifetime at 1150°C could be due to the surface sealing effect, as shown in Figure 6. Figure 6 is a polished cross section of a specimen tested at 1150°C/250 MPa with 1000 h of test time. Observations showed the formation of SiO_2 due to oxidation of MI SiC matrix completely sealed the cracks developed on the tensile surface. Note that very limited or no oxidation of both BN fiber coating and fiber was observed in the region ~50 µm below the surface region (Fig. 6b). The original BN fiber coating remains intact without oxidation. The surface sealing by a SiO_2 layer would inhibit further ingression of oxygen into the material and, thus, substantially decrease the oxygen concentration available for oxidation of BN coating and fibers. The rapid oxidation reaction of dense MI SiC matrix at 1150°C in air was able to readily seal the cracks developed on the tensile surfaces and protect the BN and fibers from oxidation, thus extending the lifetimes of the composite. When the test temperature ≥ 1200°C the Hi-Nicalon fiber exhibited a substantial strength degradation and thus lifetime performance decreased with increasing temperature.

The lifetime limiting processes for the composites with C fiber coating have been well studied and described previously [9,13,14,25]. Thus, a brief description of the process will be only reported in the present study. At low temperatures (≤ 700°C), the oxidation of graphite was the primary process detected by thermogravimetric analysis studies [29]. However, oxidation of the Nicalon fibers was also observed by transmission electron microscopy examination in the ORNL Nic/SiC tested at 425°C in air [13]. The oxidation of the Nicalon fiber was evidenced by a formation of a skin layer of SiO_2 on the fiber surface. In this case, the strength of the Nicalon fibers will continuously decrease with time because of oxidation reaction. The final failure occurs as the applied stress exceeds the residual fracture strength of the composite. At high temperatures ($\geq900°C$), the rapid removal of C fiber coating and subsequent oxidation of fiber and SiC matrix would then result in a strong bond and embrittlement of the composites.

SUMMARY

Lifetime studies in four-point flexure were performed on Hi-Nicalon™ fiber-reinforced MI SiC matrix composite at temperatures ranging from 700° to 1300°C°C in air as a function of number of harness and type of fiber coating. Lifetime results indicated that the composite exhibited a stress-dependent lifetime at stress levels above an apparent fatigue limit under the test conditions employed in the present study. The obtained lifetime response of Hi-Nicalon/MI SiC composite was similar to the trend observed in various CG-Nicalon™ fiber-reinforced SiC matrix composites. At $\leq 950°C$, the lifetimes of Hi-Nicalon/MI SiC composites decreased with increasing applied stress levels and temperatures. Also, the lifetimes were extended as test temperature increased from 950 to 1150°C as a result of surface sealing due to SiO_2 formation by the oxidation of MI SiC matrix. The lifetime limiting processes were attributed to the progressive oxidation of BN fiber coating and formation of glassy phase at the interfaces, which developed a strong interfacial bond between fiber and matrix, resulting in embrittlement of the composite with time. At temperature $\geq 1200°C$, the creep of the Hi-Nicalon fiber dictated the mechanical strength and thus lifetime performace of the composites. Results also showed that the lifetime of Hi-Nicalon/MI SiC composites could be enhanced via the use of a high number harness of satin weave (8 versus 5), due to the higher load carrying capability of the reinforcing fibers. The lifetime of the composites could be also substantially extended via the use of BN fiber coating, due to its better oxidation resistance as compared with C coating in oxidizing environments. The lifetimes of the Hi-Nicalon/MI SiC composites above the threshold stress were controlled by the progressive removal of the carbon fiber coating and subsequent oxidation of fibers and matrix.

ACKNOWLEDGEMENTS

Research sponsored by the U.S. Department of Energy, Assistant Secretary for Energy Efficiency and Renewable Energy, Office of Industrial Technologies, Industrial Materials of the Future Program, under Contract DE-AC05-00OR22725 with UT-Battelle, LLC.

REFERENCES

(1). N. Frety and M. Boussuge, "Relationship Between High-Temperature Development of Fiber-Matrix Interfaces and the Mechanical Behavior of SiC-SiC Composites," *Comp. Sci. and Tech.,* 37, 177-89 (1990).

(2). D. Singh and J. P. Singh, "Effect of High-Temperature Loading on Mechanical Properties on Nicalon Fibers and Nicalon Fiber/SiC Matrix Composites," *Ceram. Eng. & Sci. Proc.,* 14 [9-10] 1153-64(1993).

(3) S. V. Nair and Y. L. Wang, "Failure Behavior of a 2-D Woven SiC Fiber/SiC Matrix Composite at Ambient and Elevated Temperatures," *Ceram. Eng. & Sci. Proc.,* 13 [7-8] 433-41 (1992).

(4). A. Chulya, J. Z. Gyekenyesi, and J. P. Gyekenyesi, "Failure Mechanisms of 3-D Woven SiC/SiC Composites Under Tensile and Flexural Loading at Room and Elevated Temperatures," *Ceram. Eng. & Sci. Proc.,* 13 [7-8] 420-32 (1992).

(5) M. A. Karnitz, D. F. Craig, and S. L. Richlen, "Continuous Fiber Ceramic Composite Program," *Am. Ceram. Soc. Bull.,* 70 [3] 430-35 (1991).

(6) D. C. Larsen, J. Adams, L. Johnson, A. Teotia, and L. Hill, "Ceramic Materials for Heat Engines," Noyes Publications, NJ (1985).

(7) P. F. Tortorelli, S. L. Riester, and R. A. Lowden, "Influence of Fiber Coatings on the Oxidation of Fiber-Reinforced SiC Composites," *Ceram. Eng. & Sci. Proc.,* 14 [1-2] 358-66 (1993).

(8) S. Raghuraman, M. K. Ferber, J. F. Stubbins, and A. A. Wereszczak, "Stress-Oxidation Tests in SiC$_f$/SiC Composites," pp. 1015-26 in Advanced in Ceramic-Matrix Composites II, Ceramic Transaction Vol. 46, J. P. Singh and N. P. Bansal, eds., American Ceramic Society, Westerville, 1995.

(9) H. T. Lin, P. F. Becher, and P. F. Tortorelli, "Elevated Temperature Static Fatigue of a Nicalon Fiber-Reinforced SiC Composite," pp. 435-440 in the MRS Symposium Proceedings Vol. 365: Ceramic Matrix Composites-Advanced High-Temperature Structural Materials, Materials Research Society, Pittsburgh, Pennsylvania (1995).

(10) H. T. Lin and P. F. Becher, "Stress-Temperature-Lifetime Envelop of Nicalon Fiber-Reinforced SiC Matrix Composites in Air," *Composites Part A: applied science and manufacturing.* 28A (1997) 935-942.

(11) H. T. Lin and P. F. Becher, "Effect of Fiber Coating on Lifetime of Nicalon Fiber-Reinforced Silicon Carbide Composites in Air," *Mater. Sci. and Eng.* A231 (1997) 143-150.

(12) S. Jacques, A. Guette, F. Langlais, R. Naslain, and S. Goujard, "High Temperature Lifetime in Air of SiC/C(B)/SiC Microcomposites Prepared by LPCVD," pp. 381-388 in the Proceedings of High-Temperature Ceramic-Matrix Composites I: Design, Durability, and Performance, Ceramic Transactions Vol. 57, A. G. Evans and R. Naslain, eds., The American Ceramic Society, OH (1993).

(13) P. F. Becher, H. T. Lin, and K. L. More, "Lifetime-Applied Stress Response in air of a SiC-Based Nicalon Fiber-Reinforced Composite with a Carbon Interfacial Layer: Effects of Temperature (300° to 1150°C)," *J. Am. Ceram. Soc.*, 81 [7] 1919-25 (1998).

(14) E. Lara-Curzio and M. K. Ferber, "Stress-Rupture of Nicalon/SiC CFCCs at Intermediate Temperatures," *J. Mater. Sci. Letters*, 16 [1] 23-26 (1997).

(15) M. A. Kmetz, J. M. Laliberte, W. S. Willis, S. L. Luib, and F. S. Galasso, "Synthesis, Characterization, and Tensile Strength of CVI SiC/BN/SiC Composites," *Ceram. Eng. Sci. Proc.* 12 [9-10], 2161-2174 (1991).

(16) M. Leparoux, L. Vandenbulcke, S. Goujard, C. Robin-Brosse, and J. M. Domergue, "Mechanical Behavior of 2D-SiC/BN/SiC Processed by ICVI," pp. 633-640 in the Proceedings of the 10th International Conference on Composite Materials, Vol. IV: Characterization and Ceramic Matrix Composites, A. Poursartip and K. Street, eds.,Woodhead Publishing Limited, England, 1995.

(17) R. A. Lowden, K. L. More, O. J. Schwarz, and N. L. Vaughn, "Improved Fiber-Matrix Interlayers for Nicalon/SiC Composites," pp. 345-352 in the Proceedings of High Temperature Ceramic Matrix Composites, R. Naslain, J. Lamon, and D. Doumeingts, eds., Woodhead Publishing Limited, England, 1993.

(18) N. S. Jacobson, "High Temperature Oxidation of the Boron Nitride Interphase in Composites," pp. 69-70 in the Proceedings of the Symposium of High Temperature Materials Chemistry VII, Electrochemical Society, Chicago, IL, Oct. 10, 1995.

(19) E. Y. Sun, H. T. Lin, and J. J. Brennan, "Intermediate-Temperature Environmental Effects on Boron Nitride Coated-Fiber Reinforced Glass-Ceramic Composites," *J. Am. Ceram. Soc.*, 80[3] 609-614 (1997).

(20) B. W. Sheldon, E. Y. Sun, S. R. Nutt, and J. J. Brennan, "Oxidation of BN-Coated SiC Fibers in Ceramic Matrix Composites," *J. Am. Ceram. Soc.* 79 [2], 539-43 (1996).

(21) K. L. More, K. S. Ailey, R. A. Lowden, and H. T. Lin, "Evaluating the Effect of Oxygen Content in BN Interfacial Coating on the Stability of SiC/BN/SiC Composites," *Composites Part A: applied science and manufacturing* 30 (1999) 463-470.

(22) D. R. Behrendt and M. Singh, "Producing Fiber Reinforced Composites Having Dense Ceramic Matrices," US Patent 5,865,922, Feb. 2, (1999).

(23) M. Takeda, J. Sakamoto, A. Saeki, and H. Ichikawa, "Mechanical and Structural Analysis of Silicon Carbide Fiber Hi-Nicalon Types," *Ceram. Eng. Sci. Proc.* 17 [4], 35-42 (1996).

(24) H. M. Yun and J. A. DiCarlo, "Thermomechanical Behavior of Advanced SiC Fiber Multifilament Tows," *Ceram. Eng. Sci. Proc.* 17 [4], 61-67 (1996).

(25) G. N. Morscher, "Tensile Stress Rupture of SiCf/SiCm Minicomposites with Carbon and BN Interphases at Elevated Temperatures in air," *J. Am. Ceram. Soc.*, 80[8] 2029-2042 (1997).

(26) K. Oda and T. Yoshio, "Oxidation Kinetics of Hexagonal Boron Nitride Powder," *J. Mater. Sci.*, **28**, 6562-6566 (1993).

(27) S. P. Randall and J. L. Margrave, "Vapour Equilibria in the B_2O_3-H_2O System at Elevated Temperatures," *J. Inorg. Nucl. Chem.*, Vol. 16, pp. 29-35 (1960).

(28) T. Matsuda, "Stability to Moisture for Chemically Vapour-Deposited Boron Nitride," *J. Mater. Sci.* 24, 2353-2358 (1989).

(29) Tortorelli, P. F., Keiser, J. R., Riester, L., and Lara-Curzio, E., in "Continuous Fiber Ceramic Composites Program Task 2 Bimonthly Progress Report for June-July 1994," Oak Ridge National Laboratory, Oak Ridge, TN, 1994, pp. 51-54.

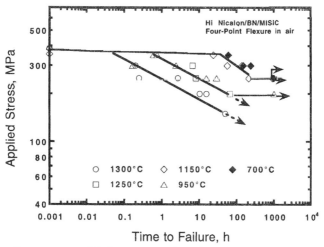

Figure 1. Stress versus lifetime curves for Hi-Nicalon fiber-reinforced melt infiltrated SiC matrix composite tested at 700°, 950°, and 1150°C in air.

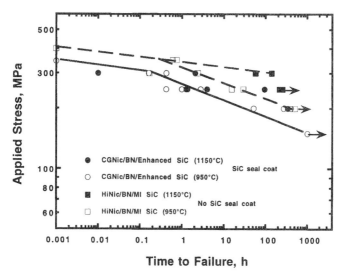

Figure 2. Stress versus lifetime curves for Hi-Nicalon/MI SiC matrix composite and commercial CG-Nicalon/Enhanced CVI SiC composites at 950° and 1150°C in air.

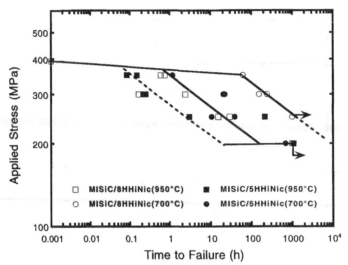

Figure 3. Stress versus lifetime curves for Hi-Nicalon/MI SiC matrix composites as a function of number of harness.

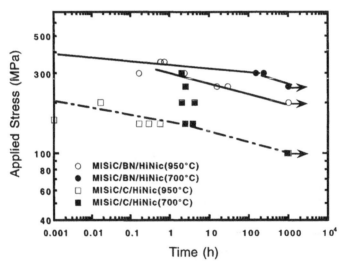

Figure 4. Stress versus lifetime curves for Hi-Nicalon/MI SiC matrix composites with BN and C fiber coating.

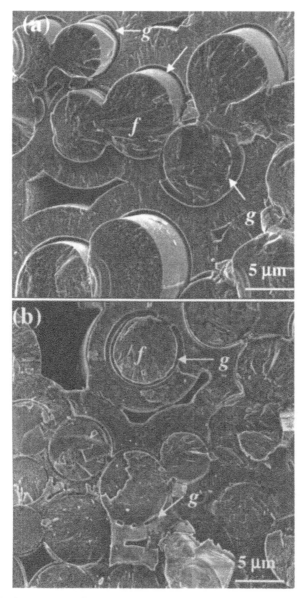

Figure 5. Fracture surface of Hi-Nicalon/MI SiC composites with BN fiber coating tested at 700°C (a) and 950°C (b) in air. Note that the *f* denotes the fiber and *g* denotes the glass phase.

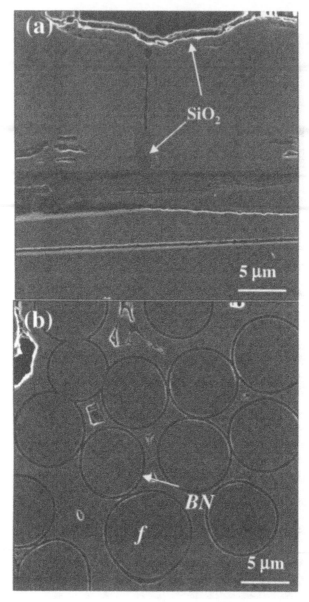

Figure 6. Polished cross section of Hi-Nicalon/MI SiC composites tested at 1150°C in air. Note that the *f* denotes the Hi-Nicalon fiber. Arrow in (a) indicates the sealing surface crack by SiO₂ due to oxidation of MI SiC.

SHEAR LAG-MONTE CARLO SIMULATION OF TENSILE BEHAVIOR OF UNIDIRECTIONAL CERAMIC MATRIX COMPOSITES

S. Ochiai and H. Okuda
International Innovation Center
Kyoto University
Sakyo-ku, Kyoto 606-8501, Japan

H. Tanaka, M. Tanaka and M. Hojo
Graduate School of Engineering
Kyoto University
Sakyo-ku, Kyoto 606-8501, Japan

K. Schulte and B. Fiedler
Technical University Hamburg-Harburg
Denickestrasse 15,
D-21071 Hamburg, Germany

ABSTRACT

The influences of interfacial debonding and residual stresses on the tensile stress-strain curve, strength and fracture morphology of unidirectional continuous fiber-reinforced ceramic matrix composites were simulated by the shear lag analysis combined with the Monte Carlo simulation method. The following features could be described. (a) With increasing interfacial strength, the fracture mode varies from the cumulative type, characterized by the gradual breakage components (fiber, matrix) and interfacial debonding in the fracture process and by the fiber pull-out in the fracture surface, to the non-cumulative one, characterized by the chain reaction mainly of breakage of components and by the overall fracture of the composite perpendicular to tensile axis without fiber pull-out. (b) In the weakly bonded composites, the damages (breakage of fiber and matrix and interfacial debonding) are accumulated intermittently, resulting in a serrated stress-strain curve. (c) In the fracture process of weakly bonded composites, the residual stresses (compressive and tensile axial residual stresses in fiber and matrix, respectively) enhance the breakage of matrix and matrix breakage-induced debonding at low applied strain, leading to reduction in strength of the composite when the fracture strain of the matrix is equal to or lower than that of fiber.

INTRODUCTION

Since the fracture toughness of the components (fiber and matrix) in fiber-reinforced ceramic matrix composites is not necessarily high, the fiber/matrix interface is controlled as to allow debonding to arrest cracks[1,2]. In such weakly bonded composites, the damages such as breakage of fiber and matrix and interfacial debonding arise and interact mechanically among each other. Such mechanical interactions play a dominant role to determine the type and location of subsequent damage and the resultant mechanical property of the composite.

One of the tools to solve the mechanical interactions among the various damage modes is the shear lag analysis. The ordinary shear lag analysis is, however, constructed based on the approximation that only fibers carry applied stress and the matrix acts only as a stress-transfer medium. Owing to such an approximation, it has the following disadvantages; (a) it can be applied only to polymer- and low yield stress-metal-matrix composites but not to ceramic-matrix ones and (b) residual stresses cannot be incorporated. To overcome such disadvantages, the authors[3-5] have recently proposed a modified method, with which the general situation (both fiber and matrix carry applied stress and also act as stress transfer media) can be described and also residual stresses can be incorporated.

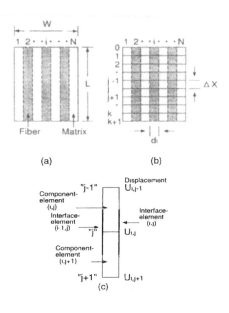

The modified method is rough as well as the ordinal one, but, due to the roughness, the mechanical interactions among the various damage modes, which vary at every occurrence of new damage, can be traced by a simple calculation. In the present work, such a modified method was combined with the Monte Carlo method, which can give spatial distribution of strength of the components, and was applied to unidirectional 2D model composite. For polymer-composites, the micro-mechanical models using the shear lag analysis + Weibull fiber statistics[6] and using the load sharing rule + Weibull fiber

Fig.1 Schematic representation of the two-dimensional model composite for simulation and notation of the displacement $U_{i,j}$, and the component- and interface-elements.

statistics + Monte Carlo method[7] have been demonstrated to be useful for description of their fracture behavior. In the present work, the contributions of matrix stress to composite one, residual stresses and residual stress-induced debonding are newly incorporated in the model, which will be applied to ceramic composites.

SIMULATION METHOD

The model composite employed in the present work is presented in Fig.1. The components (fiber, matrix) were numbered as 1,2...i,... to N from left to the right side. Each component was regarded to be composed of k+1 short component elements of length Δx, and the position at the top end was numbered as 0 and then 1, 2, 3,...j... k+1 downward, in step of Δx. The "i" component from x=(j-1) Δx to jΔx was labeled as the (i,j)-component-element and the interface from x=(j-1/2)Δx to (j+1/2)Δx between "i" and "i+1" components as the (i, j)-interface-element. The displacement of the (i,j)-component-element at x=jΔx was denoted as $U_{i,j}$. The interfacial parameter $\alpha_{i,j}$ (=1 or 0 when the (i,j)-interface is bonded or debonded, respectively) and the component parameter, $\gamma_{i,j}$ (=1 or 0 when the (i,j)-component-element is not broken or broken, respectively) were introduced to incorporate the damage map into the program. From the spatial distribution of debonded interface-elements with $\alpha_{i,j}$=0 and broken component-elements with $\gamma_{i,j}$=0, the damage map was generated.

Under the assumption that the composite has no residual stress at fabrication temperature T_1 and it is cooled down to the test temperature T_2, the residual axial stresses of fiber and matrix at T_2 were, to a first approximation, calculated by $(\alpha_m-\alpha_f)E_fE_mV_m\Delta T/(E_fV_f+E_mV_m)$ and $(\alpha_f-\alpha_m)E_fE_mV_f\Delta T/(E_fV_f+E_mV_m)$, respectively, where ΔT is the temperature difference (=T_1-T_2<0), α, E and V are the coefficient of thermal expansion, Young's modulus and volume fraction, respectively, and the subscripts f and m refer to the fiber and matrix, respectively.

The equation for stress balance is expressed by[3-5]

$$B1(i,j)U_{i,j-1}+B2(i,j)U_{i-1,j}+B3(i,j)U_{i,j}+B4(i,j)U_{i,j+1}+B5(i,j)U_{i+1,j}=B6(i,j) \quad (1)$$

where B1(i,j) to B6(i,j) are parameters including the elastic constant, cross-sectional area, volume fraction and residual stress of each component, applied strain ε_c and damage map (spatial distribution of $\alpha_{i,j}$- and $\gamma_{i,j}$ –values). As the values of B1(i,j) to B6(i,j) vary with varying damage map, they were determined at each occurrence of new damage. The stress of each component- and interface element was calculated from the $U_{i,j}$-values obtained by Eq.(1).

As the criterion for debonding, the shear stress criterion (debonding occurs when the shear stress exceeds the shear strength τ_c) and energy release rate criterion (debonding occurs when the energy release rate exceeds the critical value $G_{d,c}$) have widely been used to describe the debonding[8-11]. It has been shown that the shear stress criterion has a one to one relation to the energy release rate criterion[8]. In the present work, both criteria were used. The detailed procedures for simulation using the shear stress- and energy release rate criteria are shown elsewhere[3,5].

The strength of each component was determined by generating a random value based on the Monte Carlo procedure using the Weibull distribution[10]. The overall strain was raised in a step of 0.01~0.025%. At each strain level, the stress distribution was calculated, from which whether a new damage occurs or not was examined. If no damage arose, the overall applied strain was raised. If a new damage arose, the stress distribution was calculated under the new damage map and whether a new damage occurs further or not was examined. Such a procedure was repeated and the next damage was identified one after another, until no more damage occurred at a given strain. After stoppage of formation of damages, the strain was raised. For the increased applied strain, the same procedure was repeated until overall fracture of the composite.

RESULTS AND DISCUSSION

(a) Influences of species of precut component and residual stresses on interfacial debonding

The simulation results of the interfacial debonding under the existence of one broken fiber-element and one broken matrix-element are shown in Fig.2. The input values are listed in the caption. The numbers 1,2,3... show the order of the debonding. When no residual stress exists ($\alpha_f =\alpha_m$), debonding from the broken fiber-ends occurs prior to that of the matrix, since the Young's modulus of the fiber is higher than that of matrix and the fiber volume fraction is high (0.5) in this example. When residual stresses exist ($\alpha_f <\alpha_m$ where the residual stresses in the fiber direction

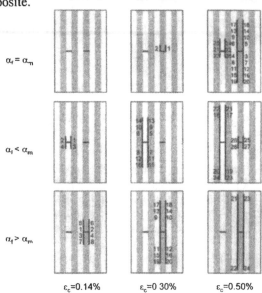

Fig.2 Influence of residual stresses on progress of debonding under the existence of one broken fiber-element and one broken matrix-element. The used values were; N=9, k=12, d_f (diameter of fiber)=12μm, V_f (fiber volume fraction)=0.5, Δx=24μm, E_f=180GPa, E_m=100GPa, G_f (shear modulus of fiber)=69GPa, G_m (shear modulus of matrix)=38GPa, τ_c=100 MPa. For the sample without residual stresses, α_f =was set to be equal to α_m. For the sample with residual stresses, α_f =1x10^{-6}/K, α_m=5x10^{-6}/K and ΔT=-1200K were input to monitor the case of $\alpha_f<\alpha_m$ and α_f =5x10^{-6}/K, α_m=1x10^{-6}/K and ΔT=-1200K to monitor the case of $\alpha_m<\alpha_f$.

are tensile and compressive for matrix and fiber, respectively), the debonding from the broken-end of the matrix section is enhanced but the debonding from the broken-end of the fiber is suppressed. When residual stresses are reversed ($\alpha_m < \alpha_f$ where the residual stresses in the fiber direction are compressive and tensile for matrix and fiber, respectively), the debonding from the broken-end of the matrix is suppressed but the debonding from the broken-end of the fiber is enhanced. Conclusively, debonding from the broken end of the component is enhanced and retarded when the broken component has tensile and compressive stresses in the fiber direction, respectively, when the magnitude of the radial and tangential residual stresses is low in comparison with that of the axial stresses.

(b) Influence of mechanical interaction among damages on debonding behavior under existence of many broken components

Figure 3 shows an example of the stress (σ_c)-strain (ε_c) curve and variation of the damage map under existence of many broken components. At $\varepsilon_c=0.21\%$, the 1st debonding occurs, whose location is determined by the mechanical interactions among the broken components. The occurrence of the 1st debonding changes the damage map and therefore the mechanical interaction, which induces the 2nd debonding. Due to the occurrence of the 2nd debonding, the damage map is renewed and the 3rd debonding is induced. In this way, the change in damage map due to the preceded debonding causes the next debonding one after another up to the 6th debonding. After the 6th debonding, overall debonding stops, since the shear stress of all interface-elements becomes lower than the interfacial shear strength under the corresponding damage

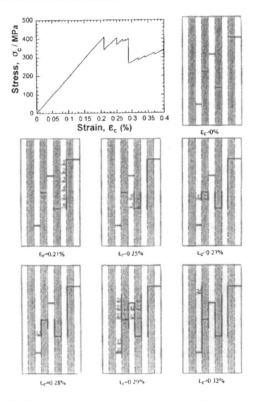

Fig.3 Stress (σ_c)-strain (ε_c) curve and variation of damage map of the composite under the existence of many broken components. The input values were; N=9, k=12, d_f =0.1mm, V_f=0.44, Δx=0.2mm, E_f=400GPa, E_m=100GPa, G_f=160GPa, G_m=40GPa, τ_c =200 MPa and α_f =α_m.

map and applied strain. Due to the progress of the 1st to 6th debonding at ε_c=0.21%, the stress-carrying capacity of composite is reduced, resulting in the drop in the stress-strain curve.

The composite stress increases again with increasing applied strain ε_c but no debonding occurs up to ε_c=0.25% since the shear stress of any interface-element does not exceed the shear strength. When ε_c reaches 0.25%, the 7th to 10th debonding occur one after another due to the mechanical interaction, resulting in loss of stress carrying capacity. After the stoppage of debonding, the composite stress again increases with increasing strain.

As shown in this micro-composite model, the overall debonding progresses intermittently with repetition of growth and stoppage, resulting in the serrated stress-strain curve. In the practical composites containing far many fibers and matrix, such a serration is not necessarily found due to the occurrence of debonding at many locations at every strain. The fundamental process obtained by the present work is superimposed in actual composites.

(c)Influence of interfacial strength on stress-strain behavior of the composite in which breakage of components and interfacial debonding occur consecutively

In order to extract just the fundamental fracture process affected by interfacial strength, a micro-composite model was used. In addition, the strength-value of each component-element was given by the same series of random values for all samples. Thus, the spatial distribution of the strength of fiber- and matrix-elements and input values other than interfacial strength were common in all samples in this part. The only difference among the samples was the magnitude of the interfacial strength. Figures 4 to 7 show the representative simulated stress-strain curves and variation of damage map for very weak, weak, intermediate and strong interfaces, respectively. The following features were found.

(1) The stress-strain curve shows also serrations due to the intermittent breakage of the components and interfacial debonding.

(2) In the sample in Fig.4 whose interface is the weakest within the present work, interfacial debonding occurs for large distance, following each breakage of the component. For instance, when the 1st damage (breakage of the matrix) occurs, follows the debonding (2nd to 9th damages at the same strain and the 10th to 17th damages at the increased strains), and when the 18th damage (breakage of matrix) occurs, it is followed by debonding (19th to 34th damages). In this way, once the weak components are broken, debonding for long distances subsequently follows it. In such a situation, the stress concentration arising from the breakage of the component approaches 1(unity)[12]. Thus the behavior of the composite tends to be similar to that of bundles of the components. In this way, the cumulative type fracture, characterized by the gradual breakage components (fiber, matrix) and interfacial debonding in the fracture process and by the fiber

pull-out in the fracture surface (Fig.4(f)), occurs in weakly bonded composites.

(3) In the sample in Fig.7 whose interface is the strongest, first the breakage of the component occurs as indicated by 1, but further damage is not induced. Then the stress increases with increasing strain and the second breakage of the component occurs as indicated by 2. Again no damage is induced by this damage. In this way, even the weakest and second weakest components are broken in the low applied strain range, the damages do not cause further damages. Namely they are isolated. However, once the 3rd breakage of the component occurs, the fracture of the components occurs one after another nearly in the same cross-section. Thus a very large drop in load arises. Interfacial debonding does not occur at this applied strain due to the strong interface. When the applied strain is raised, debonding occurs, which results in overall fracture. In this way, the non-cumulative type fracture occurs in strongly bonded composite, characterized by the chain reaction mainly of breakage of components at a certain strain after a formation of isolated damages and by the overall fracture of the composite nearly perpendicular to tensile axis without fiber pull-out (Fig.7(f)).

(4) With increasing interfacial strength, the debonding caused by the breakage of components tends to be minor and the breakage of components tends to cause breakage of other components, as shown in Figs.5 and 6 whose interfacial strengths are in the middle of very weak (Fig.4) and strong (Fig.7) ones.

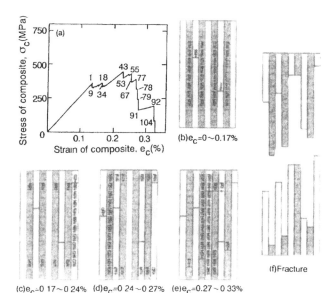

(c)e_c=0 17~0 24% (d)e_c=0 24~0 27% (e)e_c=0.27~0 33%

Fig.4 Stress-strain curve and variation of damage map of the composite with very weak interface ($G_{d,c}$=5 J/m^2). The input values were: N=9, k=12, d_f=15µm, V_1 =0.3, Δx=30µm(=2d_f), E_f=410GPa, E_m=180GPa, G_f=164GPa and G_m=72 GPa, the Weibull's scale parameter=10(fiber) and 5(matrix) and the Weibull's shape parameter =2 GPa (fiber) and 0.8GPa(matrix), which were common for Figs.4 to 7.

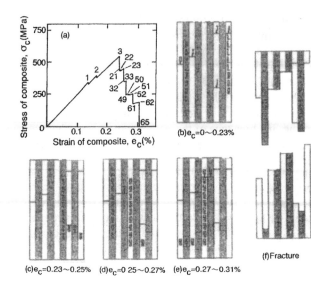

Fig.5 Stress-strain curve and variation of damage map of the composite with weak interface ($G_{d.c}=20$ J/m^2).

(b)$e_c=0\sim0.23\%$

(c)$e_c=0.23\sim0.25\%$ (d)$e_c=0\ 25\sim0.27\%$ (e)$e_c=0.27\sim0.31\%$

(f)Fracture

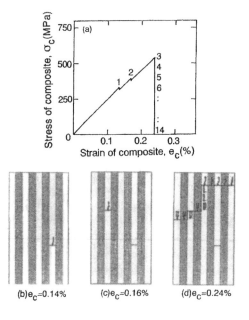

Fig.6 Stress-strain curve and variation of damage map of the composite with intermediate interfacial strength ($G_{d.c}=80$ J/m^2)

(b)$e_c=0.14\%$ (c)$e_c=0.16\%$ (d)$e_c=0.24\%$

(e)Fracture

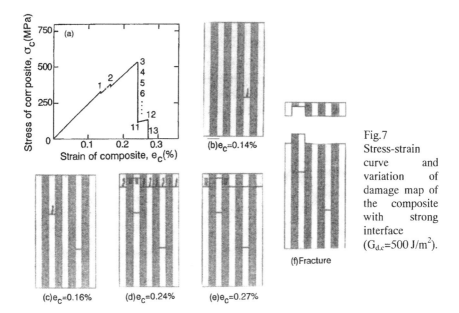

Fig.7
Stress-strain curve and variation of damage map of the composite with strong interface $(G_{d.c}=500 \text{ J/m}^2)$.

(a)

(b)e_c=0.14%

(c)e_c=0.16%   (d)e_c=0.24%   (e)e_c=0.27%

(f)Fracture

(5) As shown above, the fracture process is different for different interfacial strength. As a result, the final fracture morphology is very sensitive to interfacial strength as shown in (f) in Figs.4 to 7. Evidently, the irregularities of the fracture surface decrease, and the overall fracture tends to occur perpendicular to the tensile axis with increasing interfacial strength in accordance with the change of fracture mode from non-cumulative to cumulative type.

(d) Influence of residual stresses on the shape of stress-strain curve and strength of the composite in which breakage of components and interfacial debonding occur consecutively

In the case of case $\alpha_f < \alpha_m$, the tensile residual stress in the matrix along the fiber direction enhances the breakage of matrix and also matrix-breakage-induced debonding. It is expected that the influence of such premature fracture and interfacial debonding on the stress-strain curve and strength of the composite arises clearly especially when the inherent fracture strain of the matrix is equal to, or lower than, that of the fiber. Thus, in the preset work, the case where the inherent fracture strain of the matrix is comparable to that of the fiber is taken up as a representative and the Weibull's parameters were chosen so as to give comparable average fracture strain for fiber and matrix. Also the common spatial distribution of the strength of fiber- and matrix- elements was given for the both cases of $\alpha_f = \alpha_m$ and $\alpha_f < \alpha_m$.

Figure 8 shows the comparison of the simulated stress-strain curves between the cases of $\alpha_f = \alpha_m$ and $\alpha_f < \alpha_m$. For reference, the number of broken fiber (N_F), matrix (N_M) and interface (N_I)–elements was measured as a function of applied strain. They were, however, quite different to each other and could not be clearly shown on the same scale. Then the normalized values with respect to the final values $N_{F,f}$, $N_{M,f}$ and $N_{I,f}$, respectively, are shown in Fig.8.

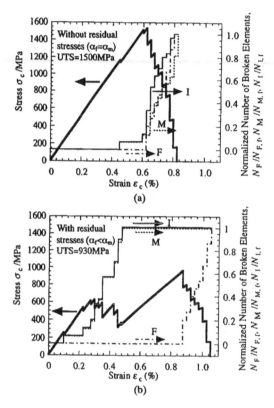

(a)

(b)

Fig.8 Examples of the stress (σ_c)-strain (ε_c) curve of the samples with and without residual stresses, together with the variation of normalized number of fractured elements of fiber ($N_F/N_{F,f}$), matrix ($N_M/N_{M,f}$) and interface ($N_I/N_{I,f}$). The signs F, M and I with arrows refer to fiber, matrix and interface, respectively. The input values were; N=31, k=29, d_f =0.14μm, V_f=0.4, Δx=0.28mm, E_f=380GPa, E_m=180GPa, G_f=150GPa, G_m=70GPa and τ_c=100 MPa, the Weibull's scale parameter=8(fiber) and 5(matrix), the Weibull's shape parameter=3GPa(fiber) and 1.4GPa(matrix) and α_f =α_m for the sample without residual stresses and α_f =5x10^{-6}/K, α_m =10x10^{-6}/K and ΔT=-1200K for the sample with residual stresses.

Comparing the variation of broken matrix-elements ($N_M/N_{M,f}$) and debonded interface-elements ($N_I/N_{I,f}$) with increasing applied strain between the samples with and without residual stresses, both $N_M/N_{M,f}$ and $N_I/N_{I,f}$ of the sample with residual stresses shift evidently to the lower strain range. It is emphasized that the matrix breakage and matrix breakage–induced debonding reach a saturation level below 0.5%, while no fiber is broken at such strain level. This means that the stress

Advanced SiC/SiC Ceramic Composites

carrying capacity of the matrix is lost at 0.5% and the composite behaves like a fiber bundle beyond 0.5% strain. In this way, the premature breakage of the matrix, followed by enhanced debonding, due to the residual stresses, causes the loss of stress carrying capacity and reduction in strength of the composite. On the other hand, in the sample without residual stresses, the matrix is broken nearly at the same strain as the fiber, and premature debonding does not occur so that both fiber and matrix can support applied load, resulting in high strength. In the present examples, due to the deleterious effect of the residual stresses stated above, the strength of the sample with residual stresses was 930MPa, which was far lower than 1500MPa of the sample without residual stresses.

CONCLUSIONS

Concerning the influence of interfacial debonding and residual stresses on tensile fracture behavior of UD ceramic matrix composites, following features could be described by the present shear lag analysis combined with the Monte Carlo simulation method.

(1) When no residual stresses exist, the debonding from the broken fiber-end occurs at lower applied strain than that from the broken matrix-end in the composite with high fiber volume fraction. When the broken component has tensile and compressive stresses along the fiber direction, the debonding from the broken component-end is enhanced and retarded, respectively.

(2) In the fundamental process of fracture obtained by the simulation using micro-composite, damages such as breakage of fiber and matrix and interfacial debonding are accumulated intermittently, resulting in a serrated stress-strain curve.

(3) The mechanical interactions among the modes of damage become different for different interfacial strength and residual stresses, and thus the order and location of the occurrence of component-breakage and interfacial debonding become different, resulting in different stress-strain behavior and strength of the composite.

(4) With increasing interfacial strength, the fracture mode varies from the cumulative type, characterized by the gradual breakage of the constituents (fiber, matrix) and interfacial debonding in the fracture process and by the fiber pull-out in the fracture surface, to the non-cumulative one, characterized by the chain reaction mainly of breakage of components and by the overall fracture of composite perpendicular to tensile axis without fiber pull-out.

(5) When the interface is weak and the fracture strain of the matrix is equal to, or lower than, that of fiber, the residual stresses (compressive and tensile axial residual stresses in fiber and matrix, respectively) enhance the breakage of matrix and matrix breakage-induced debonding. As a result, the stress carrying capacity of the composite is reduced, resulting in low composite strength.

ACKNOWLEDGEMENT

The authors wish to express their gratitude to JSPS (Japan Society for Promotion of Science) and DFG (Deutsche Forshungsgemeinschaft) for the support of the present cooperative work between Japan and Germany.

REFERENCES

[1] A. G. Evans, "The mechanical properties of reinforced ceramic, metal and intermetallic matrix composites", *Materials Science and Engineering*, **A107** 227-41(1989).

[2] A. G. Evans and F. W. Zok, "The physics and mechanics of fibre-reinforced brittle matrix composites", *Journal of Materials Science*, **29** 3857-96(1994).

[3] S. Ochiai, I. Okumura, M. Tanaka, M. Hojo and T. Inoue, "Influences of residual stresses, frictional shear stress at debonded interface and interactions among broken components on interfacial debonding in unidirectional multi-filamentary composites", *Composite Interfaces*, **5** 363-81(1998).

[4] .S. Ochiai, M. Hojo and T. Inoue, "Shear lag simulation of progress of interfacial debonding in unidirectional composites", *Composite Science Technology*, **59** 77-88(1999).

[5] S. Ochiai, M. Hojo, K. Schulte and B. Fiedler: "Nondimensional simulation of influence of toughness of interface on tensile stress-strain behavior of unidirectional microcomposite", *Composites Part A*, **32** 749-61(2001).

[6] C. M. Landis, I. J. Beyerlein and R. M. McMeeking: "Micromechanical simulation of the failure of fiber-reinforced composites", *Journal of Mechanics and Phyysics of Solids*, **48** 621-48(2001).

[7] M. Lienkamp and P. Schwartz: "A Monte Carlo simulation of the failure of a seven fiber microcomposite", *Composite Science and Technology*, **46** 139-46(1993).

[8] B. Lauke, W. Beckert and J.Singletary, "Energy release rate and stress field calculation for debonding crack extension at the fibre-matrix interface during single -fiber pull-out", *Composite Interfaces*, **3** 263-73(1996).

[9] M. He and J. W. Hutchinson, "Crack deflection at an interface between dissimilar elastic materials", *International Journal of Solids Structures*, **25** 1053-67(1989).

[10] H. C. Cao, E. Bischoff, O. Sbaizero, M. Ruehle and A. G. Evans, "Effects of interfaces on the properties of fiber-reinforced ceramics", *Journal of American Ceramic Society*, **73** 1691-99(1990).

[11] L. M. Zhou, J. K. Kim, Y. M. Mai, "Interfacial debonding and fiber pull-out stresses", *Journal of Materials Science*, **27** 3155-66(1992).

[12] S. Ochiai, K Schulte and P. W. M. Peters, "Strain concentration factors for fibers and matrix in unidirectional composites", *Composite Science and Technology*, **44** 237-56(1991).

A NEW STANDARD TEST METHOD FOR THE TENSILE EVALUATION OF CERAMIC FIBERS

Edgar Lara-Curzio
Metals & Ceramics Division
Oak Ridge National Laboratory
Oak Ridge, TN 37831-6069

ABSTRACT

In 1997 the jurisdiction of the former ASTM Standard Test Method D3379 "Standard Test Method for Tensile Strength and Young's Modulus for High-Modulus Single-Filament Materials" was transferred from ASTM committee D30 on *High Modulus Fibers and Their Composites* to committee C28 on *Advanced Ceramics*. One year later this document was withdrawn, in part because it was demonstrated that using the average of the cross-sectional area of a collection of fibers for the calculation of individual fiber strengths, as recommended by the standard, was inappropriate. This paper reviews the draft of a new ASTM test method for the tensile evaluation of single fibers, which incorporates provisions for addressing the deficiencies of the previous ASTM standard test method. At the time of this publication this test method was currently evolving through the internal ASTM balloting process on its way to becoming a standard.

Keywords

Fiber, testing, standard, strength, Weibull, distribution, Monte Carlo. ASTM, ceramic, composite, CFCC

Background

The development of continuous fiber-reinforced ceramic matrix composites (CFCCs) has been possible, in part, by the availability of strong ceramic fibers. The principal role of the fibers in CFCCs is to bridge the wake of advancing cracks in the matrix thus providing a *graceful mode of failure*. Ultimately, the tensile strength of CFCCs is determined by the distribution of tensile strengths of the reinforcing fibers. For example the tensile strength of a unidirectional CFCC can be estimated from [1]:

$$\sigma_u = v_f \sigma_c \left(\frac{2}{m+2} \right)^{1/(m+1)} \frac{m+1}{m+2} \tag{1}$$

where v_f is the fiber volume fraction, m is the fiber Weibull modulus and σ_c depends on the fiber characteristic strength and on the interfacial shear stress. Expressions that depend on the parameters of the distribution of fiber strengths have also been derived to predict the reliability of CFCCs when subjected to stress-rupture conditions [2].

The parameters of the distribution of fiber strengths can be estimated by obtaining the monotonic tensile strength of several fibers that belong to a subset that is representative of the entire fiber population, followed by a statistical analysis of the results. For many years, ASTM D3379 *Standard Test Method for Tensile Strength and Young's Modulus for High-Modulus*

Single-Filament Materials [3] was the accepted standard test method for the determination of fiber strength. According to this document, the tensile strength of a fiber is determined by the ratio of the breaking load and the average cross-sectional area of a sample of fibers that is assumed to be representative of the fiber population. However, it has been demonstrated that the calculation of individual fiber strength values according to this formula can lead to serious errors in the estimation of the parameters of the distribution of fiber strengths, particularly when the dispersion of fiber diameters is large [4-5].

To determine the magnitude of the error in the estimates of the parameters of the distribution of fiber strengths when individual fiber strength values are determined according to the former ASTM D3379 test method, let us consider the following analysis. Let us consider a collection of N fibers with similar length and uniform circular cross-sectional area. Let us assume that the diameters of these fibers are distributed according to a normal distribution with known mean fiber diameter Φ_μ and standard deviation Φ_{sd} (Figure 1).

Figure 1. Distribution of fiber diameters in a bundle with different standard deviation. $15 \pm 3 \mu m$.

Let us also assume that the fiber strengths are distributed according to a two-parameter Weibull distribution with known parameters m and σ_o. In this case, the probability that a fiber will fail when subjected to an applied tensile stress, σ is going to be:

$$P_f = 1 - \exp\left(-\frac{V}{V_o}\left(-\frac{\sigma}{\sigma_o}\right)^m\right) \tag{2}$$

where V is the fiber gauge volume and V_o is the characteristic volume associated with the characteristic strength σ_o. Let us carry out a Monte Carlo simulation according to the diagram in Figure 2. The diameter of each one of the N fibers is obtained by generating random numbers between 0 and 1 and finding the corresponding diameter for that random number (probability) according to the parameters of the distribution of fiber diameters[1]. Then, for each fiber diameter ϕ_i, the corresponding fiber cross-sectional area, A_i, is calculated using Equation (3),

$$A_i = \pi \frac{\phi_i^2}{4} \tag{3}$$

---

[1] Using function **NORMINV(probability, mean, standard_deviation)** in the computer application Microsoft Excel 98.

Similarly, the strength of each fiber is determined by generating random numbers between 0 and 1 and finding the corresponding strength for that random number (i.e.- probability of failure) according to the parameters of the distribution of fiber strengths, and the volume of that particular fiber. Once the dimensions and tensile strength of each fiber are known, the associated distribution of breaking loads is determined considering that

$$L_i = A_i \, \sigma_i \tag{4}$$

where L_i is the breaking load for the i-th fiber with strength σ_i and cross-sectional area A_i. If we were to follow the instructions of the former ASTM 3379 standard test method, we would calculate each fiber strength as the ratio of the breaking load and the average of the fiber cross-sectional areas, i.e.,

$$\sigma_i^* = \frac{L_i}{A_\mu} \tag{5}$$

where

$$A_\mu = \frac{\pi}{4} \, \Phi_\mu^2 \tag{6}$$

To differentiate between the actual tensile strength of a fiber and the strength that would be obtained by following the instructions of ASTM D3379, the latter, σ^*, will be hereafter referred as "modified fiber strength". If we assume that the "modified fiber strengths" are distributed according to a Weibull distribution, then the parameters of this distribution can be estimated by ranking the "modified fiber strengths" in ascending order and assigning them probabilities of failure according to the following estimator [5]

$$P_i^* = \frac{i - 0.5}{N_f} \tag{7}$$

Then, the parameters m^* and σ_o^* of the distribution of "modified fiber strengths" can be estimated by the maximum likelihood estimation method, or by linear regression.

To determine the effect of the dispersion in the distribution of fiber diameters on the magnitude of the error incurred in the estimate of the parameters of the distribution of fiber strengths when following D3379, the procedure outlined above was repeated ten times for a different value of the standard deviation of the distribution of fiber diameters. Then, the parameters of the distribution of "modified fiber strengths" were estimated for each case. The following arbitrary values of $\sigma_o = 3.0$ GPa, $\Phi_\mu = 12$ μm were maintained constant, while the standard deviation of the distribution of fiber diameters was assigned the following values: 1 μm, 2 μm, 3 μm, 4 μm and 5 μm. Also, calculations were carried out for different values of the Weibull modulus (3, 5, 10 and 15).

Figure 3 is a plot of $ln \, ln \, (1/(1-P_f^*))$ versus $ln \, \sigma^*$ for the various values of Φ_{sd}, along with the original distribution of fiber strengths. Note that as the standard deviation of the distribution of fiber diameters increases, the slope of the curves decreases, i.e., the spread of the distribution of "modified fiber strengths" becomes larger than that of the original fiber strength distribution.

Figure 4 is a plot of the Weibull modulus of the distribution of "modified fiber strengths" normalized by the value of the original fiber strength distribution, as a function of the value of the standard deviation of the distribution of fiber diameters. The symbols in Figure 4 represent the average of ten simulations while the error bars represent the standard deviation about the mean value. Note that results were obtained for different values of the original Weibull modulus. As it was indicated, the Weibull modulus of the distribution of "modified fiber strengths" decreases with increasing Φ_{sd}. Figure 5 shows the characteristic "modified strength" normalized by the original characteristic strength and indicates that its value increases with Φ_{sd}. Note that for the numerical values used in this exercise, which can be considered to be representative of many ceramic fibers, the Weibull modulus can be underestimated by as much as a factor of four, although the characteristic strength is overestimated by only 25%.

The results of this analysis can be rationalized as follows. When the breaking loads are divided by the average of the fiber cross-sectional areas, the intrinsic relationship between breaking loads and fiber size is lost. In general, large-diameter fibers are more likely to fail with large rather than with small loads. However, when a large load is divided by a cross-sectional area value that is smaller than its corresponding cross-sectional area, it results in a "modified fiber strength" that is much larger than the actual strength. Similarly when a small load is divided by a cross-sectional area value that is larger than its corresponding cross-sectional area, it results in a "modified fiber strength" that is much smaller than the actual strength. The overall effect of these operations is to produce a new distribution ("modified fiber strengths") that is much wider than the original strength distribution. Furthermore, this effect is magnified when the spread in the distribution of fiber diameters increases, since for the same fiber strength distribution it results in a wider distribution of breaking loads and hence in a wider distribution of "modified fiber strengths". These results have been confirmed by others [6-7]

The results from this analysis suggest that to obtain accurate fiber strength values it is necessary to determine the diameter (cross-sectional area) of each fiber. However, because the diameter of many fibers varies along their length (Figure 6) [8-10], measuring the diameter of the fiber at a single location is not enough. Specifically, if the diameter (cross-sectional area) of a fiber is determined at a single location prior to performing the tensile test, then there is a good probability that such location will not coincide with the failure location.

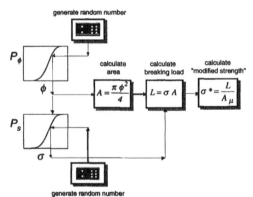

Figure 2. Schematic of Monte Carlo simulation to determine effect of fiber diameter variability on the parameters of the distribution of fiber strength.

Advanced SiC/SiC Ceramic Composites

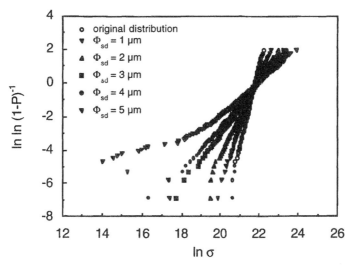

Figure 3. Weibull plot for the original distribution of fiber strengths, and for the distributions of "modified fiber strengths" obtained for different values of the standard deviation of the distribution of fiber diameters.

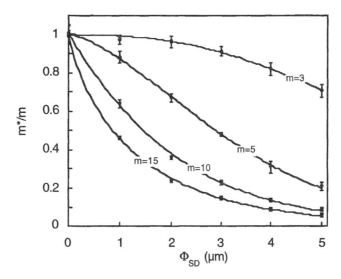

Figure 4. Effect of the standard deviation of the distribution of fiber diameters on the Weibull modulus of the distribution of "modified fiber strengths", normalized by the value of the Weibull modulus of the original distribution of fiber strengths. The error bars represent the standard deviation of ten repetitions.

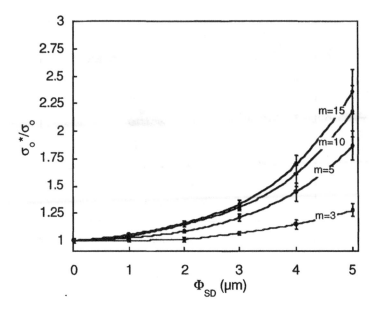

Figure 5. Effect of the standard deviation of the distribution of fiber diameters on the characteristic strength of the distribution of "modified fiber strengths", normalized by the value of the characteristic strength of the original distribution. The error bars represent the standard deviation of ten repetitions.

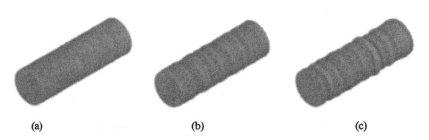

Figure 6. Profiles for 15 μm-diameter fibers with random diameter variation along their length (25 mm). Standard deviations correspond to (a) ±0.1 μm; (b) ±0.3 μm, and (c) ±0.5 μm.

To determine the magnitude of the error that is introduced in the determination of individual fiber strength values when the fiber dimensions are determined at a location that doesn't coincide with the fracture location and the fiber diameter varies along the fiber length, let us consider the following Monte Carlo simulation (Figure 7). Without losing generality let us assume that the diameter of the fibers is measured always at the same location (e.g.- at the middle of the gauge length) before the tensile test is performed. The simulation was carried out as follows: a fiber is subdivided into n segments of equal length, λ. Assuming that the parameters of the distributions of fiber strengths and average fiber diameters (among fibers in a bundle) are known, then diameter and strength values are assigned to each one of those n segments. It is also assumed that the diameter of each fiber varies randomly along its length.

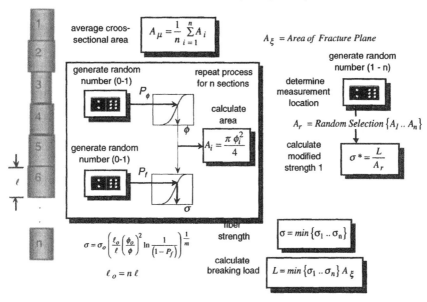

Figure 7. Schematic of Monte Carlo simulation to determine effect of variability of fiber diameter along its length on the determination of fiber strength, when the location of fiber diameter measurement doesn't coincide with the fracture location.

The value of the diameter for each segment was obtained using a random number generator and the procedure described in the previous section, whereas the strength of each fiber segment was determined according to

$$\sigma = \sigma_o \left(\frac{\ell_o}{\ell} \left(\frac{\phi_o}{\phi} \right)^2 \ln \frac{1}{\left(1 - P_f\right)} \right)^{\frac{1}{m}}$$
(8)

$$\sigma_{th} = min\{\sigma_1 .. \sigma_n\}$$
(9)

and the cross-sectional area of the fracture plane, A_ζ, is that of the segment with the smallest strength. Thus the breaking load, L, is determined as

$$L = A_\xi\, \sigma_{th} \tag{10}$$

The "modified strength", σ^*, is defined as the ratio of the breaking load L, and the cross-sectional area of the location where the fiber diameter was measured, let's say at the i-th of the n fiber segments. The error in the determination of the strength is computed as follows:

$$\text{error} = \left| \frac{\sigma_i - \sigma^*}{\sigma_i} \times 100 \right| \tag{11}$$

Figure 8 shows a plot of the average error, calculated according to Equation 11, as a function of the variability of fiber diameters along the length of the fiber. The data points in Figure 8 represent the average of 1000 repetitions using fibers 25-mm long (subdivided into 25 segments), with average diameter of 15 μm, and Weibull modulus of 5. The error bars are associated with the smallest and largest errors among the 1000 repetitions. The results in Figure 8 indicate that the error in the determination of fiber strength increases with the magnitude of the variability of fiber diameter along the fiber length when the strength is calculated using the cross-sectional area of the fiber that doesn't correspond to the cross-sectional area of the fracture plane. It has been found that for some fibers, the variability in fiber diameter along their length can be as large as ± 10% of their diameter [2-5] and therefore significant errors can occur when the cross-sectional area used for the calculation of strength doesn't correspond to the cross-sectional area of the fracture plane.

The new test standard

As a result of the analyses presented in the previous sections, a task group was organized within sub-committee C28.07 of ASTM to draft a new standard test method for the tensile evaluation of single fibers. The scope of this new document is the determination of the tensile strength of a single fiber. The user of the standard is referred to ASTM C1239 [12] in order to perform a statistical analysis to estimate the parameters of the distribution of fiber strengths. The task group is currently focused on developing a standard procedure for determining the cross-sectional area of single fibers. Considering the importance of determining the cross-sectional area of the fiber at the failure location, the test method suggest means for capturing the fracture surfaces after the test. The need to recover the fracture surfaces is also consistent with the need to do a fractographic analysis to identify the strength-limiting flaws.

The new standard test method is comprised of 12 sections: 1. *Scope*; 2. *Referenced Documents*; 3. *Terminology*; 4. *Summary of Test Method*; 5. *Significance and Use*; 6. *Interferences*; 7. *Apparatus*; 8. *Precautionary Statements*; 9. *Procedure*; 10. *Calculations*; 11. *Reporting*; 12. *Precision and Bias*. The test method addresses various techniques for preparing and mounting single fibers but recommends one technique that is widely used in the community, which consists in mounting the fibers onto cardboard holders as indicated in Figure 9. Mechanical loads can be transferred to the cardboard holder either through pins, or by clamping the cardboard holder on its ends. Section 10. *Calculations*, provides guidance for determining fiber strains and the fiber Young's modulus through determination of the cross-head displacement and machine compliance corrections (Figure 11). This procedure is based on plotting the inverse of the slope of load versus cross-head displacement curves as a function of fiber gauge length. The slope of the resulting line will be inversely proportional to the fiber's Young's modulus, whereas the intercept of the line with the vertical axis (i.e.- extrapolation to zero gauge length) will correspond to the value of the machine compliance.

Advanced SiC/SiC Ceramic Composites

This document has undergone through both sub-committee and main committee balloting. During the summer of 2002, the inclusion into the document of a precision and bias statement will be balloted and it is expected that the document will become a society-approved standard test method at the end of 2002.

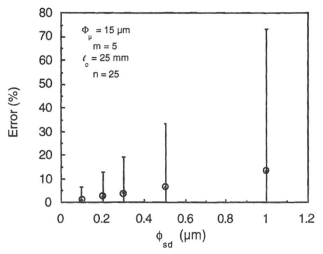

Figure 8. Magnitude of error in the determination of single-fiber strength values when the failure location doesn't coincide with the diameter measurement location, as a function of the variability in fiber diameter along the fiber length (ϕ_{sd}). Data points correspond to the average of 1000 repetitions. Error bars represent the largest and smallest values among those 1000 repetitions.

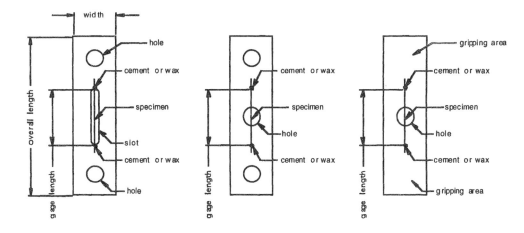

Figure 9. Typical fiber gripping techniques.

Figure 10a. Picture of typical fiber tester.

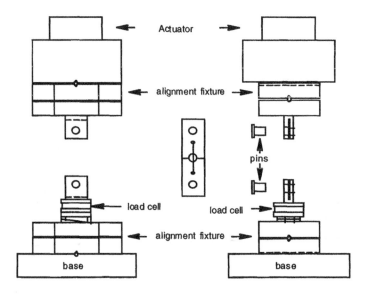

Figure 10b. Schematic of typical fiber tester

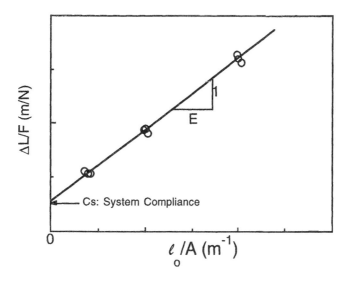

Figure 11. Method for determining the compliance of the system.

Summary

A task group within sub-committee C28.07 of ASTM has drafted a new standard test method for the tensile evaluation of single fibers. The drafting of a new standard was prompted by deficiencies of the former ASTM D3379. In particular, it was demonstrated, through Monte Carlo simulations, that large errors could be incurred in the estimation of the parameters of the distribution of fiber strengths when the strength of individual fibers is determined as the ratio of the breaking force and the average of the cross-sectional area of several fibers. It was also found that the magnitude of the error increased with increasing dispersion in the distribution of fiber diameters. For the case of typical ceramic fibers it was found that the Weibull modulus can be underestimated by as much as a factor of four while the characteristic strength is overestimated by 25%. Also, it has been demonstrated that determining the dimensions of each fiber is not enough to obtain accurate strength measurements, unless the fiber cross-sectional area is determined at the fracture plane. This is an important requisite considering that the diameter of many ceramic fibers varies along their length, and therefore, there is a large probability that measurements of the fiber dimensions prior to the tensile tests would be obtained at a location of the fiber that doesn't coincide with the fracture plane.

It is expected that the new standard test method will be officially adopted by ASTM by the end of 2002.

ACKNOWLEDGEMENTS

This work was sponsored by the U.S. Department of Energy, Assistant Secretary for Energy Efficiency and Renewable Energy, Office of Industrial Technologies, Industrial Energy Efficiency Division, Continuous Fiber-reinforced Ceramic Composites Program under contract, DE-AC05-96OR22464 with UT-Battelle, LLC.

REFERENCES

1. W. A. Curtin, "Theory of the Mechanical Properties of Ceramic-Matrix Composites," *J. Am. Ceram. Soc.*, **74** [4] 2837 (1991)
2. E. Lara-Curzio," Stress-Rupture of Nicalon™/SiC Continuous Fiber Ceramic Composites in Air at 950°C, " *J. Am. Ceram. Soc.*, **80** [12] pp. 3268-72 (1997)
3. ASTM D3379 "Standard Test Method for Tensile Strength and Young's Modulus for High-Modulus Single-Filament Materials," American Society for Testing and Materials, West Conshohocken, PA 19428
4. E. Lara-Curzio and C. M. Russ, "Why it is Necessary to Determine Each Fiber Diameter When Estimating the Parameters of the Distribution of Fiber Strengths," *Ceram. Eng. and Sci.*, **20**, (1999) pp. 681-688
5. E. Lara-Curzio and C. Russ, "On the Relationship Between the Parameters of the Distributions of Fiber Diameters, Breaking Loads and Fiber Strengths," *J. Mater. Sci. Letters*, **18**, 24 (1999) pp. 2041-2044
6. T. A. Parthasarathy, "Extraction of Weibull Parameters of Fiber Strength from Means and Standard Deviations of Failure Loads and Fiber Diameters," *J. Am. Ceram. Society*, 83, 3 (2001) pp. 588-592
7. Wilson
8. N. Lissart and J. Lamon "Statistical analysis of failure of SiC fibers in the presence of bimodal flaw populations," *Journal of Materials Science* , **32**, 22 pp. 6107-6117 (1997)
9. T. Morimoto, J. Goering, and H. Schneider, "A New Method for Measuring Diameter Distribution Along Ceramic Fibers," *Ceram. Eng. Sci. Proc.*, **19**, 3 (1998) pp. 47-54
10. G. E. Youngblood, C. R. Eiholzer, C. A. Lewinsohn, and R. H. Jones, and A. Hasegawa, and A. Kohyama, "Fiber Diameter Variation-Sample Preparation and Analysis," *Ceram. Eng. Sci. Proc.*, **20**, 3 (1999) pp. 481-486
11. T. Tanaka, H. Nakayama, A. Sakaida, and N. Horikawa, "Estimation of Tensile Strength Distribution for carbon Fiber with Diameter Variation Along Fiber," *Materials Science Research International*, **vol. 5**, No. 2 (1999), pp. 90-97
12. ASTM C1239-95 "Standard Practice for Reporting Uniaxial Strength Data and Estimating Weibull Distribution Parameters for Advanced Ceramics", American Society for Testing and Materials, West Conshohocken, PA 19428

TENSILE STRENGTH OF CHEMICAL VAPOR INFILTRATED ADVANCED SiC FIBER COMPOSITES AT ELEVATED TEMPERATURES

T. Nozawa, K. Hironaka, Y. Katoh and
A. Kohyama
Institute of Advanced Energy,
Kyoto University
Gokasho, Uji, Kyoto 611-0011, Japan

T. Taguchi and S. Jitsukawa
Japan Atomic Energy Research
Institute,
Tokai, Ibaraki 319-1195, Japan

L.L. Snead
Metals and Ceramics Division,
Oak Ridge National Laboratory
Oak Ridge, TN37831-6069

ABSTRACT

SiC/SiC composites composed of high-crystalline, near-stoichiometric SiC fiber, like Hi-Nicalon™ Type-S and Tyranno™-SA, are promising structural materials for fusion and other high-temperature applications, because of their excellent chemical, physical and mechanical stabilities at high-temperature. In order to explore performance of recently developed advanced SiC/SiC composites under severe environment and to identify the key issues for the further material development, room- and high-temperature tensile tests were conducted under the mild oxidizing environment at 1300°C for Tyranno™-SA fiber reinforced SiC matrix composites with two kinds of the fiber and matrix (F/M) interphase: pyrolytic carbon (PyC) and SiC/PyC, fabricated by the forced-flow/thermal-gradient chemical vapor infiltration (F-CVI) method. Tensile strength of both composites was significantly stable to high-temperature exposure up to 1300°C in mild oxidizing environment, with no clear difference for two interfacial structures. Also, there was no dependence of PyC thickness on tensile strength for both interphase systems. In other words, the rough surface of Tyranno™-SA fiber had the same role with the SiC pre-coating.

INTRODUCTION

SiC/SiC composites are considered one of the promising materials for applications in advanced energy industries, because silicon carbide (SiC) has inherent mechanical property stability at high-temperature, low-induced activation, after heat, and excellent corrosion resistance [1]. High-crystalline and near-stoichiometric SiC fibers like Hi-Nicalon™ Type-S and Tyranno™-SA are, in particular, stable to oxidation at high-temperature and to severe neutron exposure, because of less impurities such as oxygen and good structural order [2]. Similarly, β-SiC matrix derived by forced-flow/thermal-gradient chemical vapor infiltration (F-CVI) process, which has a highly-crystalline structure, would show good stability of strength against neutron exposure [3-5]. From these reasons, SiC/SiC composites with high-crystalline and near-stoichiometric SiC fiber and matrix are considered to have excellent physical and mechanical properties under these severe conditions. Therefore, many investigations on F-CVI process have been enthusiastically carried out at Oak Ridge National Laboratory (ORNL), as a part of Japan-US collaborations. This study focused on tensile properties of SiC/SiC composites with recently developed new SiC fibers, for the optimization of F-CVI process.

In case of the design of ceramic matrix composites, interfacial materials such as pyrolytic carbon (PyC) are, in general, formed between fiber and matrix in order for the improvement of toughness. The role of the interphase is to deflect the main cracks at the weak fiber and matrix (F/M) interface and to cause many fiber pullouts. Interfacial shear strength and friction cause the ductile fracture behavior and produces high fracture toughness. However, there has been a concern that cracks propagated along the smooth surface of the fibers such as Hi-Nicalon™ would have insufficient interfacial friction for load transfer. A rough SiC layer was designed to deposit on the fiber surface to promote the crack deflection within the interphase [6]. In addition, multi-layered SiC interphase has been developed in order to make the crack path more complex and to prevent the outer reaction gases from flowing into the crack paths.

The objective of this study is to investigate mechanical performance of recently developed stoichiometric SiC fiber reinforced F-CVI SiC matrix composites and also to identify the key implementation for the improvement of F-CVI technique. In particular, high-temperature tensile properties and the effect of several interfacial structures were discussed.

EXPERIMENTAL
Materials
All the composite disks with a 3-inch diameter and a half-inch thickness were fabricated by F-CVI method at ORNL (Table I). Plane-woven (P/W) sheets of Si-Al-C fiber: Tyranno™-SA (Ube Industries), which were stacked in the [0°/90°] direction, were used as reinforcements. Then, β-SiC matrix was deposited by using methyltrichlorosilane (MTS) carried with hydrogen.

Pyrolytic carbon (PyC) interphase and SiC/PyC interphase were deposited on

Table I Test Materials

ID	Composite Structure	Aimed Interlayer Thickness [nm]	Fiber Volume Fraction [%]	Density [Mg/m³]	Porosity [%]
1264	Tyranno-SA/PyC/FCVI-SiC	150	35.4	2.61	23.3
1265	Tyranno-SA/PyC/FCVI-SiC	300	35.3	2.72	18.0
1266	Tyranno-SA/PyC/FCVI-SiC	75	35.2	2.62	18.1
1267	Tyranno-SA/(SiC/PyC)/FCVI-SiC	100/150	38.8	2.74	15.7
1268	Tyranno-SA/(SiC/PyC)/FCVI-SiC	100/150	38.8	2.69	18.0
1269	Tyranno-SA/(SiC/PyC)/FCVI-SiC	100/300	38.8	2.71	17.2

the surface of each fiber before the F-CVI densification. Three kinds of thickness of PyC interphase; 75, 150 and 300 nm, were chosen for the evaluation of influences of their thickness on tensile properties. In order to fabricate a uniform interphase in the F/M interface, deposition of PyC was carried out in two steps: initial infiltration followed by turning the specimens upside down and infiltrating again. All the composites had a small gradient of the interfacial thickness with a maximum at the center of the upstream side. More details were discussed elsewhere [7].

Tensile Test

Tensile tests were conducted by an electromechanical testing machine (Instron Japan Co. Ltd.) on the basis of ASTM C1275 and C1359. Miniature edge-loaded tensile specimen was used (Fig. 1). All the tests were conducted at the crosshead speed of 0.5 mm/min. The step-loading tests were performed for the room-temperature tension in air for the precise evaluation of the damage accumulation around

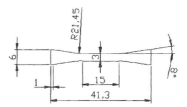

Fig. 1 Schematic illustration of the miniature edge-loaded tensile specimen

proportional limit. Simple monotonic tests were performed for high-temperature tension at 1300°C in a flow of commercial argon with about 0.1 Pa of oxygen in partial pressure. High-temperature tests were carried out following a 20 min ramp to the test temperature and a subsequent equilibration time of about 10 min. More details were described elsewhere [8].

After the tensile tests, fracture surfaces of all the specimens were examined by using scanning electron microscopy (SEM). Besides, porosity, fiber volume fraction and thickness of the PyC interphase were also measured.

Analysis

Tensile properties were analyzed by using the normalized value as shown in equation (1), which takes into consideration the large scatter of porosity among as-received materials. It was revealed that elastic modulus and proportional limit stress were in roughly inverse proportion to the porosity [8]. In this equation, stress was calculated as applied force divided by the true cross-sectional area excluding the area occupied by pores.

$$\text{Normalized Value} = \frac{\text{Original Value}}{1 - \text{Porosity}} \qquad (1)$$

It is noted that, in this analysis, it was assumed that porosity was distributed equally in any cross-section.

RESULTS AND DISCUSSION

High-Temperature Tensile Fracture Behaviors of Advanced SiC/SiC Composites

Tyranno™-SA/ FCVI-SiC had good stability in tensile properties under the high-temperature exposure regardless of the interfacial structures (Fig. 2). There was only a minor degradation of tensile strength at 1300°C in mild oxidizing environment.

SiC easily formed into SiO_2 in air by the oxidation and, even if in inert environment, SiC is oxidized due to the reaction with oxygen included as the impurity [10, 11]. The former is well known as passive

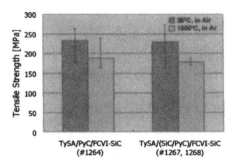

Fig. 2 High-temperature tensile strength of advanced SiC/SiC composites

oxidation and the latter is referred to as active oxidation. SiC fiber and matrix used in this study were high-crystalline and near-stoichiometric composition, and hence there were few impurities. Also, it is reported that there was no significant degradation of tensile strength in Tyranno™-SA fiber itself after the heat treatment below 1300°C in inert environment [12]. These good stabilities of each component made tensile strength of composites stable to the oxidizing attack. However, slight degradation of tensile strength should not be ignored. In air, PyC might be easily burned out by oxidation. However, under the mild-oxidizing environment, PyC oxidation seemed quite small. It might be due to the severe degradation of the in-situ fiber strength or interfacial shear properties at 1300°C.

Effect of the PyC Thickness on RT/HT Tensile Properties – Single PyC Interphase

It is reported that Tyranno™-SA/SiC composite had maximum flexural strength for a 200 nm PyC thickness. For thinner PyC thickness, there existed

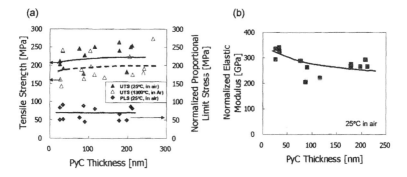

Fig. 3 PyC thickness dependencies on (a) tensile strength, normalized PLS and (b) normalized elastic modulus (single PyC interphase)

significant large stress drops [13]. However, in tension, there was no clear dependence on the PyC thickness. Tensile strength and proportional limit stress (PLS) were nearly constant over the wide range of the PyC thickness (Fig. 3). On the contrary, tensile modulus tended to decrease with the PyC thickness.

Similar to the results at room temperature, there was no effect of the PyC thickness on the tensile strength at 1300°C in Ar, although the magnitude of tensile strength decreased about 20 %. Tensile strength at high-temperature in inert environment was nearly constant in the broad range of the PyC thickness. According to the microscopic observation of the fracture surface (Fig. 4), PyC was burned out only near the surface of the composite, and, macroscopically, PyC degradation was never critical to the reduction of tensile strength. The reason for the degradation of the tensile strength at 1300°C in Ar might be the in-situ degradation of tensile strength of fiber itself.

On considering the application to neutron irradiation, thin PyC interphase also becomes an advantage because of less degradation by neutron damage. Also excellent mechanical performance in the thinner PyC is very important fact for the design of the multi-layered interphase, which is structured of the sequence of thin SiC and PyC layers.

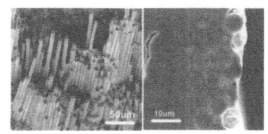

Fig. 4 Typical fracture surface appearances of advanced SiC/SiC with single PyC interphase after tension at 1300°C in mild oxidizing environment

Effect of the SiC Pre-Coating on the Fiber

Fig. 5 shows the PyC thickness dependency on the room-temperature tensile properties of SiC/SiC with SiC/PyC interphase. It was the same behavior as that in the case of using the single PyC interphase. In addition, the fracture behavior at high-temperature in inert environment was also very similar. There was about 20 % degradation of tensile strength and PyC burned out partially (Fig. 6). This is because the rough surface of Tyranno™-SA fiber played the same function of the rough SiC pre-coating, which was formed for the promotion of the crack path within the first PyC interlayer. Possibly, rough surface of Tyranno™-SA fiber promoted the crack deflection within the first PyC interlayer.

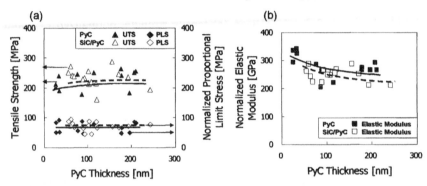

Fig. 5 PyC thickness dependencies on (a) tensile strength, normalized PLS and (b) normalized elastic modulus (at room temperature in air)

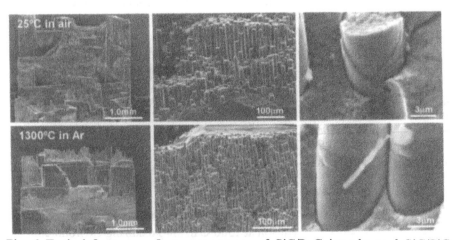

Fig. 6 Typical fracture surface appearances of SiC/PyC inter-layered SiC/SiC composites. There existed special gaps at the F/M interface after HT-tension

Advanced SiC/SiC Ceramic Composites

CONCLUSIONS

In order to investigate the performance of recently developed high-crystalline, near-stoichiometric F-CVI SiC/SiC composites, room- and elevated-temperature tensile tests were conducted by using small specimen test technique. In particular, roles of the various PyC based interphase; single PyC and SiC/PyC were identified. Key conclusions are summarized as follows.

1. High-crystalline, near-stoichiometric F-CVI SiC/SiC composites showed the excellent mechanical performance and they were characteristic in high strength, high modulus in tension.
2. Tensile strength and PLS were independent of the PyC thickness in the broad range. However, elastic modulus depended on it. It tended to decrease as the PyC thickness increasing.
3. There was no clear difference in tensile behavior between single PyC and (SiC/PyC) inter-layered composites. The PyC thickness effect on Tyranno™-SA/(SiC/PyC)/SiC composites was quite similar to that on Tyranno™-SA/PyC/SiC composites. Rough surface of Tyranno™-SA fiber had the same function with the SiC pre-coating.
4. Advanced F-CVI SiC/SiC composites with single PyC and SiC/PyC interphase showed the excellent high-temperature performance, respectively. Both types of SiC/SiC composites could maintain over 80% of their tensile strength at 1300°C in Ar. In-situ fiber strength and interfacial shear properties might be degraded, because of no significant difference between the two different interphase types at elevated temperature.

ACKNOWLEDGMENTS

The authors would like to express their sincere appreciation to Mr. J.C. McLaughlin and Dr. T.M. Besmann, Oak Ridge National Laboratory for the fabrication of materials. This work was performed as a part of collaborations at Japan-US Program for Irradiation Test of Fusion Materials (JUPITER-II) and US-DOE/JAERI Collaborative Program on FWB Structural Materials in Mixed-Spectrum Fission Reactors, Phase IV. This study was also supported by Core Research for Evolutional Science and Technology (CREST).

REFERENCES

[1] R.H. Jones, L.L. Snead, A. Kohyama, P. Fenici, "Recent Advances in the Development of SiC/SiC as a Fusion Structural Material," Fusion Engineering and Design," *Fusion Engineering and Design*, 41 15-24 (1998).

[2] L.L. Snead, Y. Katoh, A. Kohyama, J.L. Bailey, N.L. Vaughm, R.A. Lowden, "Evaluation of Neutron Irradiated Near-Stoichiometric Silicon Carbide Fiber Composites," *Journal of Nuclear Materials*, **283-287** 551-555 (2000).

[3]R.J. Price, "Effects of Fast-Neutron on Pyrolytic Silicon Carbide," *Journal of Nuclear Materials*, **33** 17-22 (1969).

[4]R.J. Price, G.R. Hopkins, "Flexural Strength of Proof-Tested and Neutron-Irradiated Silicon Carbide," *Journal of Nuclear Materials*, **108-109** 732-738 (1982).

[5]M.C. Osborne, J.C. Hay, L.L. Snead, D. Steiner, "Mechanical- and Physical-Property Changes of Neutron-Irradiated Chemical-Vapor-Deposited Silicon Carbide," *Journal of the American Ceramics Society*, **82** 2490-2496 (1999).

[6]T. Hinoki, W. Yang, T. Nozawa, T. Shibayama, Y. Katoh and A. Kohyama, "Improvement of Mechanical Properties of SiC/SiC Composites by Various Surface Treatments of Fibers," *Journal of Nuclear Materials*, **289** 23-29 (2001).

[7]T. Taguchi, N. Igawa, T. Nozawa, K. Hironaka, L. L. Snead, T. Hinoki, Y. Katoh, S. Jitsukawa, A. Kohyama and J. C. McLaughlin, "Optimizing the Fabrication Process for Superior Mechanical Properties in the Stoichiometric SiC Fiber Reinforced FCVI SiC Matrix Composite System," *Fusion Materials*, **DOE/ER-0313/31** 47-56 (2001).

[8]T. Nozawa, K. Hironaka, T. Taguchi, N. Igawa, L.L. Snead, Y. Katoh, S. Jitsukawa and A. Kohyama, "Tensile Properties of Stoichiometric Silicon Carbide Fiber Reinforced F-CVI Derived Silicon Carbide Matrix Composites," *Fusion Materials*, **DOE/ER-0313/31** 40-46 (2001).

[9]T. Nozawa, T. Hinoki, Y. Katoh, A. Kohyama, E. Lara-Curzio, "Specimen Size Effects on Tensile Properties of 2-D/3-D SiC/SiC Composites," *ASTM STP*, 1418 (2002), to be published.

[10]C. Labrugere, A. Guette, R. Naslain, "Effect of Ageing Treatment at High Temperatures on the Microstructure and Mechanical Behavior of 2D Nicalon/C/SiC Composites.1: Ageing under Vacuum or Argon," *Journal of the European Ceramics Society*, **17** 623-640 (1997).

[11]C. Labrugere, L. Guillaumat, A. Guette, R. Naslain, "Effect of Ageing Treatment at High Temperatures on the Microstructure and Mechanical Behavior of 2D Nicalon/C/SiC Composites.2: Ageing under CO and Influence of a SiC Seal-Coating," *Journal of the European Ceramics Society*, **17** 641-657 (1997).

[12]R.E. Tressler, "Recent Developments in Fibers and Interphases for High Temperature Ceramic Matrix Composites," *Composites A*, **30** 429-437 (1999).

[13]W. Yang, H. Araki, T. Noda, Y. Katoh, J. Yu and A. Kohyama, "CVI Tyranno-SA/SiC Composites with PyC and SiC Multi-interlayers," *Ceramics Engineering & Science Proceedings*, **22** 481-488 (2001).

THERMOMECHANICAL PROPERTIES OF SILICON CARBIDE FIBER-REINFORCED SIALON-BASED COMPOSITES FABRICATED BY REACTIVE MELT INFILTRATION

Satoshi Kitaoka and Naoki Kawashima
Japan Fine Ceramics Center
2-4-1 Mutsuno, Atsuta-ku, Nagoya
456-8587, Japan

Yoshiyuki Sugahara
Waseda University
3 Ohkubo, Shinjuku-ku
Tokyo, 169-8555, Japan

Yuji Sugita, Yukio Kagiya and Takahiro Banno
Chubu Electric Power Company
20-1 Kitasekiyama, Odaka, Midori-ku, Nagoya, 459-8522, Japan

ABSTRACT

BN coated SiC (Hi-Nicalon) fiber-reinforced ß'-SiAlON based composites have been fabricated by reactive melt infiltration. ß'-SiAlON was produced in-situ by the reaction of ß-Si_3N_4 and AlN reactants in the preforms with Y-Al-Si-O molten glass. The wettability of the fibers to the molten glass was improved by infiltration and pyrolysis of perhydropolysilazane. The reaction between the fiber and molten glass was depressed using a controlled N_2 atmosphere. The mechanical properties of the composites at room temperature were significantly influenced by the composition of the nitride reactants introduced into the preforms. The increase of the amount of ß'-SiAlON formed in the matrix probably increased the residual tensile radial stress between the fiber and matrix, resulting in the improvement of the mechanical properties. The high temperature mechanical performance could be improved by increasing the packing of nitride reactants into the fabrics before melt infiltration of the glass because of the decrease in the amount of residual glass.

INTRODUCTION

Continuous fiber reinforced ceramic matrix composites (CFCCs) have received considerable attention for use in gas-turbine applications, because they are expected to have both excellent thermal stability and damage tolerance. Infiltration of molten glass is an alternative method for making CFCCs. It is capable of producing complex shaped composites with fully dense matrices easily, and is a high precision and low cost technique.[1] The densification of the matrix improves corrosion resistance of the composites and leads to a wider design of products because of an increase in the elastic limit. Glass-ceramic matrices provide the unique capability of allowing the composites to densify in the glassy state with subsequent crystallization of the matrix, giving good high-temperature resistance. However, the use of composites with devitrified glass ceramic matrices is limited to low temperature.

ß'-SiAlON ($Si_{6-z}Al_zO_zN_{8-z}$, where z = 0–4.2) is also a candidate for use as the matrix material in CFCCs because of its high strength, excellent oxidation resistance and thermal shock properties.[2] This material can be produced by a reactive melt infiltration process, because ß'-SiAlON is commonly prepared by reaction sintering. For example, a molten glass containing aluminum oxide and/or silicon oxide reacts with a mixture of ß-silicon nitride and aluminum nitride powders in porous preforms and transforms to ß'-SiAlON based ceramics according to

$$(4-z)Si_3N_4 + 2zAlN + zSiO_2 \rightarrow 2Si_{6-z}Al_zO_zN_{8-z} \tag{1}$$

$$(6-z)Si_3N_4 + zAlN + zAl_2O_3 \rightarrow 3Si_{6-z}Al_zO_zN_{8-z} \tag{2}$$

Even if a ß'-SiAlON layer is formed during the reactive melt infiltration, the oxide reactants are expected to diffuse easily through the product until the high solid solution limit is reached. Because Si_3N_4 and SiAlON ceramics are wet very well by Y-Al-Si-O glass containing lanthanide oxide, the glass constituents may also move along the ceramics' grain boundaries. By adding yttrium, the composition of the residual glass during formation of ß'-SiAlON becomes enriched with Y_2O_3, so that compounds with high melting point such as YAG and Y-silicate will be produced, resulting in CFCCs with excellent thermomechanical properties. If a large amount of the unreacted nitrides are present in the preform compared with the molten glass, the amount of residual glass will be able to be significantly reduced.

In order to achieve a fully dense matrix and complete reaction, infiltration should be performed at as high a temperature as possible. SiC fibers with excellent thermal stability,

Hi-Nicalon (Nippon Carbon Co., Japan), are the most suitable for use in high temperature applications. It has been found that polymer infiltration and pyrolysis of a Si-N precursor (perhydropolysilazane) lead to improved wetting of the fibers by an Y-Al-Si-O molten glass.[3] Moreover, the reaction between the fibers and the molten glass could be inhibited by the molten glass infiltration under an N_2 pressure of 10^6 Pa. As a result, Hi-Nicalon fiber reinforced ß'-SiAlON based ceramic composites with a fully dense matrix and minimal from fiber degradation can be fabricated by reactive melt infiltration.[3] The matrix composition of the composites depends significantly on the amount of the Si_3N_4 and AlN reactants introduced into woven preforms of the fibers. The purpose of the present study is to investigate the effect of the amount of the reactants introduced into the preforms on the thermomechanical performance of the composites.

EXPERIMENTAL PROCEDURE

Orthogonal 3D woven fabrics of BN coated SiC fibers (Hi-Nicalon) were used as preforms for complex shaped components. The preform size and the number ratio of the fiber bundles are listed on Table 1. Volume % of the fibers in the preforms was approx. 30. Slurry containing a mixture of ß-Si_3N_4 (NP-500, Denki Kagaku Kogyo Co., Japan) and AlN (Dow Chemical Co.) reactant powders, solvent (ethanol or iso-buthanol) and polyethylenimine (ACROS Co.) as a dispersing agent was introduced into the preform. The mixture ratios of the reactants were changed at the slurry infiltration stage. However, only a small amount of either the Si_3N_4 or AlN powders entered into the bundles during slurry infiltration in spite of addition of the dispersing agent. It is well known that molten Y-Al-Si-O glass easily wets both powders, but not BN. Therefore, when either the Si_3N_4 or AlN powders are absent and the reaction of the SiC fibers with the molten glass is depressed, the molten glass also hardly infiltrates the bundles during the melt infiltration process.

Because perhydropolysilazane (N-N410, Clariant K. K., Japan) is a liquid at room temperature, it can penetrate into the bundles. Moreover, it can be pyrolyzed and converted to Si_3N_4 at temperatures below the melt infiltration temperature of 1773 K. The Si_3N_4 produced in the bundles makes for easy infiltration of the molten glass. Therefore, after slurry infiltration, perhydropolysilazane (N-N310, Clariant K. K., Japan), which was concentrated to about

Table I. Preform Size and Number Ratio of the Fiber Bundles.

Dimension	Preform size, mm	Fiber bundle ratio
x (l)	50	22
y (w)	30	10
z (t)	4	3

90 wt%, was introduced and pre-pyrolyzed at 873 K under N_2 at atmospheric pressure. Precursor infiltration and pyrolysis are referred to hereafter as PIP. Next, melt infiltration of Y-Al-Si-O glass, which was previously prepared by melting a mixture of 34 mass% Y_2O_3, 22 mass% Al_2O_3 and 44 mass% SiO_2 at 1648 K, was performed at 1773 K for 3 hours under an N_2 pressure of 10^6 Pa. Molten glass infiltration is hereafter referred to as MGI. After MGI, the excess glass covering the preforms was removed by grinding. The preforms were annealed at 1573 K for 20 hours under an N_2 pressure of 10^6 Pa to complete the SiAlON formation and crystallization of the residual glass in the preforms.

Samples of dimensions 40 mm×10 mm×1 mm were cut from the preforms with the same fiber alignment. Three-point flexure tests were performed at room temperature and 1773 K in argon using these samples with a span of 30 mm under a constant cross-head speed of 0.5 mm/min, and a tensile stress applied in the x-direction. The maximum stress, σ_{max}, was determined at the maximum applied load during the three-point flexure test. Work of fracture (WOF) was obtained by calculating the area under the load-displacement curve and dividing by the apparent cross-sectional area of the specimen, when either the specimen broke, or the maximum displacement of 2.5 mm (i.e., when the specimen made contact with the bottom of the bending jig) had been achieved.

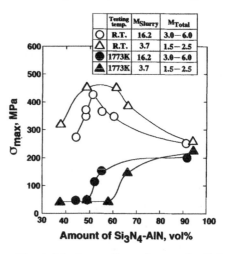

Fig. 1 Maximum flexural strength of the specimens as a function of the total amount of Si_3N_4-AlN reactants introduced into the preforms. M_{Slurry} and M_{Total} are AlN/Si_3N_4 molar ratios at the slurry infiltration stage and after PIP stage, respectively.

RESULTS AND DISCUSSION

X-ray diffraction analysis of all the composites found several crystalline phases in the matrix, such as ß'-$Si_{6-z}Al_zO_zN_{8-z}$ and YAG, etc. Figure 1 shows a plot of maximum flexural strength, σ_{max}, of the specimens as a function of the total amount of Si_3N_4 and AlN reactants, V_r, introduced into the preforms during slurry infiltration and PIP. All the composites were fabricated by slurry infiltration, PIP and MGI under high N_2 pressure. M_{Slurry} and M_{Total} are the AlN/Si_3N_4 molar ratios at the slurry

Advanced SiC/SiC Ceramic Composites

infiltration stage and after the PIP stage, respectively. Because the Si-N precursor was used at the PIP stage, the values of M_{Total} in the preform were smaller than those of M_{Slurry}. The values of σ_{max} at room temperature for the composites show maxima at 50–60 vol% of V_r for both values of M_{Total}. The values for the composites fabricated with the smaller M_{Total} are greater than those for the larger M_{Total}. The values of σ_{max} at 1773 K suddenly increase with increasing V_r, but become saturated at about 200 MPa. The values of V_r corresponding to the onset of the saturation for the composites fabricated with smaller M_{Total} are larger than those with the larger M_{Total}. As mentioned above, we used the Si-N precursor to improve the wettability of the fiber with the molten glass. This precursor was introduced into the fiber bundles and transformed into Si_3N_4 powders. The absence of Al-N species in the bundles, therefore, depressed SiAlON formation, resulting in a small amount of residual glass with a low melting temperature in the bundles. In order to overcome this problem, it may be desirable to use an Si-Al-N precursor instead of the Si-N precursor.

Figure 2 shows WOF of the specimens as a function of V_r. The dependence of V_r on WOF is similar to that of σ_{max} shown in Fig.2. The WOF values for the composites fabricated with the smaller M_{Total} are greater than those with the larger M_{Total}, especially for composites containing a lot of reactants.

Figure 3 shows scanning electron micrographs of typical fracture surfaces of specimens tested at room temperature for composites fabricated when M_{Slurry} is 16.2. The fracture surfaces corresponding to the smaller WOF, as shown in Fig.2, are relatively flat and fiber pullout is hardly observed. However, for the specimen with $V_r = 50$ vol% corresponding to the larger WOF, the pullout length of the fibers is very large. BN films with flat surfaces remain on the pullout fibers. This suggests that debonding between fibers and matrix preferentially occurs at the interface between the BN layer and the matrix.

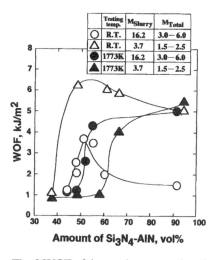

Testing temp.	M_{Slurry}	M_{Total}
○ R.T.	16.2	3.0–6.0
△ R.T.	3.7	1.5–2.5
◑ 1773K	16.2	3.0–6.0
▲ 1773K	3.7	1.5–2.5

Fig. 2 WOF of the specimens as a function of the total amount of Si_3N_4-AlN reactants introduced into the preforms.

The reason why the mechanical performance and fracture morphology of the composites depend on the reactant compositions in the preforms will be discussed from the viewpoint of interfacial residual stress, which is produced by mismatch of the thermal expansion coefficients between the fiber and matrix. Radial residual stresses on the interface are estimated on the basis of the calculated matrix compositions. Figure 4 shows the calculated amount of ß'-SiAlON produced in the matrix as a function of V_r. The amount of ß'-SiAlON was determined on the assumption that a contribution ratio of reactions (1) and (2) of ß'-SiAlON formation is proportional to the concentration ratio of Al_2O_3 and SiO_2 in the molten glass and each fraction reacted is 100%. When the matrix contains an excess amount of the Y-Al-Si-O glass reactant compared with that of the Si_3N_4 and AlN reactants, it is assumed to consist of ß'-SiAlON and a remaining glass to simplify the later residual stress calculations. The composition of the glass becomes richer in yttrium with increasing V_r. In contrast, when excess amount nitride reactants are present, the matrix consists of ß'-SiAlON and yttrium-rich glass in addition to the nitride reactants. As shown in Fig. 4, the amount of ß'-SiAlON increases with increasing V_r, passes through a maximum, and then decreases. Each maximum value is shifted to a larger V_r with decreasing z value, i.e.,

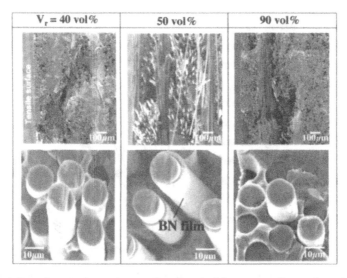

Fig. 3 Scanning electron micrographs of typical fracture surfaces of specimens tested at room temperature for composites fabricated when M_{slurry}=16.2. V_r is the total amount of Si_3N_4-AlN reactants introduced into the preforms.

decreasing M_{Total}. The matrix consists of ß'-SiAlON and the glass on the left side of each maximum, at the opposite side, ß'-SiAlON, yttrium-rich glass and the nitride reactants.

Figure 5 shows the estimated residual radial stress on the interface between the fiber and matrix at room temperature as a function of V_f. The radial stress, σ_r, on the fiber is given by [4]

$$\sigma_r = \frac{-qE_mE_f}{E_f(1+v_m)+E_m(1-v_f)}\left(\Delta\alpha\Delta T+\frac{A}{r}\right) \qquad (3)$$

where E_m, v_m and E_f, v_f are the elastic moduli and Poisson's ratios of the matrix and fiber, respectively. ΔT (1173 K) is relative to the softening temperature of Y-Al-Si-O glass, which is between its glass transformation and meting temperatures.[5] $\Delta\alpha$ is the mismatch between the coefficients of thermal expansion of the matrix and the fiber. q is a parameter which is equal to unity for the infinite matrix case considered here. A is the amplitude of fiber roughness, and r is the fiber radius. The fiber is assumed to be perfectly smooth and the roughness equal to zero. The elastic modulus and thermal expansion coefficient of the matrix in the equation were determined by simple rules of mixtures for the matrix,[6,7] whose composition was calculated from Fig.4. The radial stress changes from compression to tension with increasing amount of reactants, passing through a maximum, followed by compression. For low M_{Total}, namely low z value of SiAlON, the stress remains tensile even for large V_f. These curve shapes are related to the amount of SiAlON produced in the matrix. Because the thermal expansion coefficient of SiAlON is smaller than that of the fiber, the larger the amount of SiAlON, the greater the increase of the tensile stress. In general, it is well known that the tensile radial stress accelerates debonding and fiber pullout,

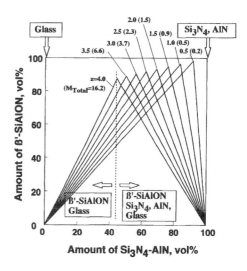

Fig. 4 Amount of ß'-SiAlON as a function of the amount of Si_3N_4-AlN reactants introduced into the preforms. z corresponds to the composition of $Si_{6-z}Al_zO_zN_{8-z}$.

resulting in excellent mechanical performance, especially WOF. As shown in Fig.5, in order to improve the mechanical properties of the composites at room temperature, the radial stress should be controlled so as to be in the tensile region. In other words, we need to lower M_{Total} and increase V_r. In this study, since the composites were fabricated with M_{Total} of 1.5 – 2.5 and 3.0 – 6.0, the radial stresses may be in the two gray regions shown in Fig. 5. The smaller M_{Total}, the wider the range of V_r that can be used to produce a tensile stress. Such a change in the radial stress is the reason why the curves of σ_{max} and WOF have peaks at 50 – 60 vol% of V_r, with the corresponding fracture morphologies shown in Figs. 1 -3.

As shown in Fig. 4, when M_{Total} is decreased, the formation of SiAlON tends to be depressed, leading to an increased amount of low melting point residual glass, in other words, suppression of yttrium enrichment of the residual glass. This probably contributes to an increasing V_r, corresponding to the onset of the saturation of σ_{max} and WOF at 1773 K shown in Figs. 1 and 2.

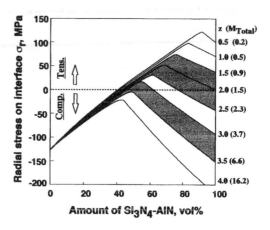

Fig. 5 Residual radial stress on the interface between the fiber and matrix at room temperature as a function of the amount of Si_3N_4-AlN reactants introduced into the preforms.

CONCLUSIONS

Continuous BN-coated Hi-Nicalon fiber reinforced ß'-SiAlON based ceramics were produced by reactive melt infiltration. An increase in the amount of ß'-SiAlON in the matrix improved the room temperature mechanical properties of the composites. This was related to the residual stress between the fiber and matrix. The high temperature mechanical performance was improved by increasing the packing of Si_3N_4 and AlN powders into the fabrics before MGI because of the decrease in the amount of residual glass.

Advanced SiC/SiC Ceramic Composites

REFERENCES

[1] J. J. Brennan, "Glass and Glass-Ceramic Matrix Composites," pp. 260-77 in *Fiber-Reinforced Ceramic Composites*, Edited by K. S. Mazdiyasni, Noyes Park Ridge, N. J. (1990).

[2] C. M. Huang, D. Zhu, Y. Xu, T. Mackin and W. M. Kriven, "Interfacial Properties of SiC Monofilament Reinforced ß'-SiAlON Composites," *Mater. Sci. and Eng.*, A201, 159-68 (1995).

[3] S. Kitaoka, N. Kawashima, T. Suzuki, Y. Sugita, N. Shinohara and T. Higuchi, "Fabrication of Continuous SiC Fiber Reinforced SiAlON Based Ceramic Composites by Reactive Melt Infiltration", *J. Am. Ceram. Soc.*, 84, 1945-51 (2001).

[4] R. J. Kerans and T. A. Parthasarathy, "Theoretical Analysis of the Fiber Pullout and Pushout Tests", *J. Am. Ceram. Soc.*, 74, 1585-96 (1991).

[5] M. J. Hyatt and D. E. Day, "Glass Properties in the Yttria-Alumina-Silica System", *J. Am. Ceram. Soc.*, 70, C283-287 (1987).

[6] Y. Kagawa and H. Hatta, "*Ceramics Matrix Composites -Tailoring Ceramic Composites-*," pp.97, Agune shofusha, Tokyo (1990).

[7] P. S. Turner, "Thermal-Expansion Stresses in Reinforced Plastics", *J. Res. N.B.S*, 37, 239-250 (1946).

FRACTURE RESISTANCE EVALUATION OF CERAMIC MATRIX COMPOSITES

Joon-Soo Park
Graduate School of Energy Science,
Kyoto University,
Gokasho, Uji, Kyoto 611-0011, Japan

Yutai Katoh and Akira Kohyama
Institute of Advanced Energy,
Kyoto University,
Gokasho, Uji, Kyoto 611-0011, Japan

Han-Ki Yoon
Dept. of Mechanical Engineering
Dong-Eui University,
24 Kaya-dong, Pusanjin-ku, Pusan 614-714, Republic of Korea

ABSTRACT

The fracture resistance of CMCs has been discussed. The effects of specimen size on fracture behavior using compact tension test specimens were investigated. Unloading-reloading sequences were used to measure the effective crack length of test specimens. The effects of the magnitude of unloading on the fracture behavior of test specimens were also investigated. 50% and 100% of unloading resulted in severe damage on the fracture surfaces of reinforcing fiber because of interfacial wear between fiber and matrix. Therefore, it was found that it is better to reduce the amount of unloading in order to reduce frictional damage on the reinforcing fibers. The crack-initiation fracture resistance of SiC fiber reinforced SiC matrix composites (SiC/SiC) and carbon fiber reinforced carbon matrix composite (C/C) were affected by test specimen size. These size effects are gradually decreased with increasing of test specimen size.

INTRODUCTION

Ceramic matrix composites (CMCs) are being developed for high temperature utilization in energy system, aerospace and other industrial applications. They offer several attractive features including high strength and modulus, improved fracture resistance, light weight, small thermal expansion coefficient and high thermal conductivity which contribute to good thermal shock resistance and high temperature stability in corrosive environments.

Although the enhancement of fracture resistance is one of key issues for the development of these materials, no standardized test methods have been established. In CMCs, several mechanisms (i.e., extensive matrix crack multiplications, fiber-matrix debonding and sliding) contribute to increase the work of fracture and to avoid catastrophic failure.[1,2] In spite of these nonlinear fracture mechanisms, many researchers have made and continued to make inappropriate use of the existing linear elastic fracture mechanics (LEFM) formalism for the analysis of fracture resistance of CMCs.[3,4] Some researchers have introduced the J-integral to fracture resistance evaluation of these materials. [5,6]However, CMCs have a persudo-ductility, J-integral is not very appropriate because of the large size of fracture process zone.[7]

Furthermore, full unloading during the application of the unloading-reloading method has been widely used to measure the effective crack length without any consideration for introducing surface damage on the reinforcing fibers.[8]

Miniaturization of the test specimen for these tests has been also identified as one of the most important issues for the evaluation of CMC because of the high cost of materials and the space constraints in experimental environments. However, little is known about the effect of test specimen size on the fracture behavior of CMCs.

The aim of the present work is to evaluate the fracture resistances of SiC/SiC and C/C composites on the basis of a global energy approach. In order to evaluate the fracture behaviors of both materials, compact tension (CT) test specimens were used and the unloading-reloading sequences were adopted to measure the effective crack length during the test. 10, 50, 100% of unloading amount and fracture test specimens with various sizes were tested to evaluate the effects of unloading amount and specimen size on the fracture behavior of CMCs and the applicability of small specimens for the fracture resistance evaluation was discussed.

EXPERIMENTAL PROCEDURES

Commercially available plain woven carbon/carbon composites (CX-31, Toyo-Ttanso, Osaka, Japan) and SiC fiber reinforced SiC matrix composites made by polymer imfiltration and pyrolysis (PIP) method were obtained for this study. The SiC/SiC composites were reinforced with Tyranno LoxM Si-Ti-C-O fiber (Tyranno-LoxM, Ube Industries, Ltd., Ube, Japan) with a surface modification in order to optimize the fiber-matrix interface.[9,10] Large amount of pores and small cracks, were observed in the matrices of both materials. The numbers of fiber bundles per unit length of SiC/SiC and C/C composites were 0.756 /mm and 1.135 /mm, respectively. The mechanical properties are listed in Table 1. CT specimens with various thicknesses and widths were used for the determination of crack growth, as shown in Fig. 1. Notch (a_0/W=0.6) was introduced using a diamond wheel with 0.4 mm thick. And then the notch tip of each test specimen was carefully sharpened by means of a razor blade with diamond paste.

Table 1 Mechanical properties of C/C and SiC/SiC composites

	SiC/SiC	C/C
Tensile modulus (GPa)	70	47
Yield strength (MPa)	143	117
Ultimate tensile strength (MPa)	199	127
Density (g/cm^3)	2.10	1.60
Number of layer in unit thickness (/mm)	4.3	4.2
Number of fiber bundle in unit length (/mm)	**0.756**	**1.135**

Two or three test specimens at the same test condition were prepared to obtain the average value of crack-initiation fracture resistance. Fracture tests were performed at a crosshead speed of 0.1 mm/min using a commercial electro-mechanical test machine (Model 5581, Instron Co., Canton, USA) in which the test specimen was repeatedly loaded and unloaded. During the tests, the fracture behavior of materials was simultaneously observed using a video-microscope. In this study, 10% of applied force was unloaded at each loading-unloading cycle. 50% and 100% unloading amounts were also applied to a set of specimens in order to study the effects of the unloading fraction.

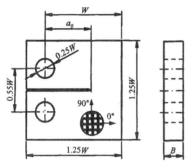

		Width W (mm)			
		40	30	20	15	10
Thickness	C/C	5	5	5~3	5~3	5~3
B (mm)	SiC/SiC	3	3	3~2	3~2	3~2

Fig. 1 The specimen geometry and dimension of C/C and SiC/SiC

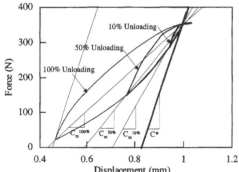

Fig. 2 Typical hysteresis loops and definition of compliance.

It is well known that a macro-crack can be defined, in the case of ceramic composites, as a specimen region within which the degree of damage is so extensive that its rigidity is negligible. Therefore, the development of a damage zone affects the specimen compliance in the same way as crack propagation.

As shown in Fig. 2, compliance

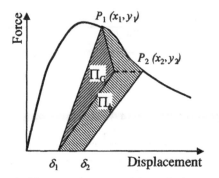

Fig. 3 Energy decomposition during crack growth from P_1 to P_2

C^* are obtained at the linear region of the unloading curve. Although, the effective crack length measured from a mean compliance C_m indicated the length from the notch tip to micro-cracks in the matrix, it is strongly depend on the nonlinear and irreversible fracture behavior (i.e. interfacial friction, crack closure and so on.) and the magnitude of unloading.

The compliance calibration curve by using C/C specimens with 20mm in width and 5 mm in thick was obtained and used for the calculation of effective crack length. The effective crack showed a good agreement with the results calculated by a relationship proposed by Paris.[11] Because of the limited number of test specimens, the effective crack length of C/C and SiC/SiC specimens with other thickness and width were calculated by that relationship described as follows.

The relationship was used to predict crack length as a function of compliance C^* for subsequent cycles j and $j+1$:

$$a_{j+1} = a_j + \left(\frac{b_j}{2}\right)\left(\frac{C^*_{j+1} - C^*_j}{C^*_{j+1}}\right)$$

where a_j = crack length at cycle j, C^*_j = compliances at cycles j, and $b_j = W - a_j$ = uncracked ligament after j cycles.

The energy consumed to propagate the crack from point 1 to point 2 is divided into two components, as shown in Fig. 3. Π_G is the energy term associated with the change in the compliance and Π_ϕ is related to the residual displacement, δ. With a sample of thickness B, the energy release rates are:

$$G(a) = \frac{\Pi_G(a)}{B\Delta(a)} = \frac{P^2}{2B}\frac{\partial C^*}{\partial a}, \quad \Phi(a) = \frac{\Pi_\phi(a)}{B\Delta(a)} = \frac{P}{B}\frac{\partial \delta}{\partial a}$$

and

$$G_R^*(a) = G(a) + \Phi(a) = \frac{P^2}{2B}\frac{\partial C^*}{\partial a} + \frac{P}{B}\frac{\partial \delta}{\partial a}$$

where P is the load.[12,13,14,15]

After the tests, the fracture surfaces of specimens were analyzed using a scanning electron microscope (JSM-6700F, JEOL, Tokyo, Japan).

RESULTS AND DISCUSSIONS

Both material systems with relatively wide specimens (W > 15 mm) fractured in a slow, stable manner and exhibited apparent nonlinear, non-elastic responses. Prior to any crack extension from the initial notch tip, distributed damage could be observed in the notch frontal in the form of matrix cracks, intra and inter-bundle delamination, which were associated with the pre-existing large pores. Subsequently, extension of a macro-crack occurred at a critical force. On the contrary, specimens with smaller width (W = 10 mm), showed a different fracture behavior. Beyond the maximum force P_{max}, compressive failure occurred and it spread from the back-face to the notch tip, probably because of the very porous nature of matrices. Thus, only the data pertinent to wide test specimen (W > 15 mm) are considered for both materials in the followings.

After the maximum force P_{max}, most structures fracture due to energy instability and the material's fracture resistance cannot be maintained. When the macroscopic crack reaches a weak spot in the material (such as a macropore, small cracks and the cross point of reinforced fiber bundles), it is reflected and rapidly extended. These kinds of multiple and rapid failures in fracture process zone lead to some peak on the force-displacement curve.

The determination of crack initiation fracture resistance is very important, CMCs will be exposed to severe service environments and therefore, the crack formation that allow the ingress of harmful gas can't be afforded because it would lead to the degradation of the internal microstructure. In this study, the crack initiation fracture resistance was defined at the maximum force P_{max} or at the first peak force, when the macroscopic crack is initiated, as shown in Fig. 4.

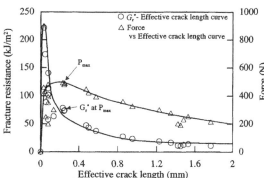

Fig. 4 Determination of crack initiation fracture resistance (C/C, W = 20 mm, t = 5 mm)

Effects of unloading amount

As shown in Fig. 5, the maximum force of test specimens decreased with increase of unloading amount and the force-displacement curves of test specimens with larger unloading amount decreased rapidly after the peak load. The area

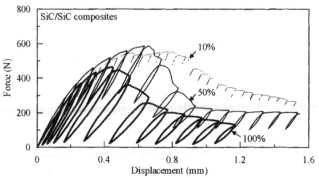

Fig. 5 Force-displacement curve of SiC/SiC test specimen with various unloading amounts

closed by unloading-reloading curve means that the irreversible energy was consumed by internal friction. The specimen with larger unloading amount had larger size of closed-area. It can be easily estimated that damage that resulted from internal friction also increases with increasing the magnitude of unloading.

Fig. 6 shows the reinforcing fibers damaged by interfacial friction during the fracture test. 50% and 100 % of unloading amounts brought about lots of straight scratches on the reinforcing fiber with strength degradation. Micro-cracking and debonding between reinforcing fiber and matrix were generated after the proportional limit. And then, the fiber surface damage of reinforcing fiber was induced by the interfacial wear, and progressively increased. The scratches were

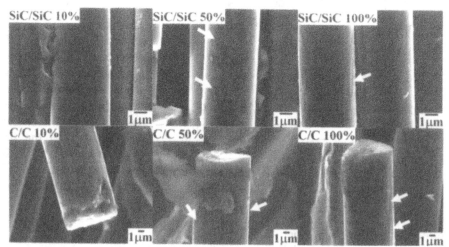

Fig. 6 SEM micrographs of the reinforcing fibers after fracture test under various unloading amount

observed mostly at the outside of fiber bundles aligned in the loading direction.

On the other hand, reinforcing fibers with 10% of unloading shows higher maximum force and less significant surface damage. The closed-loop area obtained from the test specimen with small unloading amount is smaller than those with larger one. It means that the irreversible energy consumes and the surface damage of reinforcing fiber is small. From these results, it is concluded that the small unloading amount of unloading-reloading sequences, which is a simple and effective measuring method for the effective crack length in the fracture test specimen, are desired for the fracture test in order to reduce the influence of in-situ fiber damage.

Effects of specimen thickness

Fig. 7 presents the effect of test specimen thickness on the crack initiation fracture resistance of 2D woven SiC/SiC and C/C composites. The crack initiation fracture resistance of SiC/SiC composites increases with increasing the thickness of the test specimen, and that of C/C composites also increases slightly.

In the case of isotropic materials under plain strain condition, the fracture toughness is an intrinsic materials property and does not change with specimen thickness. This opposite tendency indicates that a certain fracture mechanism concerned with test specimen thickness contributes to increasing fracture resistance of layered CMCs. The matrix of SiC/SiC composites was fabricated by the polymer ifiltration and pyrolysis (PIP) method and includes lots of pores and small cracks, which bring about the lower compression and shear strength of matrix. Additionally, its interlaminar bonding strength is very low. C/C composites also have low compression, shear and interlaminar bonding strength. It could be observed that delamination and frictional sliding between layered fabrics occurred in the fracture process zone of both materials. The layered fabrics

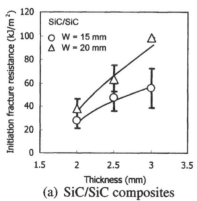 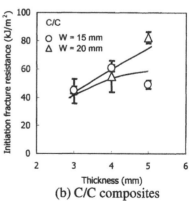

(a) SiC/SiC composites   (b) C/C composites

Fig. 7 Effects of specimen thickness on the crack initiation fracture resistance

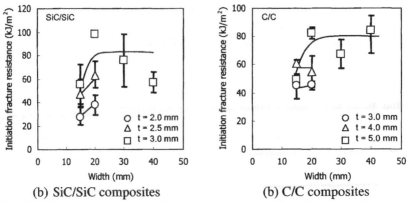

(b) SiC/SiC composites   (b) C/C composites

Fig. 8 Effects of specimen width on the crack initiation fracture resistance

of both materials were easily delaminated by the strain mismatch. It was assumed that interlaminar frictional sliding between layered fabrics is one of main reasons for the larger fracture resistance of thicker specimens.

Effects of specimen width

Fig 8 presents the effect of specimen width on the initiation fracture resistance. The initiation fracture resistance of small specimens (W=10mm) was overestimated because of a compressive failure, as we mentioned above. Initiation fracture resistance of both materials has a tendency to increase with increasing specimen width. There is a strong width-dependence in both composites. It means that the matrix crack multiplication and macroscopic crack extension are affected by the specimen width and the condition of fabric structure. Width-dependence of both composites is gradually moderated with the increasing width, because the size of fracture process zone in wide specimens (W " 20 mm) was relatively small

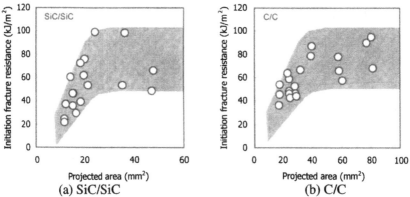

(a) SiC/SiC   (b) C/C

Fig. 9 Effect of projected area on the crack initiation fracture resistance

and localized in a vicinity of notch tip.

Size effects

The dependence of test specimen size on the fracture resistance is summarized in Fig. 9. The projected area means the product of thickness and the distance from notch tip to the back face of the specimen. It is considered that the fracture resistances of both materials are significantly affected by the specimen thickness and the size of fracture process zone as well as other fracture mechanism such as fiber bridging, fiber pull-out and matrix crack multiplication. These size effects can be reduced by the adoption of large size of test specimen and the fine fabric structure. It can be concluded that the initiation fracture resistances of both materials saturated at 20 mm in width in this case. However, further work is necessary in order to correlate the saturation level of specimen size with the condition of fabric structure.

SUMMARY

The influences of unloading amount and specimen size on the fracture resistance of SiC/SiC and C/C composite materials were investigated. The fracture tests with the unloading-reloading sequence were conducted with CT test specimens with different thicknesses and widths. The fracture test with 50 and 100% of unloading amount promoted the wear at fiber-matrix interface, which causes the in-situ fiber damage and therefore a smaller unloading amount appeared more appropriate. The crack initiation fracture resistance of SiC/SiC composites increased with increasing test specimen thickness. That of C/C composites also increased slightly. It seems that interlaminar frictional sliding is one of the energy dissipating mechanism in 2D woven CMCs. The crack initiation fracture resistance of both materials increased with increasing specimen width. It can be concluded that the size of fracture process zone changed with specimen width, and the size of fracture process zone was saturated at a certain width.

REFERENCES

[1] A. G. Evans, "Perspective on the Development of High-Resistance Ceramics," *Journal of the American Ceramics Society*, **73** [2], 187-206, (1990).
[2] H. C. Cao, E. Bischoff, O. Sbaizero, Manfred Rühle, and Anthony G. Evans, "Effect of Interfaces on the Properties of Fiber-reinforced Ceramics," *Journal of American Ceramic Society*, **73** [6] 1691-1699 (1990).
[3] H. Kobayashi, A. Todoroki and H. Nakamura, "Fracture Toughness Determination of SiC/SiC Fabric Composites", High temperature Ceramic-matrix Composites I, 431-436 (1995)

[4] S. V. Nair and Y. L. Wang, "Failure Behavior of a 2-D Woven SiC Fiber/SiC Matrix Composite at Ambient and Elevated Temperatures", *Ceramic Engineering and Science Proceedings*, **13** 433-441 (1992)

[5] S. V. Nair and Y. L. Wang, "Toughening Behavior of a Two-Dimensional SiC/SiC Woven Composite at Ambient Temperature : II, Stress-Displacement Relationship in the Crack Process Zone", *Journal of American Ceramics Society*, **81** 1149-1156 (1998)

[6] J. Y. She, J. P. Hirth, F. W. Zok and J. A. Heathcote, ,"Effect of Notch Root Radius on the Initiation Toughness of a C-Fiber/SiC-Matrix Composite" *Scripta Materialia*, **38**, 15-19 (1998)

[7] C. Droillard, J. Lamon, "FractureToughness of 2-D Woven SiC/SiC CVI-Composites with Multilayered Interphases", *Journal of American Ceramics Society*, **79** 849-858 (1996)

[8] E. Vagaggini, J. M. Domergue, and A. G. Evans, "Relatioships between Hysteresis Measurements and the Constituent Properties of Ceramic Matrix Composites I, Theory," *Journal of American Ceramics Society*, **78** 2709-2720 (1995)

[9] I. J. Davies, T. Ishikawa, M. Shibuya, T.Hirokawa and J. Gotoh, "Fibre and interfacial properties measured in situ for a 3D woven SiC/SiC-based composite with glass sealant," *Composites*, **A30** 587-591 (1999).

[10] I. J. Davies, T. Ishikawa, M. Shibuya and T. Hirokawa, "Optical microscopy of a 3-D woven SiC/SiC-based composite," *Composites Science and Technology*, **59** 429-437 (1999).

[11] ASTM E813-89, "Standard Test Method for J_{IC}, A Measure of Fracture Toughness", Annual Book of ASTM Standards, 03.01 732-746 (1989)

[12] M. Sakai, K. Urashima, and M. Inagaki, "Energy Principle of Elastic-Plastic Fracture and Its Application to the Fracture Mechanics of a Polycrystalline Graphite", *Journal of the American Ceramic Society* **66** (12) 868-874 (1983)

[13] Y. W. Mai, M. I. Hakeem, "Slow Crack Growth in Cellulose Fibre Cements," *Journal of Materials Science* **19** 501-508 (1984)

[14] M. Gomimna, D. Themines, J. L. Chermant and F. Osterstock, "An energy evaluation for C/SiC composite materials," *International Journal of Fracture* **34**, 219-228 (1987)

[15] K. R. Stull and A. Parvizi-Majidi, "Fracture Toughness of Fiber-Reinforced Glass Ceramic and Ceramic Matrix Composites," *Ceramic Engineering and Science Proceedings*, **12** [7-8] 1452-1461 (1991)

EFFECTS OF INTERLAYERS ON INTERFACIAL SHEAR STRENGTH AND FLEXURAL PROPERTIES OF Tyranno-SA FIBER-REINFORCED CVI-SiC/SiC COMPOSITES

Wen Yang, Hiroshi Araki, Tetsuji Noda, Quanli Hu, and Hiroshi Suzuki
National Institute for Materials Science, CREST-ACE
1-2-1 Sengen, Tsukuba, Ibaraki 305-0047, Japan

Yutai Katoh and Akira Kohyama
Institute of Advanced Energy, Kyoto University, CREST-ACE,
Gokasho, Uji, Kyoto 611-0011, Japan

ABSTRACT

Several SiC/SiC composites reinforced with a highly crystalline β-SiC fiber, Tyranno-SA, were fabricated using the chemical vapor infiltration (CVI) process. The flexural properties and fracture behaviors were investigated using three-point bending tests. Various pyrolytic carbon (PyC) or multi-interlayers were deposited in the composites using an isothermal CVI process to modify the interfacial bonding and fiber sliding resistance. The bending test results revealed a close PyC layer dependence of the flexural strength. The ultimate strength showed quick increase with the increasing of the PyC layer thickness up to ~100nm, beyond which slight increase of the ultimate strength via the PyC layer thickness till ~200nm was noticed. The interfacial shear strengths (ISS) of these composites were extracted with single fiber pushout tests, which revealed very large ISS from ~195MPa to over 633MPa depending on the thickness of the PyC layer.

INTRODUCTION

There is a strong and increasing interest in the R & D of continuous SiC fiber reinforced SiC matrix composites (SiC/SiC composites) for a variety of high-temperature, high-stress applications in aerospace, hot engine and energy conversion systems [1]. They are also quite attractive candidates for blanket first wall structures in nuclear fusion power systems due mainly to their inherent low induced radioactivation, radiation resistance and chemical stability at elevated temperatures [2,3]. For SiC/SiC composites, fiber/matrix interface is one of the key factors [4] that determine the materials performance. The deposition of a thin carbon coating layer on the fibers results in improved strength and crack tolerence of the composites. Recently, an alternating multiple interlayers, $(PyC-SiC)_n$, have been developed [5]. However, the understandings already established are mostly for the old-generation SiC-based fibers such as Nicalon-CG. A new advanced fiber, Tyranno-SA, which is predominantly β-SiC crystals with a near stoichiometric C/Si atomic ratio, has been developed (Ube Industry Ltd. Japan) [6]. This fiber exhibits excellent mechanical properties, coupled with much improved thermal conductivity and thermal stability. Compared with the old-generation Nicalon-CG or Hi-Nicalon[TM] fibers, the Tyranno-SA fiber possesses quite different surface characteristics such as near stoichiometric SiC surface chemistry with rough surface [7], which may have significant effects on the interfacial bonding and fiber sliding, and therefore, on the mechanical properties of the composites.

In this study, several CVI-SiC/SiC composites reinforced with the Tyranno-SA fiber were fabricated with various PyC or multi-interlayers. The flexural properties and the interfacial shear strength were investigated. The main objectives are to get understandings of the mechanical performances of these kinds of composites and the effects of the various interlayers.

EXPERIMENTAL
Composite Processing
Fibrous preforms were fabricated by stacking 11 layers of 2D plain-woven Tyranno-SA fiber cloths in 0/90°. The preforms were compressed to keep a fiber volume fraction of ~43% using a set of graphite fixtures. The normal size of the preforms was 40mm in diameter and 2.0mm in thickness. The preforms were pre-coated with various PyC or SiC/PyC multi-layers using an isothermal CVI process through the thermal decomposition of CH_4 and CH_3SiCl_3 (MTS), respectively. MTS was carried by hydrogen. The pre-coated preforms were finally densified with SiC matrix by an isothermal-forced flow CVI process at 1273K and 14.7kPa. Detailed fabrication process can be found elsewhere [8]. The densification process generally required ~15 hours.

Three-Point Bending Tests and Single Fiber Pushout Tests
Three-point bending tests (with a support span of 18mm) were performed at room temperature. Bending specimens were cut parallel to one of the fiber bundle directions of the fabric cloth using a diamond cutter and both the tensile and compression surfaces were carefully ground. Three tests were conducted for each composite. The dimension of the specimen was $30^L \times 4.0^W \times 1.5^T$ mm^3. The crosshead speed was 0.0083mm/s. The load/displacement data was recorded.

Single fiber push-out tests were carried out to extract the ISS of each composite using a load controlled micro-indentation testing system with a triangular diamond pyramidal indenter. The detailed test procedure was described elsewhere[8]. About 20 pushout tests were conducted for each specimen.

The fracture surfaces were observed using scanning electron microscope (SEM), JOEL JIM-6700F, with interfacial debonding and fiber pullouts.

RESULTS AND DISCUSSION
Flexural Properties

In total, seven composites were fabricated and investigated. The interlayer structures and thickness and composite densities/porosities are given in Table 1.

Figure 1 shows the typical load-displacement curves of the composites (the curves were shifted for clarity). Composite NL exhibited low load maximum and displayed brittle failure mode, with no signs of toughening. This is consistent with the flat fracture surface. Improved toughness was indicated in the curves for the interlayered composites. Both the load maximums and displacement at load maximums were increased. Interfacial debonding and fiber pullouts were observed at the fracture surfaces of all the interlayered composites, as typically shown in Figure 2 for composite 3L-220. Initially, the load increased linearly with the increasing of displacement, reflecting the elastic response of the composites. Deviation of the curves to the linearity occurred at certain loads, followed by a nonlinear domain of deformation until the load maximum, due mainly to the matrix cracking, interfacial debonding and fiber sliding, and individual fiber failures. Finally, the composites failed owing to the failures of the fibers. Significant inter-fabric layers delaminating occurred for two of the three bending specimens from composite SL-50 during the bending tests, primarily due to its much lower density (see as in Table 1) compared with those of the others. This might be the main reason causing the low load maximum (see as in Figure 1) for this composite.

Table I . Interlayer structure, density/porosity and mechanical properties

Composite I.D.	Interlayer structure and thickness* (nm)	Density** (kg/m^3)	Porosity (%)	PLS** (MPa)	UFS** (MPa)	ISS** (MPa)
NL	F/M	2780(10)	10.6	275(48)	281(60)	>633
SL-C50	F/C^{50}/M	2410(30)	20.4	257(5)	410(92)	331(140)
SL-C100	F/C^{105}/M	2750(10)	11.0	371(30)	567(26)	295(72)
SL-C200	F/C^{200}/M	2610(30)	14.6	339(18)	549(58)	195(51)
2L-C150	F/SiC^{150}/C^{150}/M	2370(50)	16.4	370(43)	495(85)	284(96)
3L-C50	F/C^{50}/SiC^{100}/C^{50}/M	2620(30)	15.1	354(53)	523(20)	434(165)
3L-C200	F/C^{220}/SiC^{350}/C^{150}/M	2630(30)	13.9	382(14)	596(5)	216(59)

*:The interlayer thickness was measured with high magnification SEM images with an estimated resolution of ~10nm.

**: Included in the parenthesis are the standard deviations.

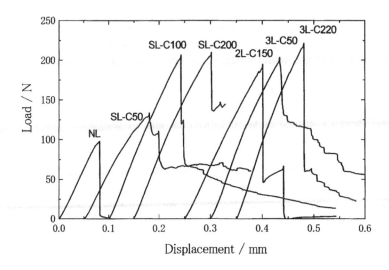

Figure 1. Representative load-displacement curves of the Tyranno-SA/SiC composites.

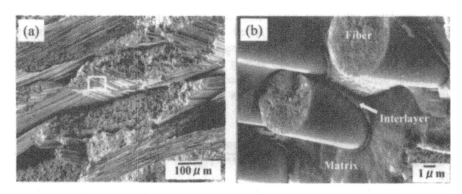

Figure 2. SEM images showing fracture surface and fiber pullouts of composite 3L-200

The proportional limit stress (PLS) and ultimate flexural strength (UFS) were extracted from the load-displacement curves and are also shown in Table 1. The PLS was the stress corresponding to 0.01% offset strain. The UFS was calculated by the simple elastic beam theory.

Effects of Composite Porosities and Interlayers

Figure 1 and Table 1 show some differences of the fracture behaviors and flexural strength among the various composites. Since the composites were fabricated by the same CVI process with similar fiber volume fraction, composite density/porosity and interlayer structure are considered to be the main reasons causing these differences. Figure 3 relates the UFS to the composite porosities. The result of a statistical study on the effects of density/porosity on the flexural strength of a SiC/SiC composite by Araki et al. [9] is also presented. The composite was fabricated by CVI process with 2D plain-woven Tyranno-SA fabric as the reinforcement, same as that of the composites in this study. All the data from Araki et al. in Figure 3 was obtained from this composite. Therefore, it is assumed that density/porosity is the main reason causing the difference of the strength (UFS) and a simple linear fitting between the strength and porosity was performed. The strength decreased, to some extent, with the increase of the porosity.

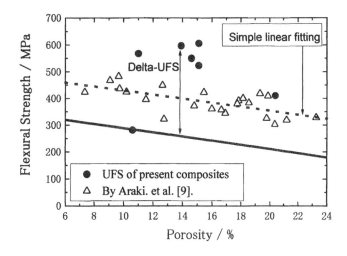

Figure 3. The UFS versus porosity for the various composites.

Assuming that the same trend of the effect of porosity on the UFS of present no-interlayered composite (solid line in Figure 3) exists and the increases of the UFSs of the present composites (Delta-UFS as noticed in the figure) were mainly owing to the deposited interlayers, a relationship between the Delta-UFS and the PyC layer thickness was obtained and is presented in Figure 4. The PyC layer thickness for the multi-layered composites is the total thickness of the PyC sub-layers in each composite. Figure 4 shows that for the PyC layered composites, the UFS increases quickly with the increase of the PyC layer thickness up to 100nm, beyond which slight increase of the UFS is obtained till ~200nm of the PyC layer. Considering the large error bars in the figure, the UFSs of those composites with multi-layers fall well in the trend between the UFS and PyC layer thickness for the PyC layered ones. This, as well as the load-displacement curves in Figure 1, indicates that the deposition of SiC sub-layer(s) either directly on the fibers or between the PyC sub-layers in the multi-layers will cause no obvious change of the strength of the Tyranno-SA fiber reinforced composites. Therefore, it is possible to improve the oxidation resistance of the interlayers, and hence, the composites, by reducing the thickness of each PyC sub-layer but keeping enough total thickness in the multi-layers without obvious change of the mechanical properties of the composites.

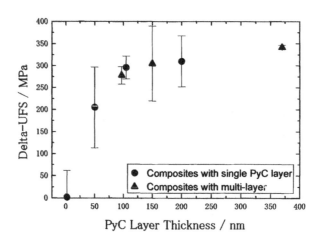

Figure 4. Effect of PyC layer thickness on the UFS

Interfacial Shear Strength and the Effect of Interlayers

The ISSs derived from the single fiber pushout tests are also shown in Table 1. Very large ISSs, from 195MPa to >633MPa, were obtained for the various composites. Large ISSs of the Tyranno-SA/SiC composites are likely due mainly to the pure SiC surface chemistry and rough surface characteristics of the Tyranno-SA fiber[13]. Figure 5 relates the ISSs to the thickness of the PyC layers, which shows close PyC layer thickness dependence of the ISS. For the PyC

Advanced SiC/SiC Ceramic Composites

layered composites, the ISS decreased from over 633MPa to 195MPa with increasing the PyC layer thickness up to 200nm. The composites with SiC sub-layers showed slightly larger ISS compared with those of PyC layered ones.

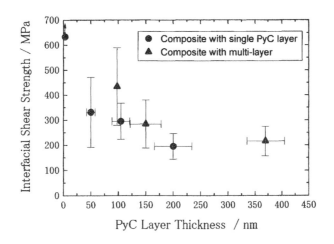

Figure 5. PyC layer thickness dependence of ISS

SUMMARY

Several Tyranno-SA fiber reinforced CVI-SiC/SiC composites with various PyC or SiC and PyC interlayers were fabricated and the flexural properties and interfacial shear strength were investigated using three-point bending test and single fiber pushout test, respectively. The flexural strengths of the composites demonstrated a close dependence on the PyC interlayer. The ultimate strength showed quick increase with the increasing of the PyC layer thickness up to ~100nm, beyond which slight increase of the ultimate strength via the PyC layer thickness till ~200nm was achieved. Large interfacial shear strengths were obtained likely due mainly to the pure SiC surface chemistry and rough surface characteristics of the Tyranno-SA fiber. For the PyC layered composites, the ISS decreased from over 633MPa to 195MPa with increasing the PyC layer thickness up to 200nm. The deposition of the SiC sub-layer in the multi-layered composites seems to cause no obvious change of the flexural strength of the materials although it is likely to increase the ISS with given total amount of the PyC layer.

ACKNOWLEDGEMENT

This work was supported by the CREST, Japan Science and Technology

Corporation.

REFERENCES
[1]A. Kohyama, M. Seki, K. Abe, T. Muroga, H. Matsui, S. Jitsukawa, and S. Matsuda, "Interactions Between Fusion Materials R&D and Other Technologies," *Journal of Nuclear Materials,* **283-287** 20-27 (2000).

[2]T. Noda, H. Araki, F. Abe, and M. Okada, "Microstructure and Mechanical Properties of CVI Carbon Fiber/SiC Composites," *Journal of Nuclear Materials,* **191-194** 539-543 (1992).

[3]L. L. Snead and O. J. Schwarz, "Advance SiC Composites for Fusion Application," *Journal of Nuclear Materials,* **219** 3-14 (1995).

[4]R. Naslain, "The Concept of Layered Interlayers in SiC/SiC": pp.23-29 in *Ceramic Transactions,* vol.58, High-Temperature Ceramic-Matrix Composites II: Manufacturing and Materials Development. Edited by A. G. Evans and R. Naslain. America Ceramic Society, Westerville, OH, 1995.

[5]T. M. Besmann, E. R. Kupp, E. Lara-Curzio, and K. L. More, "Ceramic Composites with Multilayer Interface Coatings," *Journal of American Ceramic Society,* **83** [12] 3014-20 (2000).

[6]T. Ishikawa, Y. Kohtoku, K. Kumagawa, T. Yamamura, and T. Nagasawa, "High-Strength Alkali-resistance Sintered SiC Fiber Stable to 2200C," *Nature,* **391** [6669] 773-75 (1998).

[7]W. Yang, A. Kohyama, T. Noda, Y. Katoh, H. Araki, and J. Yu, "Interfacial Characterization of of CVI-SiC/SiC composites," Submitted to *Journal of Nuclear Materials.*

[8]W. Yang, in: *Development of CVI Process and Property Evaluation of CVI-SiC/SiC Composites,* Doctoral thesis, Institute of Advanced Energy, Kyoto University, 2002

[9]Hiroshi Araki, Tetsuji Noda, Wen Yang, Quanli Hu, Hiroshi Suzuki, and Akira Kohyama, "Flexural Properties and Weibull Modulus of Several SiC/SiC Composites Prepared by CVI Method," Submitted to *Ceramic Transactions.*

FLEXURAL PROPERTIES OF SEVERAL SIC FIBER-REINFORCED CVI-SIC MATRIX COMPOSITES

Hiroshi Araki, Tetsuji Noda, Wen Yang, Quanli Hu, and Hiroshi Suzuki
National Inst. for Materials Science
1-2-1,Sengen, Tsukuba, Ibaraki 305-0047, Japan

Akira Kohyama
Institute of Advanced Energy, Kyoto University, CREST-ACE
Gokasho, Uji, Kyoto 611-0011, Japan

ABSTRACT

Several SiC/SiC composites with a disk shape of ϕ 120mm X 3-4mm were fabricated by chemical vapor infiltration (CVI) process. The reinforcements were 2D plain-woven Nicalon-CG, Hi-Nicalon, and Tyranno-SA fiber cloths. The flexural properties and fracture behaviors were investigated using three-point bending tests and scanning electron microscopy. About 30 bending tests were conducted for each composite for a statistical study on the effect of the density (and the spatial homogeneity) on the flexural strength. The Hi-Nicalon/SiC composite showed an average fracture strength of 665MPa, with the highest individual specimen strength of 815MPa. Both the composites reinforced with Nicalon-CG and Tyranno-SA fibers showed lower strength coupled with near brittle failure behaviors. Much improved flexural strength of the Tyranno-SA/SiC composite with sound fiber pullout may be expected when appropriate fiber/matrix compliant interlayer(s) is deposited.

INTRODUCTION

Matrix incorporation via chemical vapor infiltration (CVI) has advantages over other methods in high purity, minimizing damage to fibers and making near-net shape products of composites and is therefore widely used for producing SiC/SiC composites[1,2]. Fabrication of high-density SiC/SiC composites with homogeneous matrix densification, which is very important for the evaluation of the performance of the materials, is one of the primary goals for current CVI process. Recent progress on the CVI process with small size samples of disk shape with 40mm ϕ made it possible to produce dense and homogeneous

SiC/SiC composites with much improved strength and toughness[3]. However, when the composite size is scaled up, the process conditions may require modification and it becomes difficult to get uniform matrix densification with high density.

In the present paper, several SiC/SiC composites reinforced with Nicalon-CG, Hi-Nicalon, or Tyranno-SA fibers were fabricated by the CVI process. The spatial homogeneities of the composites were investigated with statistical density measurements. The flexural properties were studied using three-point bending tests. The main purposes of this investigation were to study the fracture behaviors/flexural strengths of these composites and the effects of the spatial homogeneity.

EXPERIMENTAL

2D plain-woven Nicalon-CG, Hi-Nicalon, or Tyranno-SA SiC fiber cloths laminated with 12–18 sheets were used as the composite preforms. The preforms were compressed, using a set of graphite fixtures, to keep a fiber volume fraction of ~30% for all the composites. The general dimensions of the preforms were 120mm ϕ × 3-4mm. The preforms were densified with SiC-matrix using the thermal decompositions of methyltriclorosilane (MTS) having a purity higher than 99 vol.%. MTS was carried by hydrogen with a reduced pressure 14.4kPa in the CVI reactor.

Bending test specimens were cut parallel to one of the fiber bundle directions of the fabric cloth from the composites after the CVI process. Both surfaces of the specimens were polished to remove the surface CVD (chemical vapor deposition)-SiC layer that was formed at the end of the CVI[3]. The resulting specimen size was 25mm in length, 4.0 mm in width and ~2.0mm in thickness. The density of each specimen was measured from its weight and volume. About 30 bending tests (with a span of 16mm) were conducted for each composite at room temperature with the load in the vertical direction to the laminates. The cross-head speed was adjusted to 0.0083mm/s. The microstructure and fracture surfaces were examined with scanning electron microscopy (SEM), using a JSM6100 microscope.

RESULTS AND DISCUSSION
Composite Densities and Spatial Homogeneity

Three kinds of composites were fabricated. The CVI process conditions and composite densities/porosities are shown in Table 1. The densities and spatial homogeneities of the composites were studied in a way as shown in Figure 1 for Nicalon-CG/SiC composite. Very high density, with the highest individual bending specimen density of over 2850kg/m^3 (corresponding to a porosity of <5%), was achieved by this composite. Figure 1 and Table 1 indicate that the composite was matrix-densified with quite good spatial homogeneity. No clear spatial dependence of the densities was found in Figure 1.

Table I. CVI process conditions, composite densities/porosities

Composite	CVI conditions		Density	Porosity
	Temperature	time		
	(K)	(ks)	(kg/m^3)	(%)
Nicalon-CG/SiC	1323	288	2770±60	7.3
Hi-Nicalon/SiC	1323	216	2550±90	16.0
Tyranno-SA/SiC	1323	252	2650±140	15.7

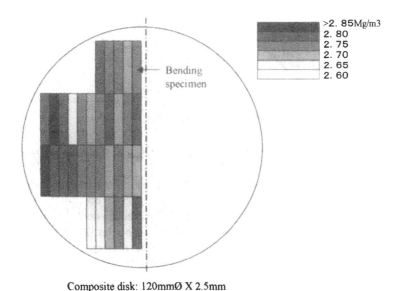

Composite disk: 120mmØ X 2.5mm

Figure 1. Density map for the Nicalon-CG/SiC composite.

The composites reinforced with Hi-Nicalon and Tyranno-SA fibers showed lower average densities with slightly increased standard deviations of the densities, as shown in Table 1. The lower density of Hi-Nicalon/SiC is considered to be due mainly to the shorter CVI time used. The Tyranno-SA fiber possesses smaller fiber diameter and much improved thermal conductivity compared with the Nicalon fibers[4]. Although rather dense Tyranno-SA/SiC composite with the average density of 2650kg/m^3 was formed in this study, further efforts on the CVI process conditions are necessary to produce high quality Tyranno-SA/SiC composites with improved density and homogenous matrix densification.

Fracture Behaviors

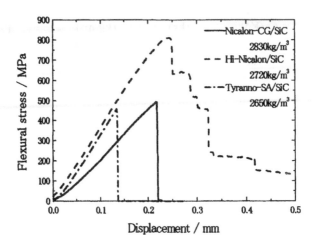

Figure 2. Representative stress-displacement curves of the three composites.

Figure 2 shows typical stress-displacement curves of the bending tests for the three composites. The specimens from Hi-Nicalon/SiC composite yielded an ultimate strength as large as 815MPa and showed a gradual decrease in stress after reaching the maximum. The stress did not become zero after the large deformation since the specimens were not torn off owing to the tolerance of fibers. On the other hand, apparent plasticity was not observed for the specimen from the Nicalon-CG/SiC composite and the ultimate strength was much lower, 493MPa, in spite of its high density as given in the figure. The fracture behavior exhibited by the specimens from Tyranno-SA/SiC composite was similar to that of the Nicalon-CG/SiC composite. Figure 3 shows the SEM images of the fracture surfaces and fiber pullouts of the three specimens. Interfacial debonding and fiber pullouts are clearly evident on the fracture surface of the specimen from the Hi-Nicalon/SiC composite (Figure 3 (a) and (b)) while the fracture surfaces of the specimens from the Nicalon-CG/SiC and Tyranno-SA/SiC composites were relatively flat and pullout of fibers were rarely found (Figure 3 (c), (d) and (e), (f)). The observation of pullout fibers at the fracture surface of the Hi-Nicalon/SiC composite specimen suggests a not very strong interfacial bonding in the materials. This is likely due to the very small roughness and the graphitic carbon shell structure [5]. Interfacial debonding and fiber pullouts yielded high strength and no catastrophic fracture behavior of the Hi-Nicalon/SiC composite.

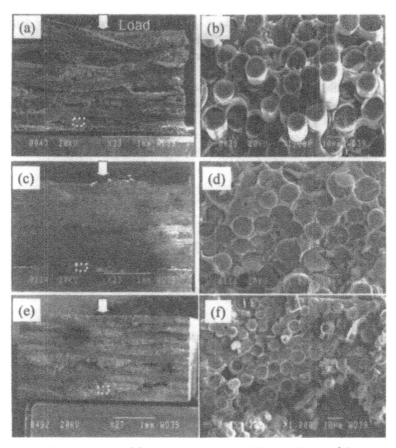

Figure 3. SEM images of fracture surfaces and fiber pullouts of the several composites ((a) and (b): Hi-Nicalon/SiC; (c) and (d): Nicalon-CG/SiC; (e) and (f): Tyranno-SA/SiC).

The Nicalon-CG fiber contains oxygen of 11.5 mass% with an amorphous structure, versus much lower value of 0.5% in the Hi-Nicalon fiber. Therefore, it is believed that some degradations of the Nicalon-CG fiber occurred during the CVI densification process which continued for 288ks at 1323K, owing to the decomposition of the fiber [6] and crystal growth, etc., and resulted in the near brittle fracture behavior of the Nialon-CG/SiC composite with low fracture strength.

Tyranno-SA is a newly developed near stoichiometric SiC fiber with predominantly β-SiC crystals structure and is very attractive for nuclear fusion application. The Tyranno-SA fiber possesses quite different surface characteristics such as near stoichiometric SiC surface chemistry with rough surface compared with the Nicalon-CG and Hi-Nicalon™ fibers [7]. Pure SiC surface chemistry and rough fiber surface result in strong interfacial bonding and fiber sliding resistance when is used as the reinforcement in SiC/SiC composites[7], and hence, result in hardly any interfacial debonding and fiber pullout upon external loading when compliant interfacial layers are not applied. Therefore, much improvement of the flexural strength of the Tyranno-SA/SiC composite with sound fiber pullout may occur when appropriate fiber/matrix interlayer(s) is deposited. The relatively small elongation (0.7%) of the fiber is also a possible reason for the near brittle type failure of the materials.

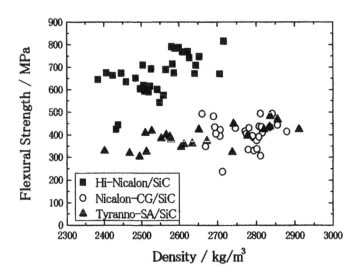

Figure 4. Relation between fracture strengths and densities of the composites.

Flexural Strength and the Effects of Densities

Figure 4 shows the relationship between fracture strength and density of the composites. The average strength of the Hi-Nicalon/SiC composite was ~665MPa. The fracture strength increased with increasing density with reduced scattering. Further increase of the fracture strength, therefore, can be expected with improved matrix densification. The strengths of Nicalon-CG/SiC and Tyranno-SA/SiC composites were 404±59 and 388±50MPa, respectively, with less density dependence compared with that of the Hi-Nicalon/SiC composite, probably due mainly to the near brittle failure behaviors.

SUMMARY

Several SiC/SiC composites reinforced with 2D plain-woven Hi-Nicalon, Nicalon-CG, or Tyranno-SA fibers were fabricated by the CVI process. The flexural properties and the effects of the space homogeneity were studied.

High density composites, with the highest individual bending specimen density of 2850kg/m^3 (corresponding to a porosity of <5%) from the Nicalon-CG/SiC composite, were fabricated with quite well spatial homogeneities.

The Hi-Nicalon/SiC composite yielded an average flexural strength of ~665MPa with inelastic fracture behavior. The strength increased with increasing the composite density with reduced scattering. Further increased strength of the composite is then expected provided improved matrix densification.

The Nicalon-CG/SiC composite showed much lower average flexural strength, 404±59MPa, with near linear elastic fracture behavior, likely due mainly to the degradation of Nicalon-CG fibers during the CVI process.

The Tyranno-SA/SiC composite exhibited similar fracture behavior as that of the Nicalon-CG/SiC composite, with the average flexural strength of 388±50MPa. Strong interfacial bonding/fiber sliding resistance is likely to be the main reason for the near brittle fracture behavior. Much improved flexural strength of the Tyranno-SA/SiC composite with sound fiber pullout may be expected when appropriate fiber/matrix compliant interlayer(s) is deposited. More efforts on the CVI process optimization are also necessary for the fabrication of high performance Tyranno-SA/SiC composite.

ACKNOWLEDGEMENT

This work was supported by the CREST, Japan Science and Technology Corporation.

REFERENCES

[1]T. Noda, H. Araki, F. Abe, and M. Okada, "Microstructure and Mechanical Properties of CVI Carbon Fiber/SiC Composites," *Journal of Nuclear Materials,* **191-194** 539-543 (1992).

[2]R. H. Jones, L. L. Snead, A. Kohyama, and P. Finici, "Recent Advancement in the Development of SiC/SiC as a Fusion Structural Materials," *Fusion Engineering Design,* **41** 15-24 (1998).

[3]W. Yang, in: *Development of CVI Process and Property Evaluation of CVI-SiC/SiC Composites,* Doctoral thesis, Institute of Advanced Energy, Kyoto University, 2002.

[4] T. Ishikawa, Y. Kohtoku, K. Kumagawa, T. Yamamura, and T. Nagasawa, "High-Strength Alkali-resistance Sintered SiC Fiber Stable to 2200C," *Nature*, **391** [6669] 773-75 (1998).

[5] Y. Katoh, A. Kohyama, T. Hinoki, W. Yang and W. Zhang, "Mechanical Properties of Advanced Sic Fiber-Reinforce CVI-SiC Composites," *Ceramic Engineering and Science Proceeding*, **21** [3] 399-406 (2000).

[6] H. Araki, H. Suzuki, W. Yang, S. Sato, and T. Noda, "Effect of High Temperature Heat Treatment in Vacuum on Microstructure and Bending Properties of SiC/SiC Composites Prepared by CVI," *Journal of Nuclear Materials*, **258-263** 1540-1545 (1998).

[7] W. Yang, A. Kohyama, T. Noda, Y. Katoh, H. Araki, and J. Yu, "Interfacial Characterization of CVI-SiC/SiC composites," Submitted to *Journal of Nuclear Materials*.

A FINITE-ELEMENT ANALYSIS OF THE THERMAL DIFFUDIVITY/ CONDUCTIVITY OF SiC/SiC COMPOSITES

R. Yamada, N. Igawa, T. Taguchi
Japan Atomic Energy Research Institute
Tokai-mura, Naka-gun, Ibaraki-ken, 319-1195, Japan

ABSTRACT

A Finite-element method (FEM) was applied for a simulation of the laser flash method for measuring thermal diffusivity. The effect of fiber and matrix thermal conductivities as well as the effect of fiber volume and fiber configuration on thermal diffusivity/conductivity of SiC/SiC composites was studied under some combinations of fibers and matrix-processing methods with changing fiber volumes of 1-D, 2-D, and 3-D fiber configurations. When using highly thermal conductive SiC fiber, the overall thermal diffusivity/conductivity of SiC/SiC composites with low matrix thermal conductivity is strongly improved with increasing the fiber volume especially for the 1-D fiber configuration. The calculation for 3-D SiC/SiC composites reveals that highly thermal conductive SiC fiber oriented to the through-the-thickness direction effectively contributes to increase the through-the-thickness thermal/diffusivity and that the fiber oriented to XY-plane has somewhat a positive effect when the fiber volume is increased due to fiber network and radially thermal diffusion in the fiber. The best X:Y:Z ratio of 3-D fiber configuration could be determined by a trade-off for obtaining better thermal and mechanical performance.

INTRODUCTION

SiC fiber-reinforced SiC composites (SiC/SiC) have been proposed for an advanced structural material for reactor-blanket modules [1-4]. For the reduction of thermal stresses in the first wall of blanket exposed to a high heat flux (~ 0.5 Wm^{-2}) from the main plasma, the thermal conductivity of SiC/SiC composites should be at least 15 $Wm^{-1}K^{-1}$ under neutron irradiation to keep thermal stresses within an allowable design stress of 200 MPa [1,2]. When the degradation of thermal conductivity due to neutron irradiation is taken into account [4-6], non-irradiated thermal conductivity of SiC/SiC composites from RT to 1200 °C should exceed at least 30 $Wm^{-1}K^{-1}$. Most data acquired so far for chemical vapor infiltrated (CVI) or polymer impregnated and pyrolized (PIP) SiC/SiC composites indicate that this requirement is quite high [3,5,6,7,8].

To achieve highly thermal conductive as well as irradiation-resistive fiber and matrix, it appears that both fiber and matrix should obtain characteristics of high density, high stoichiometry, and high crystallinity with large grain size. The recently developed sintered SiC fiber, which holds the above characteristics, exhibits a high value of thermal conductivity [9-11]. Recent matrix processing methods, such as Reaction Bonding (RB), Melt Infiltration (MI), and Liquid Phase Sintering (LPS) can produce high matrix density, and show good matrix thermal conductivity [12-14].

In this paper, we have performed 2-D thermal analysis by using a finite element method (FEM) to study how sintered SiC fiber as well as other SiC fibers affects the overall thermal diffusivity/conductivity of 1-D, 2-D, 3-D SiC/SiC composites fabricated with CVI or PIP method.

CALCULATION

A 2-D thermal analysis was carried out by using a FEM code, ABAQUS (Hibbitt, Karlsson & Sorensen, Inc.) to simulate the laser flash thermal diffusivity measurement (see Fig. 1). A rectangular geometry (10x3mm) was meshed with 8-node quadrilateral elements. The boundary conditions of the front of the specimen geometry were set to receive the flat-top laser power of 10^8 Wm^{-2} for 300 μs with the full width at half maximum (FWHM) of 500 μs, as depicted in Fig. 2, experimentally equivalent to about 4 J laser power incident on a 10 mm diameter disc. The heat flux at the rear boundary was governed by radiation whose emissivity was postulated to be 0.5. At the two side boundaries, adiabatic or radiative conditions were selected. However, no noticeable differences were exhibited

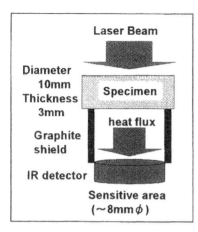

Fig. 1. The schematic diagram of the experimental setup of a specimen for thermal diffusivity measurement.

by either condition. The value of thermal diffusivity was calculated from the $t_{1/2}$-value derived from a rear surface temperature rise curve [15]. The curve was obtained by at each time step of FEM calculation plotting the temperature averaged between +-4 mm regions from the center along the rear boundary, corresponding to the sensitive area of the IR detector (see Fig. 2).

Figure 3 depicts 1-D, 2-D, 3-D fiber configurations. The incident laser power penetrated from the front to the back of the geometry, and the top surface of the rectangular solids were meshed for 2-D FEM calculations.

For the combinations of fibers and matrixes, three types of SiC fibers, i.e., Tyranno SA [10,11], Hi-Nicalon, and Hi-Nicalon Type S [16], and two types of matrix-processing, i.e., PIP and CVI, were selected. The PIP and CVI were chosen as representatives for low and high thermal conductive matrixes, respectively, and Tyranno SA and Hi-Nicalon were chosen for high and low thermal conductive fibers, respectively. Hi-Nicalon Type S may be considered as a medium thermal conductive fiber. In this calculation no interface layer between fibers and matrixes was considered.

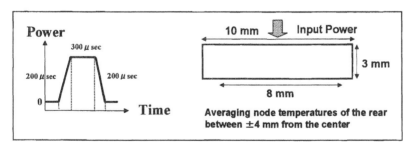

Fig. 2. The input power history and specimen's dimension for FEM calculation.

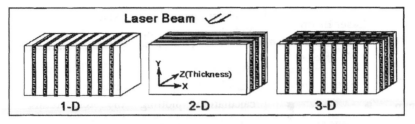

Fig. 3. Schematic diagram of 1-D, 2-D, and 3-D fiber configurations.
Fiber configurations were handled as sheets of SiC fiber, simulating SiC fiber
bundles or textures. The laser beam hit the front of the solids.

The input data of thermal diffusivity/conductivity of fibers and matrixes
used for FEM calculations are shown in Fig. 4. The CVI matrix thermal
conductivity was chosen from the values of CVD SiC, and the PIP matrix
thermal conductivity was tentatively deduced from our experimental and
calculated results [17].

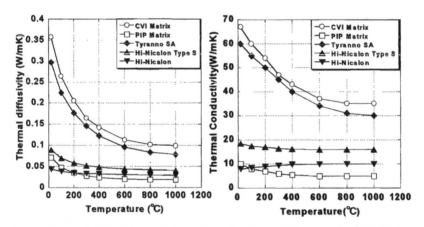

Fig. 4. Thermal diffusivity and thermal conductivity of Tyranno SA, Hi-Nicalon
Type S, Hi-Nicalon fibers, and those of PIP and CVI matrixes for FEM
calculation. The data of SiC fibers, PIP and CVI matrixes were interpolated and
extrapolated from RT and high temperature data obtained from literatures
[10,11,16,17].

RESULTS AND DISCUSSION

Figure 5 shows the calculated results of thermal diffusivity of CVI and PIP SiC/SiC composites at 900 °C for 1-D, 2-D fiber configurations (Fig. 3).

For the CVI case, the increase in the fiber volume of Tyranno SA had a minor effect on the overall composite thermal diffusivity, whereas for Hi-Nicalon and Type S fibers the thermal diffusivity was strongly reduced with increasing fiber volume. The 2-D (XY) sheets reduced the thermal conductivity more than the 1-D (Z) fibers, and when the 2-D sheets were made of a lower thermal conductive fiber, Hi-Nicalon, the Z-directional (through-the-thickness) thermal diffusivity was more reduced. It appears to be reasonable because the Z-directional thermal diffusion would largely reduced by increasing the total cross-sectional area of lower thermal conductive fibers perpendicular to the Z-direction.

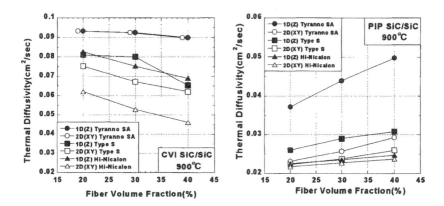

Fig. 5. Calculated results of through-the-thickness thermal diffusivity at 900°C for 1-D and 2-D CVI and PIP SiC/SiC composites vs. fiber volume of three different SiC fibers: Tyranno SA, Hi-Nicalon Type S, and Hi-Nicalon.

For the PIP case, when the matrix thermal conductivity was chosen as low as 5 $Wm^{-1}K^{-1}$ at 900 °C (see Fig. 4), not only highly thermal conductive Tyranno fiber but also Hi-Nicalon and Hi-Nicalon Type S fibers could increase the through-the-thickness thermal diffusivity with increasing fiber volume. The increase in 1-D (Z) Tyranno SA fiber volume was especially effective for increasing the Z-directional thermal diffusivity.

Figure 6 shows the calculated results for thermal diffusivity of CVI and PIP SiC/SiC composites at 200 °C for 1-D and 2-D fiber configurations. The composite's thermal diffusivity dependence on fiber type and fiber volume was basically similar to that at 900 °C, except that values of thermal diffusivity at 200 °C were larger and that the increase in thermal diffusivity of Hi-Nicalon PIP composites was nearly negligible.

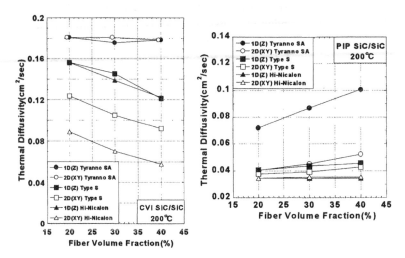

Fig. 6. Calculated results of through-the-thickness thermal diffusivity at 200°C for 1-D and 2-D CVI and PIP SiC/SiC composites vs. fiber volume of three different SiC fibers: Tyranno SA, Hi-Nicalon Type S, and Hi-Nicalon.

Since the results of 200 and 900 °C were similar and the effect of highly conductive thermal SiC fiber was prominent for the PIP case, we focused the PIP case and investigated how 3-D fiber configuration affected the through-the-thickness thermal diffusivity of PIP SiC/SiC composites at a medium temperature of 600 °C between 200 and 900 °C, as shown in the left-hand side of Fig. 7. It is clearly shown that the thermal diffusivity increased with increasing the total of fiber volume of Tyranno SA fiber as well as increasing Z-directional fiber ratio. To see the effect of Z-directional fiber contribution, we plotted the thermal diffusivity of 3-D SiC/SiC composites divided by that of 2-D ones, which is noted as through-the-thickness thermal diffusivity improvement ratio against the ratio of the fiber volume for Z-direction to the fiber volume summed for X and Y directions. This plot is shown in the right-hand side of Figure 7.

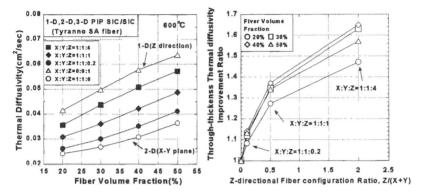

Fig. 7. Calculated results of through-the-thickness thermal diffusivity at 600 °C for 1-D, 2-D, 3-D PIP SiC/SiC composites vs. Tyranno SA fiber volume (the left-hand side), and through-the-thickness thermal diffusivity improvement ratio vs. Z-directional fiber volume divided by X- and Y- directional fiber volume (the right-hand side).

It is seen that the through-the-thickness thermal diffusivity improvement ratio was increased when the fiber configuration changed from the 2-D to the 3-D configurations. Even the smallest Z-directional fiber volume (X:Y:Z=1:1:0.2) contributed to increase the Z-directional thermal diffusivity. It is noticed that when the fiber volume was 20 %, the improvement ratio was smaller due to the direct effect of a small amount of highly thermal conductive fiber volume and that when the fiber volume was 50 %, the improvement ratio appeared to slightly diminish mostly due to the increase in the denominator, i.e., the through-the-thickness diffusivity of the 2-D (X-Y plane) as discussed later.

It is interesting to see the heat flux distribution in a specimen after the heat load of a laser shot. Typical examples of the X- and Z-directional heat flux distributions in Tranno SA fiber-reinforced PIP SiC/SiC composites are shown in Fig. 8. The specimen temperature was 600 °C before laser shot.

The strong heat flow toward the Z-direction can be seen in highly thermal conductive SiC fiber (Tyranno SA) regions configured to the Z-direction (Fig. 8A). The X-directional heat flux is also dominant in Tyranno SA fiber regions configured to both XY-plane and Z-direction (Fig. 8B), which suggests that the X-directional heat flux in the XY plane with highly thermal conductive fiber bundles would quickly flow into the Z-directional fiber bundles. This XY-planar and Z-directional fiber network would efficiently enhance the through-the-thickness thermal diffusivity. In addition, since radial thermal diffusivity of Tyranno SA fiber was assumed to be the same as the axial one, the increase in the total fiber volume also contribute to increase the through-the-thickness thermal diffusivity via radial (Z-directional) heat flow in the fiber bundles of XY plane, which would be the main reason for the result of the 2D case in Fig. 7.

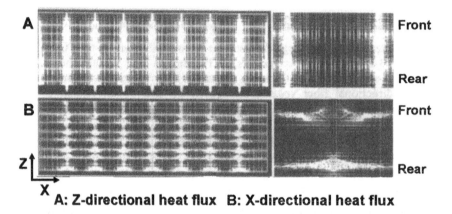

A: Z-directional heat flux   B: X-directional heat flux

Fig. 8. Heat fluxes of Z- and X-directions in 3-D PIP SiC/SiC composites (X:Y:Z=1:1:0.2,Vf=40 %) with high thermal conductive fiber, Tyranno SA, at 600 °C. The heat flux vectors are drawn by white arrows at integration points of each finite element. The figures of the right-hand side are the enlarged of the central portions of the left-hand side figures.

Finally, it should be mentioned that when carbon interface between fiber and matrix exists, the interface will reduce thermal diffusivity transverse to the fiber direction mainly due to the imperfect interfacial thermal contact [18]. Therefore, the effect of X- and Y-oriented highly thermal conductive SiC fibers for increasing the through-the-thickness thermal conductivity would be reduced by the carbon interface but this adverse effect should be small for the Z-oriented fibers because the cross-sectional area of the interface layer normal to Z-direction is much smaller than the fiber area.

It is known that for more increasing mechanical strength, more fiber should be aligned to the load direction. Therefore, the excess alignment of Z-directional fibers would reduce the tensile strength of a plate-type specimen. The present study suggests that since the small amount of Z-directional fiber is effective for increasing the through-the-thickness thermal conductivity, the amount should be determined by optimizing the best thermo-mechanical performance.

SUMMRY

In the case of low thermal conductive matrixes, such as PIP or porous low density CVI, both Hi-Nicalon and Type S fibers have no negative effect on the composite thermal diffusivity. When using highly conductive SiC fiber, such as Tyranno SA fiber, the overall thermal diffusivity and conductivity of PIP SiC/SiC composites are strongly improved with increasing the fiber volume especially for the 1-D fiber configuration.

In the case of high thermal conductive matrixes, such as dense CVI, RB, MI, low thermal conductive SiC fibers, such as Hi-Nicalon fiber, worsens· the overall composite's thermal diffusivity and conductivity with increasing the fiber volume. Even Hi-Nicalon Type S fiber behaves similarly as Hi-Nicalon fiber, whereas Tyranno SA fiber does not worsen the composite thermal diffusivity and conductivity.

A calculation for 3-D SiC/SiC composite reveals that Tyranno SA fiber oriented to the through-the-thickness direction effectively contributes to increase the through-the-thickness thermal/diffusivity and that the fiber oriented to XY-plane has somewhat a positive effect when the fiber volume is increased due to fiber network and high radial thermal diffusivity of the fiber. The best X:Y:Z ratio for 3-D fiber configuration could be determined by taking account of both thermal and mechanical performance.

REFERENCES

[1]S. Nishio, S. Ueda, I. Aoki, R. Kurihara, T. Kuroda, H. Miura, T. Kunugi, Y. Seki, T. Nagashima, M. Ohara, J. Adachi, S. Yamazaki, I. Kawaguchi, T. Hashimoto, K. Shinya, Y. Murakami, H. Takase, T. Nakamura, "Improved Tokomak Concept Focusing on Easy Maintenance", Fusion Engineering and Design **41** 357-364 (1998).

[2]S. Ueda, S. Nishio, Y. Seki, R. Kurihara, J. Adachi, S. Yamazaki, DREAM Design Team, "A Fusion Power Reactor Concept Using SiC/SiC Composites", J. Nucl. Mater. **258-263** 1589-1593 (1998).

[3]G.W. Hollenberg, C.H. Henager, Jr., G.E. Youngblood, D.J. Trimble, S.A. Simonson, G.A. Newsome, E. Lewis, "The Effect of Irradiation on The Stability and Properties of Monolithic Silicon Carbide and SiC$_f$/SiC Composites up to 25 dpa", J. Nucl. Mater. **219** 70-86(1995).

[4]L. L. Snead, R. H. Jones, A. Kohyama, P. Fenici, "Status of silicon carbide composites for fusion", J. Nucl. Mater. **233-237** 26-36 (1996).

[5]D. J. Senor, G. E. Youngblood, C. E. Moore, D. J. Trimble, G. A. Newsome, J. J. Woods, "Effects of Neutron Irradiation on Thermal Conductivity of SiC-based Composites and Monolithic Ceramics", Fusion Technology **30** 943-955 (1996).

[6]R. H. Jones, D. Steiner, H. L. Heinisch, G. A. Newsome, H. M. Kerch, "Radiation Resistant Ceramic Matrix Composites", J. Nucl. Mater. **245** 87-107 (1997).

[7]R. Yamada, T. Taguchi, J. Nakano, N. Igawa, "Thermal Conductivity of CVI and PIP Composites ", Ceram. Eng. Sci. Proc. **20**[3] 273-280 (1999).

[8]R. Yamada, T. Taguchi. N. Igawa, "Mechanical and thermal properties of 2D and 3D SiC/SiC composites", J. Nucl. Mater. **283-287** 574-578(2000).

[9]T. Ishikawa, Y. Kohtoku, K. Kumagawa, T. Yamamura, T. Nagasawa, "High-Strength Alkali-resistant Sintered SiC fibre stable to 2,200 °C", Nature **391** 773-775 (1998).

[10]T. Nakayasu, M. Sato, T. Yamamura, K. Okamura, Y. Katoh, A. Kohayama, "Recent Advancement of Tyranno/SiC composites R&D", Ceram. Eng. Sci. Proc. **20** (4) 301-308 (1999).

[11]T. Ishikawa, S. Kajii, T. Hisayuki, "High Heat-Resistant SiC-Polycrystalline Fiber and Its Fiber-Bonded Ceramic", Ceram. Eng. Sci. Proc. 21 (4) 323-330 (2000).

[12]A. Sayano, C. Sutoh, S. Suyama, Y. Itoh, S. Nakagawa, "Development of a

Reaction-Sintered Silicon Carbide Matrix Composite", J. Nucl. Mater. **271-272** 467-471 (1999).

[13]T. Taguchi, N. Igawa, R. Yamada, M. Futakawa, S. Jitsukawa, "Mechanical and Thermal Properties of Dense SiC/SiC Composite Fabricated by Reaction-Bonding Process", Ceram. Eng. Sci. Proc. **22**[3] 533-538 (2001).

[14]Y. Katoh, S-P. Lee, M. Kotani, S-M. Dong, A. Kohayama, "Development of SiC/SiC composites by Various Matrix Densification Process", CREST Int. Sym. on SiC/SiC Composite Materials R&D and its Application to Advanced Energy Systems", May 20-22, 2002, Kyoto, Japan (to be published in the same issue of Ceramic Transaction by American Ceramic Society).

[15]W. J. Parker, R. J. Jenkins, C. P. Butler, G. L. Abbott, "Flash Method Thermal Diffusivity, Heat Capacity, and Thermal Conductivity", J. Appl. Phys. 32 (1961) 1679.

[16]A. Urano, J. Sakamoto, M. Takeda, Y. Imai, H. Araki, T. Noda, "Microstructure and Mechanical Properties of SiC fiber "Hi-Nicalon Type S" Reinforced SiC Composites", Ceram. Eng. Sci. Proc. 19 (3) (1998) 55.

[17]R. Yamada, N. Igawa, T. Taguchi, S. Jitsukawa, "Highly Thermal Conductive, Sintered SiC Fiber-reinforced 3D-SiC/SiC Composites: Experiments and Finite-element Analysis of the Thermal diffusivity/ conductivity", Proc. of 10th Int. Conf. on Fusion Reactor Materials, Baden –Baden, 2001, to be published in J. Nucl. Mater..

[18]G. E. Youngblood, D. J. Senor, R. H. Jones, "Optimizing the Transverse Thermal Conductivity of 2D-SiC/SiC Composites, I. Modeling", Proc. of 10th Int. Conf. on Fusion Reactor Materials, Baden –Baden, 2001, to be published in J. Nucl. Mater..

MICROSTRUCTURE EVOLUTION IN HIGHLY CRYSTALLINE SiC FIBER UNDER APPLIED STRESS ENVIRONMENTS

Tamaki Shibayama, Yutaka Yoshida*, Yasuhide Yano* and Heishichiro Takahashi
Center for Advanced Research of Energy Technology, Hokkaido University
Sapporo 060-8628, Japan

ABSTRACT

Because they exist high specific strength, fracture and toughness at elevated temperatures compared with monolithic ceramics, silicon carbide matrix composites reinforced by continuous SiC fibers are currently being considered for high temperature applications such as aerospace components, gas turbine energy conversion systems and nuclear reactors. It is important to evaluate the creep properties under tensile loading of a SiC fiber to determine their usefulness for structural components. However, we have little knowledge on microstructure of crept specimens, especially at the grain boundary.

Recently, a simple fiber bend stress relaxation (BSR) test was introduced by Morscher and DiCarlo. Interpretation of the fracture mechanism at the grain boundary is also essential to improve mechanical properties.

In this paper, effects of applied stress by BSR test on microstructural evolution in advanced SiC fibers are described and discussed with our results of microstructure analysis on an atomic scale by using advanced microscopy.

*Graduate Student, Graduate School of Engineering, Hokkaido University, Sapporo 060-8628, Japan

INTRODUCTION

SiC/SiC composites are currently being considered for high temperature applications such as fusion reactors, because of high specific strength at elevated temperatures, improved fracture toughness compared with monolithic ceramics and low induced radioactivity.[1] Since the SiC fibers will experience thermal and mechanical stresses at elevated temperatures, it is important to evaluate long-term phase stability and creep properties of SiC fibers. Interpretation of the fracture mechanism at the grain boundary between matrices and fibers is also essential to improve the mechanical properties. However it is hard to obtain creep properties under tensile loading of small diameter SiC fibers by conventional equipments. Recently, a simple fiber bend stress relaxation (BSR) test has been developed by Morscher and DiCarlo[2] which can evaluate several different specimens at one time.

In this study, the BSR test was applied to several kinds of SiC fibers including Tyranno-SA.[3] The BSR test was conducted to evaluate the stress relaxation ratio as a function of testing time and temperature. TEM specimens of the SiC fibers before and after BSR test were prepared by Focused Ion Beam (FIB) method. HR-TEM analysis was carried out by means of JEOL 2010F with GIF 200. Tyranno-SA SiC fiber showed higher creep resistance property rather than the other SiC fibers.

In this paper, the mechanism of grain boundary stabilization under high temperature creep by small additional Al is introduced and effects of applied stress by BSR test on microstructural evolution in advanced SiC fibers are described and discussed.

EXPERIMENTAL PROCEDURE

The BSR test was applied to several kinds of SiC fibers including Tyranno, Hi-Nicalon and Tyranno-SA at elevated temperatures in the range of 900°C to 1500°C under 1MPa Ar flow for 1 h. For Tyranno-SA, ultra-high purified Ar ($O_2 < 0.1$ ppb) and high purity Ar ($O_2 < 0.1$ ppm) were also used to eliminate oxygen partial pressure effects.[4] For each test condition, an 8mm loop diameter applied to each SiC fibers. The SiC fibers placed in graphite apparatus and boats. The BSR test was applied to Tyranno SA SiC fibers to evaluate the stress relaxation as a function of time, temperature and environment. It can be determined by the simple relation;

$$m(t, T) = 1-R_0/Ra \tag{1}$$

where m is BSR ratio, t is testing time, T is testing temperature, R_0 is initial loop diameter and Ra is crept and relaxed fiber diameter at R.T..[2]

TEM specimens of the SiC fibers before and after BSR test were prepared by Focused Ion Beam (FIB) method. HR-TEM analysis was carried out by means of JEOL 2010F with GIF 200.

RESULTS AND DISCUSSION

Fig. 1 shows a cross sectional TEM micrograph of Tyranno-SA by FIB method. In general, SiC fibers displayed uniform isotropic microstructures. However, Tyranno-SA showed non-uniform microstructures. Tyranno-SA was produced by a polymer derived process similar to the other SiC fibers. But Tyranno-SA was sintered at a temperature greater than 1500°C following ceramization around 1000°C. The decomposition gas including SiO was generated during the sintered process. Therefore several voids formed in the matrices of Tyranno-SA fibers.

As shown in fig.1 by arrows, there are several voids in the inner one third mid region the fibers. The density of Tyranno-SA should be lower than that of the former version of Tyranno SiC fibers. However this fiber showed higher mechanical properties compared to the other polymer derived SiC fibers.[3] Moreover Tyranno-SA shows higher creep resistance compared to the other SiC fibers. This result suggested that the small amount of Al might improve the mechanical properties, especially high temperature creep resistance.

Fig. 1. Cross sectional TEM micrograph of Tyranno-SA.

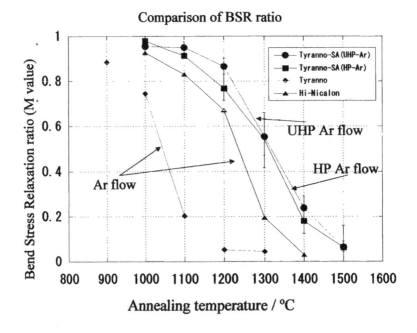

Fig. 2. Stress relaxation ratio comparison for several SiC fibers tested under Ar flow for 1h.

Fig. 2. shows stress relaxation ratio comparison for Tyranno-SA and former SiC fibers in Ar flow for 1 h test. From a BSR point of view, Tyranno-SA has highest high temperature creep resistance in this study. The probable mechanism for stress relaxation at high temperature under applied stress was grain boundary sliding. The microstructure of the different fibers are as follows: Tyranno is amorphous, Hi-Nicalon is nano-crystalline in amorphous matrix and Tyranno-SA is highly crystalline structure but not homogenous. The differences of BSR ratio observed in this study could be strongly correlated with their microstructure. In addition, microstructure and chemical composition of the grain boundary can reveal the mechanism of the strengthening of sintering SiC fibers such as Tyranno-SA.

Fig 3. shows EDS analysis results of the grain boundary of Tyranno-SA before and after BSR tests.

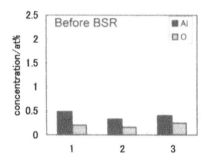 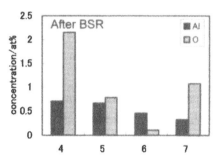

Fig. 3. EDS analysis at grain boundary of Tyranno-SA before and after BSR tests.

As shown in Fig. 2, Al contents did not change before and after BSR tests. However O contents increased after BSR tests. This result suggested that oxidation was occurred at the grain boundary under applied stress at high temperatures. This is probably related to the formation of complex oxide at grain boundary. Fig. 4 shows HRTEM images before and after BSR test. There is no amorphous layer between the grains before BSR test.

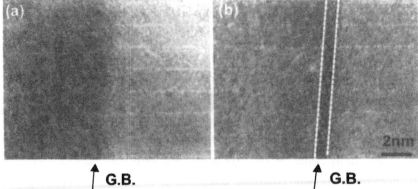

↑ **G.B.**   ↑ **G.B.**

Fig. 4. HRTEM of the grain boundary in Tyamno-SA (a) before and (b) after BSR test.

However amorphous (granular contrast) can be observed at the grain boundary after BSR tests. This result suggested that the formation of amorphous layer should be related to grain boundary sliding and to stabilize each SiC grain under applied stress condition. As far as we know, it is a first concept of self healing property for SiC fibers. To control the additional elements and it's quantity, SiC fibers could show super-plastic property in near future.

SUMMARY

(1) Tyranno-SA SiC fiber displayed non-uniform microstructures and Tyranno-SA SiC fiber showed higher creep resistance property rather than the other SiC fibers.

(2) HRTEM images showed that there was no amorphous layer between the grains before BSR test and EDS analysis showed Al segregation at the grain boundary. On the other hand, highly distorted contrast observed at the grain boundary and oxygen also increased at the grain boundary after BSR test.

(3) This result suggested that the formation of amorphous layer should be related to grain boundary sliding by creep deformation

REFERENCES

[1] P. Fenici, A. Rebelo, R. Jones, A. Kohyama and L. Snead, "Current status of SiC/SiC composites R&D," *Journal of Nuclear Materials*, **258-263** 215-225 (1998).

[2] G. N. Morscher and J. A. DiCarlo, "A Simple Test for Thermomechanical Evaluation of Ceramic Fibers," *J. Am. Ceram. Soc.*, **75** [1] 136-149 (1992).

[3] Toshihiro Ishikawa, Yasuhiko Kohtoku, Kiyoshi Kumagawa, Takemi Yamamura and Toshio Nagasawa,"High-strength alkali-resistant sintered SiC fiber stable to 2,200°C," *Nature*, **391** [6669] 773-775 (1998).

[4] T. Shibayama, G. W. Hei, H. Takahashi, M. Kawasaki and A. Kohyama, "Interface structure analysis of SiC fibers reinforced SiC matrix composites by energy filtering TEM," *J. of Electron Microscopy*, 48[6] 893-897 (1999).

Joining Technologies and Advanced Energy Applications

HIGH TEMPERATURE BRAZING FOR SiC AND SiC$_F$/SiC CERAMIC MATRIX COMPOSITES

B.Riccardi
Associazione EURATOM-ENEA, ENEA CR Frascati, PB 65- 00044 Frascati (Rome), Italy

C.A.Nannetti
ENEA CR Casaccia, 00060 S.Maria di Galeria (Rome), Italy

J.Woltersdorf and E.Pippel
Max-Planck-Institut für Mikrostrukturphysik, Weinberg 2, D-06120 Halle/Saale, Germany

T.Petrisor
Technical University of Cluj-Napoca, Romania, ENEA Consultant

ABSTRACT

The paper presents the results of the development of a brazing technique for monolithic SiC and SiC$_f$/SiC composites. This brazing technique is based on the use of Si-16Ti (at. %) and Si-18Cr (at %) eutectic alloys. The brazing temperature of the used alloys allows to avoid the degradation of the fibre/matrix-interfaces in the composite materials at least for advanced stoichiometric SiC fibre composites. All the joints showed excellent adhesion and no discontinuities and defects at the interface, while the brazing layer revealed a fine eutectic structure. In particular, in the composite joints the brazing layer appeared well adherent both to the matrix, the fibres and the fibre-matrix interphase, and the brazing alloy infiltration looked sufficiently controlled. The brazing alloys were characterized by X ray diffraction and scanning electron microscopy (SEM). All the brazed joints were analysed by SEM and the Si-16Ti joints were also characterized by high resolution transmission electron microscopy investigations of the microstructure and of the nanochemistry (HREM, EELS, esp. ELNES). These analyses revealed atomically sharp interfaces without interdiffusion or phase formation at the interface leading to the conclusion that direct chemical bonds are responsible for the adhesion.

Shear tests performed at room temperature and 600°C on lap joint specimens gave remarkable results: the samples manufactured with monolithic SiC exhibit a high shear stress level and mainly cracked in the SiC bulk (150 MPa at RT for Si-18Cr joints), while the composite samples exhibited a RT shear strength up to 70 MPa for Si-16Ti and up to 80 MPa for Si-18 Cr with failure occurring mainly in the base material.

INTRODUCTION

Silicon carbide and SiCf/SiC ceramic matrix composites are attractive materials for energy application because of their chemical stability and mechanical properties at high temperature [1]. Nevertheless, in order to manufacture complex components the availability of suitable joining techniques is necessary. Among several joining techniques under development [2,3,4,5,6], the brazing is one of the most promising[7]. The requirements of a suitable brazing material are: chemical compatibility and wettability with SiC substrate, thermal expansion coefficient similar to that of the SiC substrate, high shear strength, and for the composite joints a brazing temperature low enough to avoid a degradation of the fibres and the fibre-matrix interface.

The possibility to use pure silicon without active metal filler as braze for silicon carbide has been assessed in previous works [8]. Characteristics of Si are a good chemical compatibility and wettability with silicon carbide, in particular at 1480°C the contact angle between liquid Si and solid SiC is 38° [9]. Moreover the thermal expansion coefficient α is similar to that of silicon carbide: $\alpha_{Si}(RT) = 3.0 \times 10^{-6} \ K^{-1}$ and $\alpha_{SiC}(RT) = 4.0 \times 10^{-6} \ K^{-1}$. Unfortunately the use of pure silicon leads to serious problems because of the high melting point (1410°C) that may degrade fibres or fibre-matrix interfaces. This aspect represents the main drawback also in the case of joints performed by reaction forming techniques employing infiltration of molten silicon into joining parts interspaced by carbon or carbon-SiC mixtures to be converted into SiC by the infiltrating silicon.

In this paper, a recently developed silicon carbide brazing technique is reviewed and discussed. The alloys used are based on eutectic compositions of silicon-titanium and silicon-chromium.

EXPERIMENTAL

The use of Si-Ti and Si-Cr eutectic alloys was proposed in order to take advantage of the lower melting point with respect to pure Si, and the presence of titanium and chromium which behave as active elements. The Si-Ti and Si-Cr phase diagrams [10] show the presence of two eutectics of interest (Si-16Ti at.% and Si-18Cr at.%) with melting points of 1330°C and 1305°C respectively. The Si-16Ti eutectic is composed of free silicon and TiSi$_2$, while the Si-18Ti is composed of free Si and SiCr$_2$. The joining cannot be performed simply by mixing Si-Ti or Si-Cr powders because in this way it is not possible to get the above-mentioned eutectics. Therefore, the alloys have to be prepared previously, by means of a melting procedure, able to produce a fine eutectic structure [11]; in particular, Si-Ti and Si-Cr mixtures were melted by a plasma torch and then re-melted several times by electron beam.

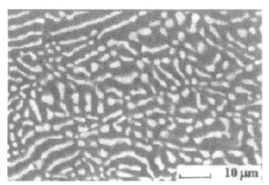

Fig. 1: Micrography of a Si-16Ti alloy before the joining process: grey zones = Si ; white zones=TiSi$_2$

Afterwards, the obtained ingots were reduced to powders by crushing and milling and finally used for the brazing experiments. Fig. 1 shows a SEM picture of the Si-16Ti alloy prior to brazing; evidencing a fine and homogeneous microstructure composed of Si and TiSi$_2$. X-ray diffraction confirmed that no other phases than Si and TiSi$_2$ and Si and SiCr$_2$ were detected in the proposed brazing alloys (Fig.2 and 3).

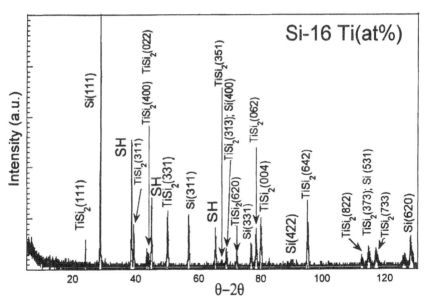

Fig. 2: XRD pattern of Si-16Ti alloy *(SH=sample holder)*

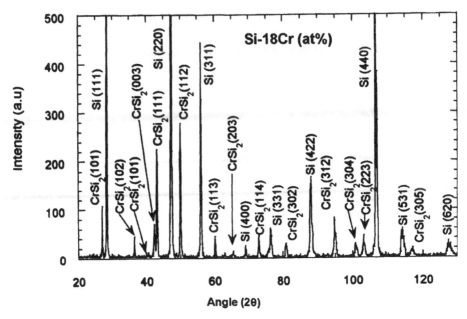

Fig. 3: XRD pattern of Si-18Cr alloy

The joining was carried out by using monolithic polycrystalline α-SiC (Hexoloy-SA ®Carborundum) and a SiC$_f$/SiC composite produced by SNECMA (CERASEP® N3-1), with the latter consisting of a pseudo tri-dimensional weave of Nicalon™ CG fibres, densified by chemical vapour infiltration (CVI) and finally SiC coated by chemical vapour deposition (CVD). The above materials have the necessary chemical stability at the brazing temperature. The typical properties of the monolithic and composite joining parts to be joined used are reported in tables 1 and 2 [12,13].

Table I. Main properties of Hexoloy-SA ®Carborundum

Property	Temperature (°C)	
Density	20	3.07 g/cm^3
Apparent porosity	20	0 %
Young's modulus	20	350 GPa
Thermal expansion coefficient	20-1000	4.02 10^{-6} 1/K
Modulus of Rupture (MOR)	20-1600	380-410 MPa

Table II. Main properties of SNECMA-CERASEP® N3-1 composite

Property	Temperature	
	20°C	1000°C
Density	2.4 – 2.5 g/cm³	--
Thermal expansion coefficient	4.0 10^{-6} 1/K	--
Tensile strength (in plane)	300 MPa	--
Tensile strain (in plane)	≥ 0.6 %	0.3-0.4 %
Trans laminar shear strength	(200 ± 20) MPa	--
Inter-laminar shear strength	40 MPa	30 MPa
4 points bending strength	600 MPa	--

The used samples were 12 x 10 x 3 mm³ plates both of monolithic SiC and SiC$_f$/SiC composite. Bulk SiC samples didn't need any surface treatment since their roughness was in the order of 1 μm.The composite specimens were ground in order to reach a surface roughness in the order of a few microns. The mentioned CVD coating (> 100 μm) was partially removed by surface preparation, thus some fibres remained uncoated and thus exposed to the brazing alloy.

After ultrasonic cleaning in acetone and the application of the brazing alloy between the two pieces to be joined, the samples were inserted in the oven and kept in contact during thermal cycle with a 1 N load. The first joinings were carried out in inert atmosphere (Ar+3% H$_2$) but then they were always performed in a vacuum furnace (10^{-6} mbar) because that facility allowed a better control of the thermal cycle. The samples were heated up to the eutectic temperature with a heating rate of 10 °C/min; the hold time at melting temperature was about 10 min; cooling down to 600°C was performed at 20°C/min followed by natural cooling down to room temperature.

MICROSTRUCTURE AND NANOCHEMISTRY

All the joints were examined to detect macroscopic defects or cracks. Afterwards cross-sections of monolithic SiC and SiC$_f$/SiC composite joints were examined by SEM equipped with Energy Dispersive X ray spectroscopy (EDX). For both alloys, the joint thickness of composites was in the range 20-30 μm, but a local variation in the thickness could be observed depending on the surface roughness . The joint thickness of monolithic specimens was generally higher due to the absence of infiltration in the impervious SiC; the values ranged from few tenth of microns up to 100 μm depending on the amount of braze deposited between the pieces to be jointed. SEM micrographs showed no discontinuities and defects at the joint interface and no unmelted particles. Moreover the eutectic structure showed a morphology comparable with that of the starting powder

(Fig.4), the Si-18Cr was generally finer than that of Si-16Ti. The duration of the brazing cycle allowed to control sufficiently the infiltration of the joints of composites. Sometimes infiltration was observed but it was limited to no more than a couple of fabric layers close to the joint interface.

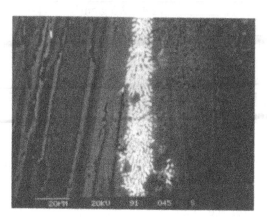

Fig. 4 SEM images of a joint performed between SiC$_f$/SiC composites (Si-18Cr)

EDX maps (Fig.5 and 6) showed absence of Ti and Cr within SiC (close to the join interface). The oxigen maps evidenced only a very slightly higher oxygen content in the brazing alloy with respect to the bulk SiC. No macroscopic reaction layers were visible at the interface.

In order to study the interface structure and the nature of the bonds between the Si-16Ti eutectic alloy and the SiC and SiC$_f$/SiC composites, investigations were performed by transmission electron microscopy (TEM) including high resolution or atomic plane imaging (HREM), and electron energy-loss spectroscopy (EELS) for chemical analysis (cf., e.g., [14,15]). The EELS method allowed to estimate the kind and concentration of the chemical elements with a spatial resolution in the order of 1-2 nm. In particular, the analysis of the near-edge fine structures (ELNES) of the relevant ionisation edges allowed to characterise the chemical bonding state of individual elements with the same resolution. Also with these analyses, no Ti diffusion into SiC and no new phases were found in the related interfaces. In some TEM images strain-contrast contours have formed along the interface (cf. Fig.7). This mechanical strain could result from the thermal expansion mismatch between the SiC bodies and the brazing alloy and hints at strong bonding in the interface.

Advanced SiC/SiC Ceramic Composites

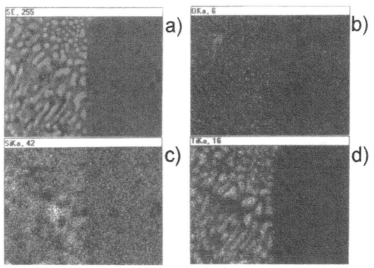

Fig. 5: Interface between Si-16Ti joint and bulk SiC (a) and O (b), Si (c) and Ti(d) mapping (EDX)

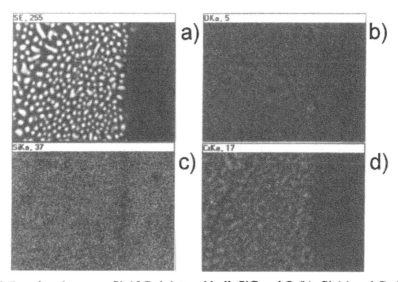

Fig. 6: Interface between Si-18Cr joint and bulk SiC and O (b), Si (c) and Cr (d) mapping (EDX)

As results both from EELS measurements and HREM images, all the analysed interfaces between the substrate and the brazing alloy could be proved to be nearly atomically sharp, i.e., there is no detectable interdiffusion or formation of new phases. Thus, the high strength macroscopically measured in these joining systems must be attributed to direct chemical Si-Si and Si-C bonds in the case of a Si/SiC interface with additional Si-Ti or Ti-C bonds at TiSi$_2$/SiC interfaces. These considerations hold for SiC$_f$/SiC and monolithic SiC joints as well.

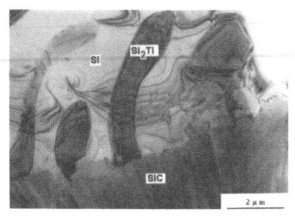

Fig. 7   TEM image of the interface area between SiC/SiC$_f$ andSi-16Ti brazing,

SHEAR STRENGTH

Among all the mechanical features of the joints, the shear strength is one of the leading properties to assess the reliability of any joint technique. Several methods to test the shear strength of joints [16] have been proposed. In this work, the shear tests were performed following an "ad hoc" modification of the ASTM D905-89 test procedure [17]. Although a pure shear strain field cannot be assured by this procedure, this method is, however, a simple one suitable to obtain a rather good estimation of shear strength and a good way for a comparative evaluation. Tests have been performed at RT (Fig. 8a) and 600°C (Fig. 8b) for Si-16Ti joints and only at RT for Si-18Cr joints. The crosshead speed was 0.6 mm/min.

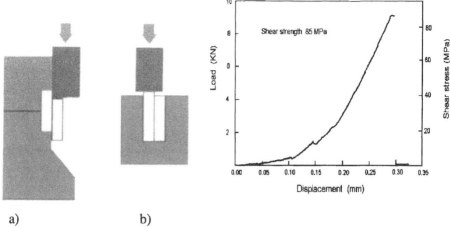

a)   b)

Fig.8: Scheme of the shear test arrangement at RT(a) and 600°C (b)

Fig.9: A shear load displacement curve (composite sample)

The results can be summarised as follows. The samples joined by using the Si-16Ti alloy and manufactured with the composite samples exhibited a shear strength of 71 ± 10 MPa at RT and up to 70 MPa at 600°C. The samples joined by using the Si-18Cr alloy and manufactured with monolithic SiC cracked at 150 MPa at RT, while the composite samples exhibited a shear strength of 80 ± 10 MPa at RT. The shear strength level measured are similar to those obtained by high performance reaction forming technique but tested with a different method [18]. Moreover, all the tests, that were carried out at least on five specimens for each typology of brazed joints, gave sufficiently reproducible results with a limited scattering. A typical shear load-displacement curve for composite samples is shown in Fig. 9. With the exception of an initial adjusting phase, the trend appears practically linear up to failure with a very limited toughness.The joint strength was slightly affected by the residual roughness and open porosity of the composite substrate. Observation of the fracture surfaces revealed that failure was always cohesive in all tested specimens. In monolithic samples, the failure occurred mainly in the substrate but sometimes started at the joint interface and propagated in the bulk SiC. Since the compressive strength of Hexoloy should be higher than 1 Gpa, failure of the substrate is likely due to not pure compressive loading and to some local stress concentrations. Consequently the joints strength is likely to be even higher than that measured.

In the composite samples, the failure always occurred in the composites (Fig. 10), leading to the conclusion that the limiting parameter of the performance was the shear strength of the composite itself that is locally further increased by the brazing alloy infiltration.

Fig.10:Typical composite specimen failure image (arrow)

CONCLUSIONS

The proposed joining technique, which employs the eutectic Si-Ti and Si-Cr alloys, appears suitable for joining of SiC and SiC$_f$/SiC composites. Joints with low residual strains and satisfactory mechanical features were obtained.
Following SEM analysis all the joints investigated did not show any defects in the brazing layer which maintained, even after joining, a fine eutectic structure. Moreover, concerning composites, the brazing alloy infiltration looked sufficiently controlled. Systematic investigations of the microstructure and of the nanochemistry (HREM, EELS, esp. ELNES) of Si-Ti joints led to the conclusion that direct chemical bonds are responsible for the adhesion. Shear tests of the joints of SiC$_f$/SiC composites showed remarkable values of the bonding strength (about 70-80 MPa) which were scarcely influenced by the testing temperature at least up to 600 °C and in the experimental conditions (quite high cross head

speed) employed, while joints of monolithic SiC (Si-18Cr) exhibited up to 140 MPa.

The technique presents some disadvantages, such as need of grinding the surfaces to joint in order to get tight tolerances and a tight joining thermal cycle, but there is no evidence of unsuitability for the joining of large pieces.

In order to assess the suitability of the alloy for energy conversion application, the chemical behaviour of the joints with respect to oxidation will be investigated in the near future.

REFERENCES

[1] R. Naslain, Materials design and processing of high temperature ceramic matrix composites: State of the art and future trends. Advanced composite materials Vol. 8, n.1, pp 3-16.

[2] M.Ortelt, F.Ruehle, H.Hald, H.Weihs, J.Greenwood, A.Pradier in "High Temeperature Ceramic Matric Composites" p. 760, Edited by W.Krenkel, R.Naslain H.Schneider, WILEY-VCH Weinheim (Germany) 2001.

[3] O.M.Akselsen, J.Mat.Sci., 27, 569-579 (1992)

[4] T.J.Moore, J.Am.Ceram.Soc, 68 [6], C151-C153(1985).

[5] M.Salvo, M.Ferraris, P.Lemoine, M.Appendino Montorsi, M.Merola, J.Nucl.Mat., 212-215, 1613-1616 (1994)

[6] J.Martinez Fernandez, A.Munoz, F.M.Varela-Feria, M.Sing. Journal European Ceramic Society Vol 20 [14-15] (2000) 2641-2648

[7] M. M.Schwartz, Ceramic joining. ASM International (1990)

[8] M. Ferraris, C. Badini, M. Montorsi, P. Appendino, H. W. Scholz, J. Nucl. Mat. 212-215 (1994) 1613-1616

[9] J. G. Li, H. Husner, Journal of Material Science Letters 10 (1991) 1275-1276

[10] ASM Handbook. Volume 3 : Alloy phase diagrams. ASM International (1992)

[11] ENEA Patent 478 (2001) –RM2001A000101

[12] Hexoloy-SA .Technical specification. Carborundum Italia s.r.l. (1998).

[13] A.La Barbera, B.Riccardi, C.A.Nannetti, A.Donato, L.F.Moreschi, J.Nucl. Mat 294 (2001) 223-231

[14] E. Pippel, J. Woltersdorf, G. Pöckel, G. Lichtenegger, Microstructure and Nano-chemistry of Carbide Precipitates, Mater. Charact. 43, 41-55 (1999)

[15]R. Schneider, O. Lichtenberger, J. Woltersdorf, Phase identificationin in composite materials by EELS fine structure analysis, J. Microsc. **183**, 39-51 (1996)

[16]S. Cordeau, C. Taffarel, Characterisation de la resistance au cisaillement de jonctions, Technical Note CEA-CEREM Grenoble (France) DEM n.38/97 (1997).

[17] B. Riccardi, A. Donato, P. Colombo, G. Scarinci : Development of homogeneous joining techniques for SiC/SiCf composites. R.Beaumont, P.Libeure, B.de Gentile, G.Tonon (Eds) Fusion Technology 1998- 20th SOFT Marseille

[18] M.Singh, E.Lara Curzio, Transactions of the ASME. Vol 123 (2001) pp288-292.

JOINING SIC-BASED CERAMICS AND COMPOSITES WITH PRECERAMIC POLYMERS

Paolo Colombo
Dipartimento DICASM, Università di Bologna, Viale Risorgimento 2, 40136 Bologna, Italy and Department of Materials Science and Engineering, The Pennsylvania State University, University Park, PA 16801
A. Donato and B. Riccardi
Associazione EURATOM-ENEA, ENEA CR Frascati, Via E. Fermi, 27, Frascati (Roma) 00044, Italy
J. Woltersdorf and E. Pippel
Max-Planck Institut für Mikrostrukturphysik, Weinberg 2, D-06120 Halle, Germany
R. Silberglitt[*] and G. Danko[$]
FM Technologies Inc., Fairfax, VA 22032
C. Lewinsohn[£] and R. Jones
Pacific Northwest National Laboratory, Richland, WA 99352

ABSTRACT
Monolithic SiC bodies (sintered SiC, reaction bonded SiC) were successfully joined using a silicone resin. Maximum values as high as 220 MPa in bending and about 50 MPa in shear tests were reached for samples joined at 1200°C. SiC/SiC$_f$ composites were also joined using a silicone resin or a polycarbosilane and various fillers. In this case, the strength of the joints was much lower, increasing only when specific fillers or reinfiltration of the joints were employed. The reason for the lower strength possessed by SiC/SiC$_f$ joints was attributed to the surface morphology of the composites.

INTRODUCTION
Silicon carbide is an important engineering ceramic because of its high strength and stability at elevated temperatures, and it is currently fabricated also in the form of a ceramic matrix composite, using SiC fibers as reinforcement. SiC/SiC$_f$

---

[*] now with RAND, Arlington, VA 22202, USA
[$] now with Pratt and Whitney, East Hartford, CT 06108, USA
[£] now with Ceramatec Inc., 2425 South 900 West, Salt Lake City, Utah 84119, USA

---

composites, in particular, have attracted a strong interest as components for the "first-wall" in a fusion energy system. In fact SiC/SiC$_f$ composites possess the thermal, mechanical, and nuclear stability necessary for the application in the field of nuclear fusion [1]. When considering joining of a SiC/SiC$_f$ composite, certain additional issues arise due to the use of composite as opposed to monolithic silicon carbide: specifically the degradation of fibers at an elevated temperature (determined by the type of fiber), the oxidation of the fiber-to-matrix interphase (in the presence of oxygen), and the surface roughness and porosity associated with the presence of the reinforcement within the ceramic matrix. Therefore, a method allowing joining of SiC/SiC$_f$ parts, but without compromising the overall properties that are needed, is required. For joining SiC-based ceramics, including composites, direct diffusion bonding, brazing, co-densification of green bodies and binders, reactive metal bonding, pressurized combustion reactions, in-situ displacement reactions, glassy interlayers, and reaction bonding techniques have been used, with various degrees of success [2, 3]. Generally speaking, diffusion bonding requires elevated temperatures (not compatible with the retention of fiber strength in composite materials) or pressures, while active metal brazing tends to create complex reaction layers which sometimes exhibit poor thermo-mechanical properties. The use of preceramic polymers for joining SiC-based ceramics has been proposed in the recent literature [4-13]. Indeed, the use of preceramic polymers for joining offers a number of attractive features, such as easy application, low processing temperatures and possibility to tailor the joint composition by adding suitable fillers to the preceramic precursor, but nevertheless several issues remain to be addressed and some specific problems still need to be solved. Preceramic polymers are organoelement polymers, generally containing silicon, which undergo a polymer-to-ceramic conversion when heated at temperatures ranging from 800 to 1400°C. In order to maintain the shape of the preformed body, the polymer, if not thermosetting itself, must be cross-linked, either by oxidation, e-beam or radiation curing. The pyrolysis of cross-linked polymers is accompanied by the formation of gaseous reaction products, high volume shrinkage and a pronounced density increase [14]. When the pyrolysis temperature is below about 1400°C, the produced material can be classified as an amorphous covalent ceramic, which usually transforms into a nanostructured crystallized ceramic at higher temperatures [15].

EXPERIMENTAL

Three types of SiC-based ceramics were used in the joining experiments: 1) hot pressed α-SiC (cylindrical specimens with a diameter of 13 mm and a thickness of 4 mm, containing 0.5-1 % of either Al or B as sintering aid and having a porosity of 3 %, fabricated by ENEA); 2) reaction bonded SiC (SiSiC, bars 76x7x6 mm; HD530, Norton Company, Worcester, MA, consisting mainly of α-SiC crystals

Advanced SiC/SiC Ceramic Composites

with a bimodal grain size distribution (approximately 10 and 100 μm) embedded in a free silicon phase (23 vol. %), with a density of 3.02 g/cm^3 and an open porosity less than 1%); 3) SiC/SiC$_f$ composites (two different SiC/SiC$_f$ composites (2D and 3D) produced by SEP, a division of Snecma (Saint-Medard-en-Jalles, France) were used. They consist of a bi-directional cloth (CERASEP N2-1) and of a tri-dimensional weave (CERASEP N3-1) of Nicalon fibers densified by chemical vapor infiltration (CVI). The monolithic ceramics were polished on one side using a 5 μm SiC paper (achieving a roughness of ~ 0.1 μm), and ultrasonically cleaned in acetone prior to joining. Original SiC/SiC$_f$ composite specimens (cut into 10x8x3 mm coupons for the joining experiments) had a rather complex surface morphology (roughness Ra ~15-25 μm, average peak-to-valley height Rz ~ 100 to 150 μm). Direct joining of as-received specimens (joining taking place on the coupons' main surface) was never successful, because the rough surfaces contain cavities in which the preceramic polymer is subjected to a large shrinkage upon pyrolysis, and the resulting defects in the ceramic layer behave as crack initiation sources [9].

In order to improve the surface quality, the composite specimens were thus mechanically polished on one side, using SiC and diamond paper, and ultrasonically cleaned in acetone prior to joining. After this procedure, the total roughness of both materials (measured on an area a few square millimeters wide) was greatly reduced. After grinding, the SiC/SiC$_f$ composite's surface was comprised of flat areas (a few hundred μm^2 wide) and of deep "valleys" where the fibers are interwoven. The surface roughness, measured only on the flat areas where joining mainly takes place, was of the order of 0.25-0.5 μm. It has to be pointed out that the SiC/SiC$_f$ composites have a thick SiC over-coat (> 100 μm) deposited by CVD, which was not completely removed by the mechanical surface preparation. Thus the fibers were not exposed and did not participate to the joint formation. The real area of contact (flat zones) is smaller than the nominal one, typically of the order of 60 to 80% for the 3D composite specimens [9, 11].

The preceramic polymers used in the joining experiments were: a methyl-hydroxyl-siloxane (SR350, General Electric Silicone Products Div., Waterford, NY) and two polycarbosilanes (PCS, Dow Corning X9-6348, and allylhydrido-polycarbosilane (AHPCS, Starfire Systems, Watervliet, NY). In some cases, the polymers were loaded with a monomodal or a bimodal powder mixture (SiC or Si nano and micro powders, Al$_{88}$-Si$_{12}$ alloy,) or with ceramic fibers and cloths (plane weave carbon or Nicalon (SiC) fiber cloth, carbon or Nicalon fibers impregnated with SR350 resin, carbon felt impregnated with SR350 resin). The polymers were dissolved in suitable solvents, and the viscosity of the solution was varied by changing the amount of solvent and of fillers. The viscous solutions were applied to the specimens to be joined using a spatula, and the samples were then overlapped obtaining a sandwich structure. In the case of reaction bonded SiC, butt joints were also produced by cutting the overlapped joined specimens. In some cases, a crosslinking step by heating at 150-200°C in air was applied to cure the preceramic polymer. Reinfiltration experiments were also performed, using a diluted polymer solution without additives. After the infiltration in a vacuum container, the specimens were pyrolyzed, and the cycle was then repeated a few times. No pressure was applied during the thermal treatment.

Pyrolysis was conducted either by annealing in a conventional tube oven or by hybrid microwave heating, processing always in an inert atmosphere. The pyrolysis cycle (heating and cooling rate, maximum temperature, dwelling time) were varied in order to assess the influence of the processing parameters on the strength and quality of the joints. The experimental details for the morphological, microstructural, compositional and mechanical analyses performed can be found elsewhere [5, 9, 10, 11, 13].

RESULTS AND DISCUSSION

The values of the fracture shear stress, measured at room temperature, for α-SiC monolithic samples joined using a polysiloxane (SR350) and a polycarbosilane (PCS) at various temperatures are presented in Fig. 1a. The thickness of the joint, measured after the 200°C heat treatment, was 20 ± 2 μm. First of all it is evident that, contrary to silicone resin, polycarbosilane does not serve an effective joining material, probably because of its different decomposition behavior upon pyrolysis (lower ceramic yield, thermosetting polymer) leading to the formation of defects in the joint. The shear strength increases with the joining temperature, and for samples pyrolyzed at 1200°C joined using the SR350 silicone resin, the most of the failures occurred not in the joint region but in the sintered α-SiC body. Since the shear strength of ceramics is significantly lower than their tensile strength it is possible that the elastic mismatch between the joint material and the substrate generates a large stress intensification in the substrates near the interface. It is also possible that defects in the joint layer propagate into the higher modulus substrate and cause failure. Other reasons that might justify this behavior are flaws generated in the SiC substrate by clamping (that activate only when the joint strength is sufficient enough to transfer stress). Among the motives that explain the increase in shear strength of the joints increase with increasing joining temperature are the decrease of residual stresses in the joint with decreasing joint thickness due to the shrinkage of the pyrolysis product, the development of more SiC-like species in the SiOC amorphous ceramic structure, and the densification of the polymer-derived ceramic material [14, 15].

The joining layer thickness strongly affected the joint strength (see Fig. 1b), in particular joint strength decreased with increasing joint thickness (thickness was measured before pyrolysis, after the crosslinking step. The joining material was

SR350). In fact, evolution of gases and shrinkage during processing are identified as critical processes that may control the presence of strength limiting flaws and residual stresses. Computed values of the shrinkage stress for different combinations of shrinkage rate and viscosity indicate that shrinkage rates greater than 1×10^{4} s^{1} and values of viscosity above 1×10^{11} Pa-s are required to generate significant values of stress (>10 MPa) that may form cracks in the joints [13].

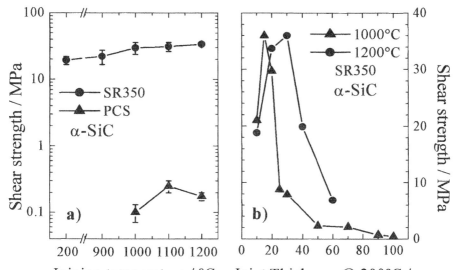

Joining temperature / °C   Joint Thickness @ 200°C / μm

Figure 1a. Shear strength as a function of joining temperature (α-SiC)
Figure 1b. Shear strength as a function of joint thickness (α-SiC)

Furthermore, the cooling rate and the joint thickness will influence the final stress state of a joint after cooling, with the residual stresses potentially leading to the formation of edge cracks, which may be avoided if the joint thickness is below a critical value [16].

Reaction-bonded SiC was also successfully bonded using SR350 polysiloxane. The flexural strength (butt joints) and shear strength, respectively, as a function of pyrolysis temperature are shown in Fig. 2a and 2b. High strength values can be obtained for processing temperatures higher than 1000°C. Due to the similarity between the thermal expansion coefficient of the polymer-derived SiOC ceramic (approximately 3.14×10^{-6} K^{-1}) and of the SiC material, no major residual stresses were observed in the joints, as shown by the casual propagation of a crack, initiated by Vickers indentation, in the joint region [10].

Optical micrographs of SiSiC joined samples at different temperatures, showing the decrease of joint thickness due to the shrinkage of the pyrolysis product with increasing processing temperature, as well as the good quality of the joint (dense, continuous, without delamination or cracks), are shown in Fig. 3.

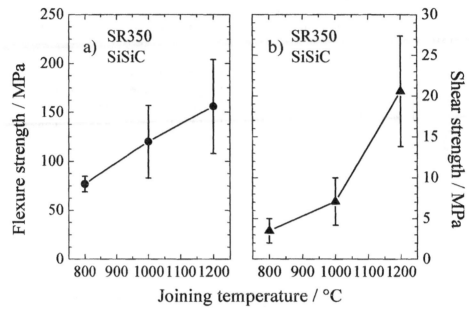

Figure 2a. Flexure strength as a function of joining temperature (SiSiC)
Figure 2b. Shear strength as a function of joining temperature (SiSiC)

SiC/SiC$_f$ specimens were also joined using SR350 polysiloxane resin. An SEM micrograph indicating that while the joint, obtained by processing at 1200°C, appears to be continuous, some voids are present due to the rough surface morphology of the ceramic composite is shown in Fig. 4a. The flat interface structure observed by high resolution electron microscopy (HREM) (Fig. 4b) and the lack of any reaction layer suggest a joining mechanism involving the direct formation of chemical bonds between the SiC/SiC$_f$ bodies and the ceramic joining material, which consists in an amorphous SiOC ceramic. Micro-chemical analyses showed no oxygen diffusion occurring between the SiOC and the SiC ceramics [9].

For all types of SiC-based ceramic substrates used, the mechanism of failure was cohesive, since traces of the pyrolyzed polymer could be found on each surface that had been joined. This indicates the formation of a strong adhesive bond between the substrate and the preceramic-derived ceramic material as well as the presence of a limited amount of residual stresses.

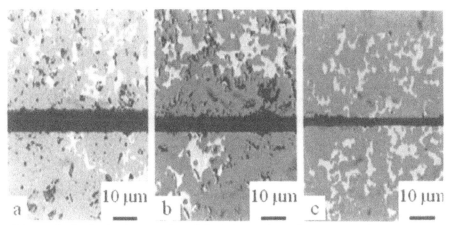

Figure 3. Optical micrographs of the joint region of SiSiC samples joined at different temperatures: a) 800°C; b) 1000°C; c) 1200°C

However, in the case of the SiC/SiC_f ceramic composite, the shear strength for the joint was not satisfactory, being on the average below 4 MPa [11]. The low strength was attributed to the presence of defects in the joining layer arising from shrinkage and densification of the preceramic polymer, concentrated within the non-flat areas of the composite surface. 2D and 3D composites, because of their different surface morphology, gave different results, with a lower strength for the samples possessing the most strongly textured surface (3D) [11].

The introduction of filler powders increased the strength and somewhat reduced the scattering of the data, due to the great decrease of shrinkage during pyrolysis, and the effective reduction defects in the layer. Average shear strength as high as 17.4 ± 3.5 MPa was achieved when using an active filler ($Al_{88}Si_{12}$ powder), that melts before reacting to give ceramic products, thus ensuring a more homogeneous distribution of the reaction products within the joining layer, instead of concentrating them in scattered particles. By reinfiltration of these joints using a diluted polymer solution without additives (up to 4 times), followed by further pyrolysis at 1200°C, the shear strength strongly increased, due to the reduction of porosity within the joint itself (see Fig. 5). The difference in effectiveness between inert and active fillers might be related to the different distribution of the filler within the joining layer after processing (scattered particles in the first case, homogeneous distribution of reaction products in the second one).

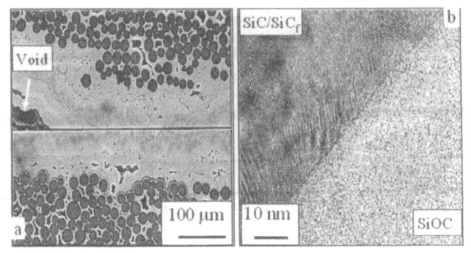

Figure 4a. SEM micrograph of a joint (SiC/SiC$_f$)
Figure 4b. HREM micrograph of a joint (SiC/SiC$_f$)

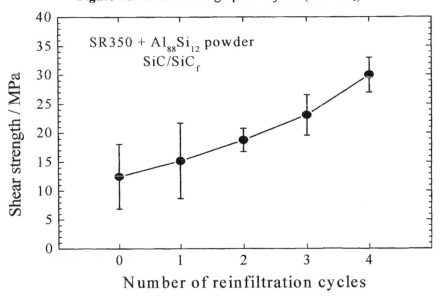

Figure 5. Shear strength as a function of the number of reinfiltration cycles, for joints between 3D composites obtained using SR350 and Al-Si powder (SiC/SiC$_f$)

SiC/SiC$_f$ composite specimens were also joined using a polycarbosilane (AHPCS) loaded with a SiC powder mixture (monomodal or a bimodal in size). Both

Advanced SiC/SiC Ceramic Composites

conventional and microwave hybrid pyrolysis of the preceramic polymer slurry produced SiC/SiC$_f$ joints that were continuous and with similar fracture strength (< 6 MPa). In general, the microwave hybrid heated specimens produced a coarser, grainier microstructure in the ceramic material, a slightly lower fracture stress and a higher Weibull modulus, both decreasing with increasing temperature. These results are consistent with a greater degree of conversion of the amorphous covalent ceramic pyrolysis product to crystalline SiC by the microwave hybrid heating process [17].

The importance of morphology of the surface in the case of joining ceramic composite samples (sandwich joint) was demonstrated using a slurry of liquid AHPCS and SiC powder (weight ratio SiC/AHPCS = 2.5) and repeating the joining procedure on the same surface of the sample for a number of times, after detaching the joined samples by mechanical testing. Indeed, the shear strength of joined 3D SiC/SiC$_f$ ceramic composite specimens increased considerably with decreasing surface roughness of the substrate (Fig. 6b). Microscopy and profilometry data show that each iteration of the polishing and testing procedure produced a smoother surface (Fig. 6a), because of the effective filling of the surface voids by the joining material ceramic residues. Since the surfaces that were joined were polished with 400 grit paper prior to each joining step, the change in surface morphology is due solely to the filling of the large inter-bundle pores present in the composite substrates, leading to a reduction in the flaw size or the number of flaws. It is also possible, however, that the surface treatment increased the contact area between the joint and the substrate and decreased the effective stress in the joining material (for the same load). Since mechanical interlocking can play an important role in the development of strong joints, a controlled study of the influence of surface treatment and contact area is therefore recommended.

Although low values of strength were obtained for joints between SiC/SiC$_f$ composites, which were unlike joints between monolithic ceramics and did not exhibit tolerable values of strength, we must point out that most of the work was carried out with the aim of satisfying criteria relevant to using the materials in a nuclear fusion reactor. Thus, very stringent limitations were enforced on the joining material and processing technique, such as the use only of low activation constituents (such as Si, C, O), and a processing temperature lower than 1200°C, in order not to damage the fibers of the composites used in the experiments.

It is quite possible that the use of novel ceramic composite materials, such as the ones obtained with the NITE process [18] possessing a very different surface morphology from the current ones, the adoption of novel generation ceramic fibers allowing a much higher service temperature (and thus higher processing temperature), or the refinement of processing methods would allow successful joining of ceramic composites using preceramic polymers.

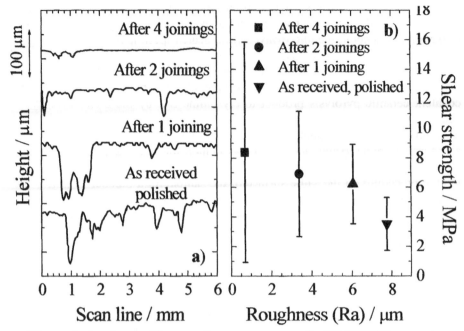

Figure 6. a) stylus profilometer line-scan patterns for 3D SiC/SiC$_f$ ceramic composites; b) shear strength (MPa) as a function of surface roughness

From the data presented, which cover various materials and experimental conditions, it is evident that the joint strength depends simultaneously on a large number of variables (composition and characteristics of the joining material, processing schedule, surface morphology of the specimens, preparation of the surface, testing methodology, ...) whose optimization requires a considerable experimental work. Furthermore, when joining ceramics using preceramic polymers, some issues specific to this technology have to be considered. These include: the preceramic precursor's yield, the setting characteristics, the rheology of the joining mixture, the amount and type of possible fillers, the maximum processing temperature, dwelling time, heating rate and the pyrolysis atmosphere. In particular, a critical area that still requires improvement in order to render more widely applicable this promising joining technology is the reduction of the shrinkage associated with the polymer-to-ceramic conversion of the precursors, resulting in the creation of strength-limiting flaws caused by stresses developing in the joint layer, and in the difficulty of producing hermetic seals (especially when joining rough surfaces).

CONCLUSIONS
Joints with up to 80% of the strength obtained by the best available joining methods [3, 19, 20, 21] have been obtained using preceramic polymer precursors as adhesives. Using preceramic polymers and fillers it is indeed possible to join SiC-based ceramics in a way that is rather simple, versatile and economical from the processing point of view. However, while the joining of monolithic ceramics has been successful, joining of ceramic composites, due to the morphology of the surfaces, has not resulted in joints with values of high strength. In the latter case, the use of fillers and, often, a careful preparation of the sample, are recommended.

ACKNOWLEDGEMENTS

Part of this work was carried out in the frame of the E.C. Fusion Technology Long-Term Programme - Materials 1995/98 SDS 3.1.1 and Advanced Materials 1999/02 ADV 1.2.1.

REFERENCES

[1] R.H. Jones, L.L. Snead, A. Kohyama and P. Fenici, "Recent advances in the development of SiC:SiC as a fusion structural material," *J.Nucl.Mater.*, **258-263** 1546-1550 (1998)

[2] A.C. Ferro and B. Derby, "Liquid phase bonding of siliconized silicon carbide," *J.Mater.Sci.* **30** 6119-35 (1995)

[3] T. Iseki, " Joining of SiC ceramics"; pp. 239-63 in *Silicon Carbide Ceramics - 1*, edited by S. Somiya and Y.Inomata, Elsevier Applied Science, London, 1991

[4] S. Yajima, K. Okamura, T. Shishido, Y. Hasegawa and T. Matsuzawa, "Joining of SiC to SiC using polyborosiloxane," *Am.Ceram.Soc.Bull.*, **60** 253 (1981)

[5] A. Donato, P. Colombo and M.O. Abdirashid, "Joining of SiC to SiC using a preceramic polymer"; pp. 471-36 in *High-Temperature Ceramic-Matrix Composites I: Design, durability and performance, Ceramic Transactions* Vol. 57. Edited by A.G. Evans and R. Naslain, The American Ceramic Society, Westerville OH, 1995

[6] I. Ahmad, R. Silberglitt, Y.L. Tian and J.D. Katz, "Microwave joining of SiC ceramics and composites"; pp 455-50 in *Microwaves: Theory and Application in Materials Processing IV*, Ceramic Transactions 80. Edited by D.E. Clark, W.H. Sutton and D.A. Lewis. The American Ceramic Society, Westerville, OH, 1997

[7] W.J. Sherwood, C.K. Whitmarsh, J.M. Jacobs and L.V. Interrante, " Joining Ceramic Composites Using Active Metal/HCPS Preceramic Polymer Slurries," *Cer.Eng.Sci.Proc.*, **18** 177-84 (1997)

[8] I.E. Anderson, S. Ijadi-Maghsoodi, Ö. Ünal, M. Nostrati and W.E. Bustamante, pp. 25-40 in *Ceramic Joining*, Ceramic Transactions 77. Edited by I.E. Reimanis,

C.H. Henager and A.P. Tomsia. The American Ceramic Society, Westerville OH, 1997

[9]E. Pippel, J. Woltersdorf, P. Colombo and A. Donato, "Structure and composition of interlayers in joints between SiC bodies," *J.Europ.Ceram.Soc.*, 17 1259-65 (1997)

[10]P. Colombo, V. Sglavo, E. Pippel and J. Woltersdorf, "Joining of Reaction-Bonded Silicon carbide using a preceramic polymer," *J.Mater.Sci.*, 33 2409-16 (1998)

[11]P. Colombo, B. Riccardi, A. Donato and G. Scarinci, "Joining of SiC/SiC$_f$ ceramic matrix composites for fusion reactor blanket applications," *J.Nucl.Mater.*, 278 127-35 (2000)

[12]J. Zheng and M. Akinc, "Green state joining of SiC without applied pressure," *J.Am.Ceram.Soc.*, 84 2479-82 (2001)

[13]C.A. Lewinsohn, P. Colombo, I. Reimanis, Ö. Ünal, "Stresses Arising During Joining of Ceramics Using Preceramic Polymers" *J.Amer.Ceram.Soc.*, 84 2240-45 (2001)

[14]P. Greil, "Active-Filler-Controlled Pyrolysis of Preceramic Polymers," *J. Am. Ceram. Soc.*, 78 835-48 (1995)

[15]R. Riedel, "Advanced Ceramics from Inorganic Polymers"; pp. 1-50 in Materials Science and Technology, A comprehensive treatment. Vol. 17B Processing of Ceramics, Part II, R. J. Brook ed. VCH Wurzburg, Germany, 1996

[16]H.P. Kirchner, J.C. Conway, Jr. and A.E. Segall, "Effect of joint thickness and residual stresses on the properties of ceramic adhesive joints: I, Finite element analysis of stresses in joints, *J.Am.Ceram.Soc.*, 70 104-09 (1987)

[17]G.A. Danko, R. Silberglitt, P. Colombo, E. Pippel and J. Woltersdorf, "Comparison of microwave hybrid and conventional heating of preceramic polymers to form silicon carbide and silicon oxycarbide ceramics," *J.Amer.Ceram.Soc.*, 83 1617-25 (2000)

[18]A. Kohyama, "Overview of CREST-ACE Program for SiC/SiC Ceramic Composites and Their Energy System Applications"; in Proceedings of CREST-International Symposium on SiC/SiC Composite Materials R & D and Its Application to Advanced Energy Systems, May 20-22, 2002, Kyoto, Japan

[19]B.H. Rabin, and G.A. Moore, "Joining of SiC-Based Ceramics by Reaction Bonding Methods," *J. Mat. Synth. & Proc.*, 1 [3] 195-201 (1993)

[20]M. Singh, "A Reaction Forming Method for Joining of Silicon Carbide-Based Ceramics," *Scripta Mater.*, 37 [8] 1151-1154 (1997)

[21]C.A. Lewinsohn, R.H. Jones, M. Singh, T. Nozawa, M. Kotani, Y. Katoh, and A. Kohyama, "Silicon Carbide Based Joining Materials for Fusion Energy and Other High-Temperature, Structural Applications,"*Ceram. Eng. & Sci.Proc.*, 22 [4] 621-625 (2001)

MODELING OF FRACTURE STRENGTH OF SiC/SiC COMPOSITE JOINT BY USING INTERFACE ELEMENTS

H. Serizawa and H. Murakawa   C.A. Lewinsohn
Joining and Welding Research Institute,   Ceramatec, Inc.
Osaka University   2425 South 900 West,
11-1 Mihogaoka, Ibaraki,   Salt Lake City, Utah 84119, USA
Osaka 567-0047, Japan

ABSTRACT
    To examine the influence of joint geometry on the strength of ceramic joints, the interface element was proposed as a simple model to represent the mechanism of failure in an explicit manner. It was applied to the analysis of the fracture strength of SiC/SiC composite specimen joined by ARCJoinT[TM], and the effect of the scarf angle on the joint strength was studied by changing the scarf angle between the composite and the joint. It was clearly shown that the joint strength was governed both by the surface energy and the bonding strength. Moreover, it was found that the strength was affected by the order of the singularity in stress field. Thus, the proposed method was considered to have a great potential as a tool to study the failure problems of various structures.

INTRODUCTION
    Silicon carbide-based fiber reinforced silicon carbide composites (SiC/SiC composites) are promising candidate materials for high heat flux components because of their high temperature properties, chemical stability and good oxidation and corrosion resistance[1-5]. For fabricating large or complex shaped parts of SiC/SiC composites, the technique of joining between simple geometrical shapes is considered to be an economical and useful method. Joints must retain their structural integrity at high temperatures and must have mechanical strength and environmental stability comparable to those of the parts being joined. As a result of R & D efforts, an affordable, robust ceramic joining technology (ARCJoinT[TM]) has been developed and constitutes one of the most suitable methods for joining SiC/SiC composites among various types of joining technologies[6].
    To establish useful design databases, the mechanical properties of joints must be accurately measured and quantitatively characterized. Among the various properties of joints, their tensile strength is one of the most important mechanical properties. Bending tests have been widely used to determine the tensile strength of SiC/SiC composites at high temperatures[7,8], but tensile tests are preferred since a pure tensile stress can be applied to the entire specimen. Recently, a new technique of tensile test using a small specimen of SiC/SiC composite has been proposed[9].
    The strength of joints is largely influenced by its geometry. In order to study this influence, the level of stress and the order of the singularity in stress field are commonly employed for the relative evaluation of strength. Although detailed information on the stress field is provided, little information on the criteria of the fracture is obtained from these types of studies. This comes from the fact that, the physics of failure itself is not explicitly modeled in these analyses.

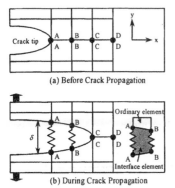

(a) Before Crack Propagation

(b) During Crack Propagation

Fig. 1　Representation of crack growth using interface element.

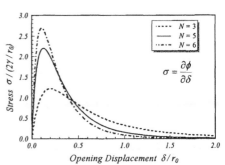

Fig. 2　Relation between crack opening displacement and bonding stress.

The interface element, which directly models the formation of fracture surfaces may have the potential of not only providing insight into the criteria of fracture but also of providing a quantitative prediction of strength.　In this research, the effect of joint shape on the magnitude of the stress singularity and strength was studied for the tensile evaluation of a SiC/SiC composite joint using the interface element.

INTERFACE POTENTIAL

　　Essentially, the interface element is the distributed nonlinear spring existing between surfaces forming the interface or the potential crack surfaces as shown in Fig. 1.　The relation between the opening of the interface δ and the bonding stress σ is shown in Fig. 2.　When the opening δ is small, the bonding between two surfaces is maintained.　As the opening δ increases, the bonding stress σ increases till it becomes the maximum value σ_{cr} at the opening δ_{cr}.　With further increase of δ, the bonding strength is rapidly lost and the surfaces are completely separated.　Such interaction between the surfaces can be described by the interface potential.　There are rather wide choices for such potential.　The authors employed the Lennard-Jones type potential because it explicitly involves the surface energy γ which is necessary to form new surfaces.　Thus, the surface potential per unit surface area, ϕ, can be defined by the following equation.

$$\phi(\delta) = 2\gamma \cdot \left\{ \left(\frac{r_0}{r_0 + \delta} \right)^{2N} - 2 \cdot \left(\frac{r_0}{r_0 + \delta} \right)^{N} \right\}, \tag{1}$$

where, constants γ, r_0, and N are the surface energy per unit area, the scale parameter and the shape parameter of the potential function, respectively.　The derivative of ϕ with respect to the opening displacement δ gives the bonding stress σ acting on the interface,

$$\sigma = \frac{\partial \phi}{\partial \delta} = \frac{4\gamma N}{r_0} \cdot \left\{ \left(\frac{r_0}{r_0 + \delta} \right)^{N+1} - \left(\frac{r_0}{r_0 + \delta} \right)^{2N+1} \right\}. \tag{2}$$

As it is seen from the above equation, the bonding stress σ is proportional to the surface energy γ and inversely proportional to the scale parameter r_0.

Advanced SiC/SiC Ceramic Composites

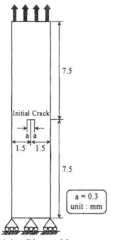

7.5

Initial Crack

a a
1.5 1.5

7.5

a = 0.3
unit : mm

Fig. 3(a)   Plate with center crack.

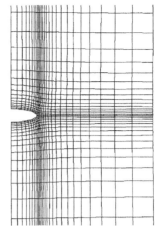

Fig. 3(b)   Mesh division near crack tip
superposed on deformation.

In the case that the crack propagation path is known, the growth of the crack under the applied load can be analyzed in a natural manner by arranging such interface elements along the crack propagation path as shown in Fig. 1.   In other cases, where the crack path cannot be expected, the crack growth behavior also can be simulated by arranging the interface elements between all the ordinary finite elements[10].   In those cases, the criteria for crack growth based on the comparison between the driving force and the resistance as in the conventional methods[11,12] is not necessary.

INFLUENCE OF PARAMETERS
    Among the three parameters involved in the interface energy function, only the surface potential γ has a clear physical meaning.   While those of the scale parameter r_0 and the shape parameter N are not very clear.   To clarify the influence of the scale parameter r_0 on the numerical results of the failure problem, the brittle fracture of a SiC/SiC composite plate with a center crack under tensile load was analyzed.   Although the latest SiC/SiC composite exhibits non-linear behavior, the analysis was conducted assuming linear elastic behavior in two dimensional plain stress in order to avoid the effects of plastic deformation on the scale parameter influences.   Young's modulus and Poisson's ratio were set to 300 GPa and 0.3, respectively.   Figure 3(a) shows the model where the plate was loaded by applying a vertical displacement at one end of the plate.   The value of the surface energy γ of SiC/SiC composite was assumed to be 30 N/m according to a result of the compact tension test[13] and the shape parameter N was set to 4 based on our previous work[14].   The value of the scale parameter was changed from 1.0×10^{-4} to 100 μm. One half of the plate was model and meshed into 27 x 50 elements, where the mesh division near the crack tip was set to fine as shown in Fig. 3(b).   In this model, the fracture load, F, is analytically obtained as the following equation according to Griffith's criterion.

$$F = A \times \sqrt{\frac{2\gamma \cdot E}{\pi a \cdot \alpha^2}} \qquad (3)$$

$$\alpha = f(\xi) = \frac{1 - 0.5 \cdot \xi + 0.370 \cdot \xi^2 - 0.044 \cdot \xi^3}{\sqrt{1-\xi}}, \quad \xi = \frac{2a}{w} \qquad (4)$$

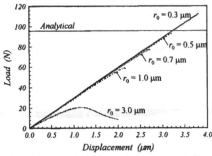

Fig. 4(a)   Load-displacement curves of plate
with center crack.

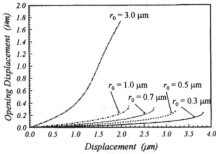

Fig. 4(b)   Applied displacement-crack tip
opening in plate with center crack.

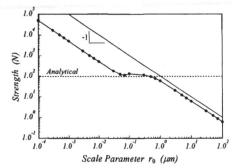

Fig. 5   Effect of scale parameter on predicted strength of plate with center crack.

Where A, E, $2a$ and w are the cross sectional area of specimen, Young's modulus, the center crack length and the specimen width, respectively.

Figure 4(a) shows the load-displacement curves in cases of r_0 = 0.3, 0.5, 0.7, 1.0, 3.0 μm and the relations between the opening displacement at the crack tip and the applied displacement are also as shown in Fig. 4(b). When the scale parameter was large such as the case with r_0 = 3.0 μm, stable crack growth was observed just after the maximum load and the computation stops due to loss of stability. On the other hand, when the scale parameter was small, the instability occurred suddenly without clear increase of the opening displacement or drop of load. The load at the loss of stability was defined as the point of fracture. In these calculations, the load step or mesh density had no influence on the transition from stable to unstable crack growth. So the effect of the scale parameter is summarized in Fig. 5 in logarithmic scale. For comparison, the fracture load computed using the analytical formula (Eq. (3)) is also shown in the figure. As it is clearly seen from these results, the curve could be divided into three parts with respect to the size of the scale parameter r_0. For 0.05 μm < r_0 < 0.5 μm, the fracture load was almost independent on the scale parameter and the value was almost the same as that predicted by Eq. (3). When the scale parameter was smaller or larger than this range, the slope of the curve became -1. This could be explained as follows. When 0.05 μm < r_0 < 0.5 μm, the process corresponds to brittle fracture of the plate with crack growth. Thus, the critical load was determined only by the surface energy γ independently of r_0. According to Eq. (2), the bonding strength and the rigidity of the interface became small with the increase of the scale parameter. Therefore, the plate broke in the separation mode without significant deformation of the plate or crack growth when the scale

Advanced SiC/SiC Ceramic Composites

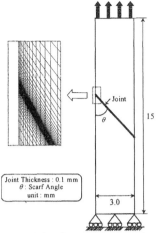

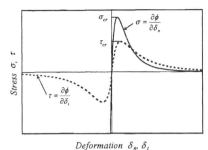

Fig. 7   Displacement-stress relation at interface
for opening and shear modes.

Fig. 6   Model and mesh division of ceramic
joint.

parameter was large.   On the other hand, when the scale parameter was small, the bonding strength became larger than the stress at the crack tip in FEM model.   In this case, the instability was not governed by the surface energy γ but by the bonding strength σ_{cr}, which means that the fracture could not be predicted by the Griffith's criterion.   These computed results clearly show that the effect of the surface energy and the bonding strength on the fracture and the stability limit depend on the combination of the deformability of the plate and the mechanical properties of the interface.

On the other hand, from our previous studies about the peeling of two bonded elastic strips, the scale parameter r_0, the shape parameter N and the mesh division had no influence on the peeling process and the process was mainly governed by the surface energy γ[15,16].

STRENGTH OF CERAMIC JOINT
Model for Analysis
As an example of the applicability of the analysis to a ceramic joint, two SiC/SiC composite specimens jointed by the ARCJoinT[TM] as illustrated in Fig. 6 was analyzed.   The length, the width and the thickness were 15 mm, 3mm and 1 mm, respectively according to the proposed tensile test[9].   The thickness of the joint was set to 100 μm, for a typical example of ARCJoinT[TM] [6,17].   Young's moduli and Poisson's ratios of SiC/SiC composite and the joint were assumed to be 300 GPa, 393 GPa, 0.3 and 0.19, respectively[18].

In order to examine the effect of scarf angle on the joint strength, the mechanical properties of the interface element need to be defined for both opening and shear modes since the mode of failure is mixed mode.   Although it is expected that there will be an interaction between the two modes, only the case when they are independent is considered in this analysis.   The effect of their interaction will be addressed in future work.   According to this assumption, the interface potential ϕ could be defined as a sum of those for the opening mode ϕ_n and the shear mode ϕ_t as in the following equations.

$$\phi(\delta_n, \delta_t) \equiv \phi_n(\delta_n) + \phi_t(\delta_t) \tag{5}$$

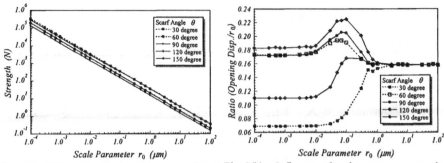

Fig. 8(a)   Effect of scale parameter on predicted Fig. 8(b)   Influence of scale parameter on ratio
             strength of ceramic joint.   of opening or shear displacement.

$$\phi_n(\delta_n) = 2\gamma_n \cdot \left\{ \left(\frac{r_{0n}}{r_{0n} + \delta_n} \right)^{2N} - 2 \cdot \left(\frac{r_{0n}}{r_{0n} + \delta_n} \right)^{N} \right\} \tag{6}$$

$$\phi_t(\delta_t) = 2\gamma_t \cdot \left\{ \left(\frac{r_{0t}}{r_{0t} + |\delta_t|} \right)^{2N} - 2 \cdot \left(\frac{r_{0t}}{r_{0t} + |\delta_t|} \right)^{N} \right\} \tag{7}$$

Where, δ_n and δ_t were the opening and the shear deformation of the interface. Due to the symmetry of shear deformation, the interface potential for the shear mode ϕ_t was assumed to be a symmetric function of the shear deformation δ_t as shown in Fig. 7.

Although the tensile strength of the composite joint was experimentally found to be different from the shear strength[6,17], the surface energies γ_n, γ_t were assumed to be the same and equal to 3 N/m according to the fracture energy of a porous ceramic layer[19]. In order to examine the influence of scarf angle θ on the joint strength, the interface elements were arranged along the lower interface between SiC/SiC composite and the joint.

Effect of Scarf Angle on Joint Strength

In order to examine the influence of the scarf angle on the joint strength, the strength was calculated through a series of computations where the scale parameters r_{0n}, r_{0t} were equally varied from 1.0×10^{-4} to 100 μm and the scarf angle θ was changed from 30 to 150 degrees. As expected from the assumption of the interface potential in Eq. (5), the mode of the failure changed at 45 or 135 degree with increasing the scarf angle. The effects of the scale parameters and the scarf angle on the joint strength are summarized in Fig. 8(a). In contrast to the case of center cracked plate discussed in the preceding section, the horizontal part shown in Fig 5 was not observed in the composite joint. This means that the joint strength was not determined by the surface energies γ_n, γ_t alone. Its magnitude was governed by both the surface energies and the bonding strength σ_{cr}. Figure 8(b) shows the relations to the scale parameter between the scale parameters and the ratio of the opening or shear displacement at the edge of the interface in the failure. As mentioned in the previous section, when the scale parameter r_0 was small or large the strength was dependent on the stress, which means that the opening displacement at the interface in the failure was also dependent on the scale parameter. Although the slope of the curves in Fig. 8(a) was almost constant, it was found that the influence of the scale parameters on the joint strength could be divided into three parts from Fig. 8(b) as same as the case of center cracked plate.

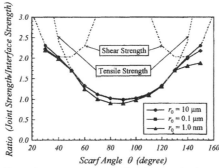

Fig. 9   Effect of scarf angle on predicted strength of ceramic joint.

Namely, the joint strength was strongly influenced by the bonding strength in the range larger than 1 μm or smaller than 0.01 μm, while the surface energy was the dominating parameter in the other region.

In order to study the influence of the scarf angle, the relation between the ratio of the joint strength to the bonding strength of the interface element σ_{cr} and the scarf angle was summarized in Fig. 9 where the scale parameters were selected to be 0.001, 0.1, and 10 μm since the effect of the scale parameter on the joint strength might change in these cases according to Fig. 8(b).   In Fig.9, two dot lines, "Tensile Strength" and "Shear Strength", means the theoretical strength of the joint, F, which can be calculated from the interfacial strength σ_{cr} as follows.

$$F = A \times \frac{\sigma_{cr}}{\sin^2 \theta} \quad \text{(Tensile Strength)} \qquad (8)$$

$$F = A \times \frac{\sigma_{cr}}{\sin \theta \cdot \cos \theta} \quad \text{(Shear Strength)} \qquad (9)$$

Where A is the cross sectional area of specimen.   When the magnitude of the scale parameter decreases, the predicted joint strength became smaller than the theoretical strength.   Since the strength of jointed materials is largely affected by the stress singularity cased by the mismatch of mechanical properties, the influence of the scarf angle on the order of the singularity was calculated.   As a result by using Dunders's parameter[20,21], the order of the singularity had a positive value at the scarf angle ranging from 51 to 95 degree, and Figure 9 shows that the difference between the predicted joint strength and the theoretical strength became larger in this area.   So, although the joint strength is not determined by the order of the singularity in stress field, this result suggested a strong relation between the order of the singularity and the joint strength.

CONCLUSIONS

The interface element was proposed as a simple model to model the mechanism of failure in an explicit manner.   It was applied to the analyses of the fracture strength of a plate with an initial crack and SiC/SiC composite specimen jointed by ARCJoinT[TM].   The conclusions can be summarized as follows.
(1)   In case of the plate with a center crack, the computed fracture load agreed fairly well with the analytical solution when the failure mode was crack growth type, namely dependent on the surface energy of crack.

(2) In case of ceramic composite joint, it was clearly shown that the strength was governed both by the surface energy and the bonding strength.

(3) It was found that the strength of ceramic composite joint was affected by the order of the singularity in the stress field.

(4) The proposed method with the interface element is considered to have a great potential as a tool to study the failure problems of various structures.

ACKNOWLEDGEMENTS

The authors are extremely grateful to Dr. M. Singh, NASA Glenn Research Center, for valuable information. This work was supported by Core Research for Evolution Science and Technology : Advanced Material Systems for Conversion of Energy.

REFERENCES

[1]P. J. Lamicq, G. A. Bernhart, M. M. Dauchier and J. G. Mace, *Am. Ceram. Soc. Bull.*, **65** [2], 336 (1986).

[2]J. R. Strife, J. J. Brennan and K. M. Prewo, *Ceram. Eng. Sci. Proc.*, **11** [7-8], 871 (1990).

[3]R. H. Jones and C. H. Henager, Jr., *J. Nucl. Mater.*, **212-215**, 830 (1994).

[4]R. H. Jones, C. A. Lewinsohn, G. E. Youngblood and A. Kohyama, *Key Engineering Materials*, **164-165**, 405 (1999).

[5]H. Serizawa, C. A. Lewinsohn, G. E. Youngblood, R. H. Jones, D. E. Johnston and A. Kohyama, *Ceram. Eng. Sci. Proc.*, **20** [4], 443 (1999).

[6]M. Singh, *Key Engineering Materials*, **164-165**, 415 (1999).

[7]C. H. Henager, Jr. and R. H. Jones, *J. Am. Ceram. Soc.*, **77** [9], 2381 (1994).

[8]C. A. Lewinsohn, C. H. Henager, Jr. and R. H. Jones, *Ceramic Transactions*, **74**, 423 (1996).

[9]K. Hironaka, T. Nozawa, T. Taguchi, Y. Katoh, L. L. Snead and A. Kohyama, *Ceram. Eng. Sci. Proc.*, **23**, (to be published).

[10]M. Shibahara, H. Serizawa and H. Murakawa, *Proc. 11th Int. Offshore and Polar Eng. Conf.*, 297 (2001).

[11]K. Honda and Y. Kagawa, *Acta Metall. Mater.*, **43**, 4, 1477 (1995).

[12]H. Serizawa, S. Sato, H. Tsunakawa and A. Kohyama, *Proc. 11th Int. Conf. Comp. Mater.*, **II**, 759 (1997).

[13]R. K. Bordia, D. H. Roach and S. M. Salamone, *Proc. 10th Int. Conf. Comp. Mater.*, **IV**, 711 (1995).

[14]M. Ando, H. Serizawa and H. Murakawa, *Ceram. Eng. Sci. Proc.*, **21** [3], 195 (2000).

[15]Z. Q. Wu, H. Serizawa and H. Murakawa, *Key Engineering Materials*, **166**, 25 (1999).

[16]H. Murakawa, H. Serizawa and Z. Q. Wu, *Ceram. Eng. Sci. Proc.*, **20** [3], 309 (1999).

[17]C. A. Lewinsohn, R. H. Jones, M. Singh, T. Nozawa, M. Kotani, Y. Katoh and A. Kohyama, *Ceram. Eng. Sci. Proc.*, **22** [4], 621 (2001).

[18]H. Serizawa, C. A. Lewinsohn and H. Murakawa, *Ceram. Eng. Sci. Proc.*, **22** [4], 635 (2001).

[19]B. F. Sorensen and A. Horsewell, *J. Am. Ceram. Soc.*, **84** [9], 2051 (2001).

[20]J. Dunders, *J. Appl. Mech.*, **36**, 650 (1969).

[21]J. Dunders, *J. Composite Materials*, **1**, 310 (1967).

DUAL-BEAM IRRADIATION EFFECTS IN SiC

H. Kishimoto, Y. Katoh, K.H. Park, S. Kondo and A. Kohyama
Institute of Advanced Energy, Kyoto University
Gokasho, Uji, Kyoto 611-0011, Japan

ABSTRACT
    Silicon Carbide (SiC) has a superior performance for the structural materials of fusion system. Dual-beam irradiation study is one of the most attractive methods for the evaluation of fusion materials. Duet facility, Institute of Advanced Energy, Kyoto University, has enough performance for the investigation of SiC and SiC/SiC composite, which is considered as the structural material of a fusion system. Dual-beam study made clear the irradiation effects due to the defect accumulation and helium deposition and established the evidence of the potential of SiC. The trend of swelling was observed up to 1673 K, and the point-defect swelling reduced with the increasing of irradiation temperature. The void formation by vacancy clustering was not observed in the present work. The helium effect on the amount of swelling was observed up to 1073 K. At temperature over 1273 K, the helium did not affect to the amount of point-defect swelling but formed the cavities on the grain boundary, resulting in an increase of the total amount of swelling.

INTRODUCTION
    Silicon Carbide (SiC) has been considered as the structural materials for a fusion system. In a fusion environment, the high-energy neutron by D-T fusion reaction induces the heavy cascade and nuclear reaction in SiC. For the research of SiC/SiC composites, very high temperature, heavy accumulation of defects and multiple structure of a composite are also important factor. The fundamental study of the effect of defect accumulation was limited, the understanding of point-defect behavior under a fusion environment was necessary [1]. Dual-beam irradiation simultaneously uses two accelerators to irradiate a heavy ion with a helium ion, which simulate the synergistic effects of irradiation induced defects and helium produced by the nuclear reaction [2]. The irradiation condition including irradiation temperature, beam flux and helium ratio were controlled flexibly and precisely in a dual-beam irradiation experiment. The upper limit of irradiation

temperature is 1773 K, which covers the temperature range of the blanket wall in a fusion system [3]. The defect accumulation changes the property and generally reduces the performance of structural materials. The detail of irradiation effects in SiC had not been well-known. Thus the objective of present work was to provide an insight into the trend of swelling, which reflects the defect accumulation. The evaluation methods were also developed for the post-irradiated materials. The microstructural investigation and the micro-indentation study were performed for the analysis of irradiation effects and helium behavior.

EXPERIMENTAL

The material used is chemical vapor deposited (CVD) high purity polycrystalline cubic (3C)-SiC produced by Morton. The CVD-SiC was cut into disk shape and its irradiation surface was polished. The dimensions of specimen were 3.0 mm in diameter and 0.25 mm in thickness. Ion-beam irradiation was carried out at the Dual-Beam for Energy Technology (DuET) Facility, Kyoto University. The specimens were irradiated with 5.1 MeV Si ions and 0.65 MeV helium ions. The irradiation experiment were carried out by the single-ion irradiation, which used only 5.1 MeV Si ion, and dual-ion irradiation, which irradiated helium ion with Si ion at the same time. Helium ions were degraded in energy and slowed down by a thin aluminum foil degrader placed in the target station. The depth profiles of displacement damage, concentrations of Si and He were calculated by TRIM-92 code. A sublattice-averaged displacement energy of SiC was assumed to 35 eV. The damage level was up to 100 dpa, and irradiation temperature was up to 1673 K. Displacement damage rate and He/dpa were 1e-3 dpa/s and 60 appm [4]. Irradiation surface was polished by diamond powders. For the investigation of irradiation effects in SiC, TEM observation, swelling evaluation and micro-indentation study were performed. The thin foil specimens for TEM investigation were prepared by a focused ion beam (FIB) processing [5, 6]. The microstructural investigation was performed with JEOL JEM-2010 conventional TEM. The swelling behavior was evaluated by the measurement of the step height between irradiated and non-irradiated surfaces with the interferometric profilometry [7]. The nano-indentation testing was carried out at the room temperature for the evaluation of mechanical property changes induced by an ion-irradiation. The hardness tests used the nano-indentation device with Berkovich diamond tip. The hardness test was performed in the load range of 0.5 to 100 gf with constant interval/resolution rate of 20/500, which leads to the 10 seconds holding time during indentation test [8].

RESULTS
Swelling Evaluation

The heavy ion-irradiation is considered to induce the heavy cascade as

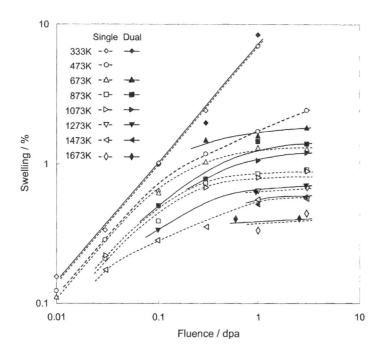

Fig.1 The temperature and dose dependences of swelling under a single- and dual-ion irradiation of SiC

similar to 14 MeV fast-neutron irradiation in a fusion environment [9]. Fig. 1 shows the result of dose rate and temperature dependence of defect accumulation induced swelling by the single-ion irradiation in SiC at temperature up to 1673 K. The point-defects accumulate with increasing of dose rate and cause the increasing of swelling at the relatively low dose region. Under the irradiation environment, the defect accumulation would simultaneously occur with the thermal annealing. At the near of room temperature, the defect accumulation increases a swelling with increasing the dose rate without saturation, and causes the amorphization at about 1 dpa [7]. The point-defect swelling obviously saturates at about 1 dpa at temperature ≥ 673 K.

In a fusion environment, the existence of helium, which is produced by the (n, α) nuclear reaction, affects the point-defect accumulation. The helium effect was studied by a dual-ion irradiation experiment. The tendency of defect accumulation

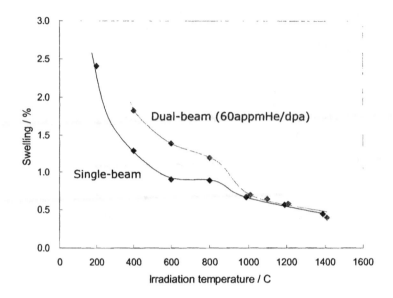

Fig.2 Temperature dependence of 'saturated' swelling in single- and dual-ion irradiated SiC.

under helium existence by a dual-ion irradiation is the same as that under non-helium existence by a single-ion irradiation. The temperature dependence of saturation swelling at 3 dpa is shown in Fig. 2. The saturation swelling continuously reduces with the irradiation temperature up to 1673 K. The helium existence enhances the swelling at relatively lower temperature up to 1073 K. There is little helium effect at temperature ≥ 1273 K.

TEM Investigation

The microstructural revolution in an irradiation environment has been studied by the neutron irradiation [10] but the precise investigation is still limited. TEM images of SiC microstructure single-ion irradiated at 1473 K are shown in Fig 3. At 1473 K, the void swelling by a vacancy clustering does not occur. There are a lot of small dislocation loops and black-dot shape defects. It is estimated that the void formation by only the defect accumulation is not easy because of the high migration energy of vacancies [11].

In a dual-ion irradiation, helium accumulation causes cavities on the grain boundaries at temperature ≥ 1273 K. At 1273 K, relatively heavy defect

Advanced SiC/SiC Ceramic Composites

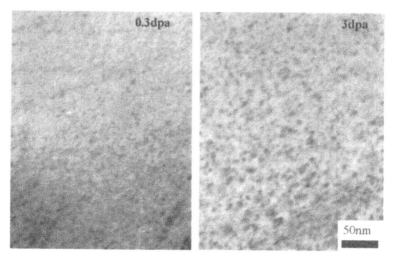

CVD-SiC 5.1MeVSi 1473K z=<110> g=[002]

Fig.3 TEM images of microstructure of SiC single-ion irradiated at 1473 K.

accumulation is necessary for the observation of cavities by TEM. Fig. 4 shows the results of properties of cavities by the analysis of TEM images [12]. The helium cavities grow with defects and helium accumulation. The swelling by helium cavities is about 0.3 % after the dual-ion irradiation at 1273 K and 100 dpa. At 1673 K, the helium cavities grow faster than that observed at 1273 K as shown in Fig. 5. Fig 5 (a) which shows the microstructure irradiated at 3 dpa and 1673 K by a single-ion irradiation. The void formation by vacancy clustering is not observed in the present condition. The helium cavities by a dual-ion irradiation are observed in Fig. 5 (b). The observations indicate that the void swelling occurs under the helium existence significant easier than that without helium.

Indentation Study of Ion Irradiated SiC
    The damaged range by an ion irradiation is limited near of the surface. The depth of damaged range depends on the energy and species of incident particle, which is generally not enough for the conventional mechanical tests. The indentation technique has been developed for the mechanical property investigation of ion-irradiated materials. The defect accumulation by an ion irradiation enhances the hardness of SiC as shown in Fig. 6. From Fig. 6 (a), the dose rate effect is apparent and has a tendency similar to the swelling by a

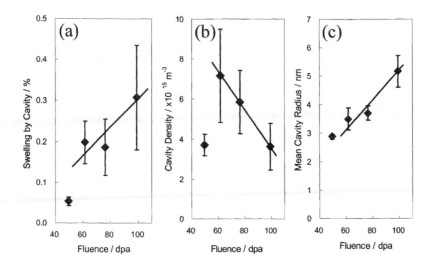

Fig.4 Analysis of the helium cavity property in SiC irradiated at 1273 K up to 100 dpa; (a) swelling by cavity, (b) cavity density, and (c) mean cavity radius.

single-ion irradiation. The hardness sensitively reflects the point-defect increase in the low dose region. The hardness enhancement saturates at the relatively lower dose rate than that observed in the case for the swelling. The other characteristic of an irradiation induced hardening in Fig. 6 (a) is the less dependence on irradiation temperature in the range of 1073 K to 1473 K. This phenomenon indicates that the dislocation loop does not significantly grow in this temperature region. Fig. 6 (b) shows the results of a dual-ion irradiated specimen. The significant helium effect to the hardening is not observed in the present study.

DISCUSSION
Defects Accumulation
    The defect accumulation by irradiation reflects to the swelling. The evaluation of saturation swelling shows that the behavior of point-defects changes at about 1073 K as shown in Fig.2. A possibility of this transfer is the onset of vacancy migration [13, 14]. At the temperature lower than 1073 K, the migration of interstitial dominates the irradiation effects. At above 1073 K, the vacancy starts to migrate, and the vacancy migration enhances the efficiency of recombination of point-defects, resulting in the swelling reduction. One important

Advanced SiC/SiC Ceramic Composites

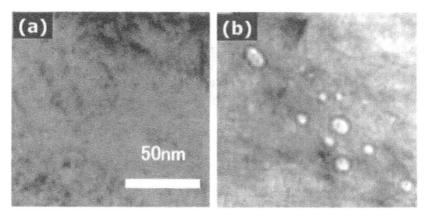

Fig.5. TEM images of SiC irradiated at 1673 K and 3 dpa; (a) single-ion and (b) dual-ion.

possibility is that void swelling by the vacancy clustering occurs at over 1073 K under an irradiation environment. However the void formation depends on the migration behavior of point-defects.

At the near of room temperature, the point-defects form small clusters, which are observed as the dot-shape by TEM [7]. At higher temperature, the density of point-defects decreases with increasing the irradiation temperature because of the enhancement of recombination efficiency of interstitials and vacancies. The other mechanism system to reduce the density of free point-defects is the growth of dislocation loops. As shown in Fig. 4, the dislocations do not form the network at 1273 K and 3 dpa. The shape of dislocation loops indicate the growth of it is slow.

The study of defect migration has been performed by ESR, TEM or MD calculation [11, 15]. The migration of carbon interstitials is considered to be easier than those of both vacancies and silicon interstitials. The result shows that only the migration of carbon interstitial is not enough to induce the rapid growth of the dislocation loops (Fig. 4). The results of swelling, as shown in Fig. 2, clearly indicate that the migration and recombination of defects must depend on the irradiation temperature. Therefore, not only the migration of defects but also the other factors are necessary to explain these phenomena.

In the discussion of a fusion material, the understanding of irradiation effects on the metals is the basis for the consideration of irradiation effects for SiC. The bond angle of atoms in SiC is limited because of a covalent crystal, which makes the displacement of atoms in SiC crystal harder than that in a metal crystal. The other consideration is the charge of defects in SiC [16]. Since the electronegativity of silicon is different from that of carbon, SiC has a partial ionic property. There is

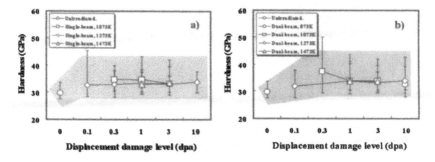

Fig.6 The irradiation effects to hardness of SiC up to 1673K; (a) single-ion and (b) dual-ion.

a possibility that a driving force exists for the defects condensation to maintain the stoichiometric or at least the electroneutrality. Under such a restriction, the high migration energy of silicon interstitial might prevent the dislocation loops rapidly growing. The vacancy also has a charge, the void formation has a possibility to be influenced by the restriction of electroneutrality.

Effects of Helium Production

In a fusion environment, the synergistic effects of defects and produced helium are estimated to affect the behavior of materials. The helium effect to swelling has two stages as shown in Fig. 2. At relatively low temperature up to 1073 K, the helium existence increases the point-defect swelling. It is possible that the helium interstitials are trapped at the vacancies to be stable, which prevents the recombination of point-defects [17]. This enhancement is not observed at over 1273 K. One considerable result, which concerns with the reduction of the helium effects to swelling at over 1273 K is shown in Fig. 4. At 1273 K, heavy helium accumulation causes the helium cavity production. At 873 K, the heavy helium accumulation does not lead the helium cavity production [7]. The helium is estimated to start to be released from a vacancy at 1073 K [18]. The helium is an electroneutral, and its charge must not affect to the clustering of helium. The helium cavity formation is significantly easier than clustering of vacancies, under the helium existence, also the void swelling by helium cavity occurs at temperature ≥ 1273 K.

CONCLUSION

The dual-ion irradiation study established the research of defect and helium behavior in SiC under the irradiation environment. The facility of dual-ion irradiation experiment and the evaluation method for the swelling, microstructure

Advanced SiC/SiC Ceramic Composites

revolution and mechanical property change were developed. The advanced ion irradiation experimental method demonstrated the capability to detect the defect behavior in SiC. The swelling of SiC reduced with the increasing of irradiation temperature. The temperature dependence of swelling behavior changes at 1073 K. This phenomenon was suggested to be due to the onset of vacancy migration. At over 1273 K, the void formation by the vacancy clustering must occur, but was not observed up to 1673 K in the present work. The microstructural investigation and the micro indentation study showed that growth of dislocation loops was relatively slow and seems not to significantly depend on the irradiation temperature. The behavior of swelling indicates that not only the migration of interstitial but the other factor was estimated to affect the defect behavior. This paper suggested the possibility that the charge of defects could prevent the rapid growth of defect clustering.

The dual-ion irradiation experiment showed the trend of helium effects to the swelling and defect behavior. At relatively low temperature up to 1073 K, the helium enhanced the swelling of SiC. This enhancement is considered to be caused by the helium trapping to the vacancies. The trapped helium prevented the recombination of point-defects and reduced the efficiency of thermal annealing. At over 1273 K, the helium effects to swelling reduced because of the helium release from the vacancies. The released helium become the clusters on the grain boundary of SiC, and the helium cavities were observed at over 1273 K. The electroneutral helium is probably not to be affected by a charge, and the clustering of helium is much easer than that of vacancies. The void swelling by helium cavity occurs at the lower temperature than that observed for void formation by the vacancy clustering.

ACKNOWLEDGEMENT

The dual-beam ion irradiation experiment was carried out with the assistance of H. Sakasegawa, T. Suzuki, H. Ogiwara and Dr. K. Jimbo, Institute of Advanced Energy, Kyoto University.

REFERENCES
[1] P. Fenici, A.J. Frias Rebelo, R.H. Jones, A. Kohyama, L.L. Snead, *J. Nucl. Mater.*, **258-263** 215 (1998).

[2] J.C. Corelli, J. Hoole, J. Lazzaro and C.W. Lee, *J. Amer. Ceram Soc.*, **66** 529 (1983).

[3] A. Kohyama, Y, Katoh, M. Ando, K. Jimbo, *Fusion Engneering and Design* **51-52** 789-795 (2000)

[4] L.L. Snead, R. H. Johns, A. Kohyama, P. Fenici, *J. Nucl. Mater.*, **233-237** 26-36 (1996).

[5] Y. Katoh, A. Kohyama, T. Hinoki, *European Conference on Composite*

Materials Science -Technologies and applications, **4** 351-357 (1998).

6. Y. Katoh, T. Hinoki, A. Kohyama, T. Shibayama and H. Takahashi, *Ceramic Engineering and Science Proceedings*, **20** [4] 325-332 (1999) .

7. H. Kishimoto, Y. Katoh, A. Kohyama and M. Ando, *Effects of Radiation on Materials*, ASTM STP 1405, American Society for Testing and Materials, (2001) 775-785.

8. K. H. Park , Y. Katoh , H. Kishimoto , K. Jimbo and A. Kohyama, " Evaluation of Dual-Ion Irradiated β-SiC by Means of Indentation Methods, " *Journal of Nuclear Materials*, Printing.

9. S. Ishino, *J. Nucl. Mater.*, **239** 24-33 (1996).

10. R.J. Price, *J. Nucl. Mate.*, **48** 47-57 (1973).

11. H. Huang, N. Ghoneim, J.K. Wong, M.I. Baskes, Modelling Simul. Sci. Eng. **3** 615-627 (1995)

12. A. Kohyama, Y. Kohno, K. Satoh, N. Igata, *Journal of Nuclear Materials*, **122-123** 619-623 (1984).

13. I.I. Geicz, A.A. Nestrov, L.S. Smirnov, *Rad. Effects,* **9** 243-246 (1971).

14. H. Itoh, M. Yoshida, I. Nashiyama, *J. Appl. Phys.*, **77** 15 (1995).

15. S.Z. Zincle, C. Kinoshita, *J. Nucl. Mater.*, **251** 200-217 (1997).

16. L.W. Hobbs, F.W. Clinald Jr., S.J. Zinkle, R.C. Ewing, *J. Nucl. Mater.,* **216** 291 (1994).

17. H. Huang, N. Ghoneim, *J. Nucl. Mater.*, **212-215** 148-153 (1994).

18. A. Hasegawa, B. M. Oliver, S. Nogami, K. Ave, R.H. Jones, *J. Nucl. Mater.*, **283-287** 811-815 (2000).

SILICON-CARBIDE AS PLASMA FACING OR BLANKET MATERIAL

T.Hino, T.Jinushi, Y.Yamauchi, M.Hashiba and Y.Hirohata
Department of Nuclear Engineering, Hokkaido University, Sapporo, 060-8628
Japan

Y. Katoh and A.Kohyama
Institute for Advanced Energy, Kyoto University and CREST-ACE, Uji, 611-0011
Japan

ABSTRACT

Silicon-carbide is a candidate material for plasma facing components and blanket components in a next generation of fusion reactors. As the plasma facing material, one of important issues to be clarified is property associated with fuel hydrogen retention, which affects fuel hydrogen recycling and in-vessel tritium inventory. For SiC and SiC/SiC composite, deuterium retention properties were investigated using an ECR ion source and a technique of thermal desorption spectroscopy. The retained amount of deuterium was observed to be comparable to that of carbon fiber composite, CFC. However, the desorption temperature was approximately 150 K lower than that of CFC. In addition, the chemical erosion was extremely smaller than that of CFC. Thus, the use of silicon-carbide has advantages in terms of fuel hydrogen retention and chemical erosion, compared to the case of graphitic materials such as CFC.

As the blanket material, SiC/SiC composite may be employed with helium gas coolant. Therefore, one of major concerns is permeation of helium gas through the SiC/SiC composite. For several SiC/SiC composite materials recently developed, the helium gas permeability was measured using a vacuum system consisting of high and low pressure chambers, and fixing structure for sample. The permeability of the SiC/SiC composite made by NITE process using nano-powder of SiC was observed to be extremely low, approximately 4×10^{-11} m^2/s. The permeability of SiC made by using only nano-powder of SiC was $(2-8) \times 10^{-13}$ m^2/s. Since the permeability is quite low, requirement for additional use of metal with SiC/SiC composite can be reduced, and then the engineering design for SiC/SiC blanket may become simple.

INTRODUCTION

Typical low activated materials to be used for a next stage of fusion reactors are ferritic steel, vanadium alloy and SiC/SiC composite[1]. As the plasma facing material, graphite such as carbon fiber composite, CFC, has been widely employed in the present large fusion devices[2,3]. In the fusion demonstration reactor, the use of low activated materials has been required for plasma facing components such as first wall and divertor wall and blanket components. Compared to ferritic steel or vanadium alloy, SiC or SiC/SiC composite as the plasma facing material has an advantage on radiation loss power determined by impurity of fusion plasma, because of its low atomic number. Since the energy confinement of fusion plasma depends fuel hydrogen recycling and impurity level due to material erosion, fuel hydrogen retention and erosion properties have to be clarified for silicon-carbide. There are little investigations on these issues for the silicon-carbide. The fuel hydrogen retention is also important for evaluation of in-vessel tritium inventory. For SiC and SiC/SiC composite, we examined the deuterium retention and chemical erosion of these silicon-carbide materials using an ECR ion source and a technique of thermal desorption spectroscopy, TDS. The obtained results were compared to the case of graphitic material, CFC.

In a case that SiC/SiC composite is used for the blanket material, the coolant is helium gas. The advantage of this blanket system is a high energy conversion efficiency because of high operation temperature, 1100 K. One of major concerns is the permeation of helium gas. If the permeation is significant, the use of only SiC/SiC composite as the blanket material may become difficult and the blanket design becomes complicated. Thus, SiC/SiC composite with low permeation is necessary for the blanket. Several SiC/SiC composite materials have been recently developed by different methods for this purpose. The helium gas permeability was measured for these materials, and then the possibility of SiC/SiC composite as the blanket material was discussed.

EXPERIMENTS

For the measurements on deuterium retention and chemical erosion, an ECR ion irradiation device[4,5] shown in Fig.1 was used. The irradiation temperature was RT and the ion energy was 1.7 keV. The fluence of deuterium ion was changed from 3×10^{21} to 5×10^{22} D/m^2 to obtain the fluence dependence of retained deuterium amount. The retained deuterium amount was measured by TDS in the same irradiation chamber after the ion irradiation. The irradiated sample was resistively heated from RT to 1273 K with a heating rate of 50 K/min. Major gas species desorbed were D$_2$, HD and CD$_4$. The retained deuterium amount was obtained by integrating desoption spectra of these species to the heating time. The desorption of CD$_4$ represents the degree of chemical erosion. Thus, the CD$_4$ desorption spectrum was also examined. As the sample, SiC in the surface of SiC

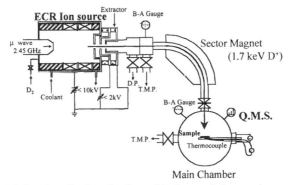

Fig.1   ECR ion irradiation device with thermal desorption spectroscopy.

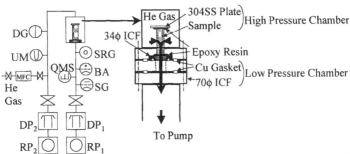

Fig.2   Vacuum system for measurement of helium gas permeability.

Converted Graphite (Toyo Tanso) and SiC/SiC composite (Ube Industries) were employed. For comparison, CFC (CX-3002U) (Toyo Tanso) was also employed.

For the measurement of helium gas permeability, a vacuum system consisting of a high pressure chamber and a low pressure chamber[6,7] shown in Fig.2 was newly established. The high or low pressure chamber corresponds to upper or down streaming side, respectively. Between two chambers, the sample was fixed using epoxy resin. By using this fixing, the vacuum leak was sufficiently suppressed. In the case that stainless steel was used as the sample, no leak was detected. The helium gas was filled into the high pressure chamber using a mass flow controller. The helium pressure was monitored by using a mercury manometer, UM. The range of the pressure was widely changed from 10^2 to 10^5 Pa. The shape of the sample was a cylinder or a flat plate. The high and low pressure chambers were evacuated by using a rotary pump and/or a diffusion pump. The pumping system was selected according to the pressure of the high or low pressure chamber. The ultimate pressure of the high or low pressure chamber was 1×10^{-5} Pa, which is sufficiently enough for the present measurement. Since the helium permeation takes place, the helium pressure of low pressure chamber increases. The rise of the helium pressure was monitored by spinning rotor gauge (SRG), Bayard-Alpert (BA) gauge and quadruple mass spectrometer (QMS). The value of SRG gives the absolute pressure, and then BA gauge and QMS were correlated by SRG. The pressure difference between two chambers gives the permeability defined by

$$K = P_L \, D \, S_{eff} \, / \, P_H \, A .\qquad(1)$$

Here P_L and P_H are pressures in the high and low pressure chambers, respectively, D and A thickness and geometric area of the sample, respectively, and S_{eff} effective pumping speed. The effective pumping speed was obtained using a standard leak of helium. The effective pumping speed was 4.5×10^{-2} m^3/s. The permeability was obtained from Eq.(1).

RESULTS
Deuterium Retention and Chemical Erosion
    After the deuterium ion irradiation at RT, the thermal desorption spectra of D_2, HD and CD_4 were obtained by TDS. Figure 3 shows the desorption spectra of D_2 for SiC and SiC/SiC composite when the fluence was 5×10^{22} D/m^2. For the case of SiC/SiC composite, the spectrum was similar to that of SiC. From this figure, it is seen that the desorption peak of the silicon-carbide is approximately 150 K lower than that of CFC. Similar result was obtained for the case of hydrogen ion irradiation[6]. The fluence dependence of retained deuterium amount was obtained by changing the fluence of deuterium ion. Figure 4 shows the retained deuterium amount as a function of the ion fluence for SiC, SiC/SiC composite and CFC. The retained deuterium amount roughly saturated at around 2×10^{22} D/m^2. In every material, the saturated amount was roughly the same, $(4-6) \times 10^{21}$ D/m^2. These results show that in the case of the silicon-carbide both the fuel hydrogen retention and in-vessel tritium inventory can be reduced by annealing with relatively low temperature, although the retention at RT is comparable to that of CFC. The desorption rate of CD_4 is shown in Fig.5. The desorption rate of CD_4 represents the degree of chemical erosion. Compared to the case of CFC, both SiC and SiC/SiC composite had an extremely low desorption rate. In the fusion reactor, the temperature of plasma facing wall is estimated as 600 – 1000 K. If the silicon-carbide is used for the plasma facing wall, the chemical erosion can be largely suppressed. The present results shows that the use of silicon-carbide has advantages for fuel hydrogen retention affecting fuel hydrogen retention and in-

Advanced SiC/SiC Ceramic Composites

vessel tritium inventory, and chemical erosion, compared to the case of graphitic materials such as CFC.

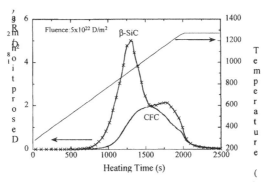

Fig.3   Thermal desorption spectra of D_2 for SiC and CFC.

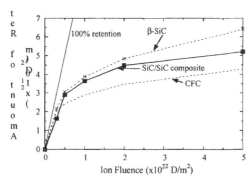

Fig.4   Fluence dependence of retained deuterium amount for SiC, SiC/SiC composite and CFC.

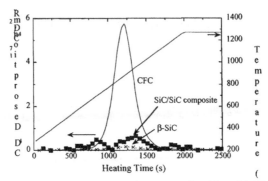

Fig.5   CD₄ desorption rate versus heating time for SiC, SiC/SiC composite and CFC.

Helium Gas Permeability

Helium gas permeability was measured for several kinds of SiC/SiC composite made by different methods, i.e., PIP (Polymer Impregnation and Pyrolysis), HP (Hot Pressing), (PIP and MI(Melt Infiltration)), and (LPS(Liquid Phase Sintering) and HP) methods. These samples were produced by Ube Industries. The shape of the sample was a cylinder or a flat plate. As the sample made by (LPS and HP) method, two samples, LPS1 and LPS2, were prepared. In both samples, Tyranno-SA was used as the fiber. In LPS1, sub-micron powder of SiC was used as the matrix. In LPS2, nano-powder of SiC was used as the matrix. For LPS2, the new process called Nano-powder Infiltration and Transient Eutectoid (NITE) process has been developed[8]. In addition, monolithic SiC made by using only nano-powder of SiC (LPS3) was prepared by (LPS and HP) method. This sample is not a SiC/SiC composite. For the samples produced by PIP, HP and (PIP and MI) methods, the permeabilities are shown in Fig.6. The permeability was roughly constant to the pressure in the high pressure chamber. Among these materials, the sample made by (PIP and MI) method had a lowest permeability, approximately 10^{-6} m²/s. The structure of the sample was observed by a scanning electron microscope, SEM. As the decrease of permeability, the structure became denser and also the pore size became smaller. For comparison, the permeability of isotropic graphite was measured. This value was approximately 10^{-5} m²/s, which is one order of magnitude larger than that of the sample made by (PIP and MI) method.

358   Advanced SiC/SiC Ceramic Composites

For the case of LPS1, LPS2 and LPS3, the permeation of helium was

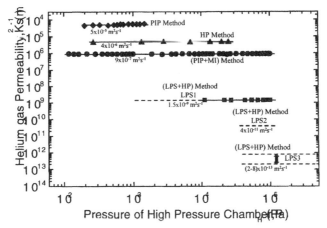

Fig.6   Helium gas permeability versus pressure of high pressure chamber for
SiC/SiC composites and bulk SiC.

extremely small and then the pressure in the low pressure chamber was not
directly measured by SRG. However, the helium pressure rise was monitored by
both BA gauge and QMS. By assuming that signal intensities of BA gauge and
QMS are proportional to the absolute pressure, the helium pressure rise in the low
pressure chamber was correlated to the absolute pressure. Then, the permeabilitis
for LPS1, LPS2 and LPS3 were estimated. As shown in Fig.6, the permeabilitis of
LPS1, LPS2 and LPS3 were roughly 1.5×10^{-9}, 4×10^{-11} and $(2-8) \times 10^{-13}$ m^2/s,
respectively. The permeability of the sample made by (LPS and HP) method was
remarkably small. In particular, the use of nano-powder as matrix is quite
effective for reduction of permeability. As the SiC/SiC composite, LPS2 has a
lowest permeability.

If the SiC/SiC composite is used for coolant pipes in the blanket, leak flux of
helium becomes too large, so that a hermetic material such as metal has to be
employed to wrap the pipes. On the contrary, the requirement for use such the
metal can be reduced if the permeability is low, and then the blanket design
becomes simple. In the present study, it was shown that the extremely low
permeability could be obtained for the case of LPS2. Therefore, it may be possible
to use the SiC/SiC composite as the blanket structure material. However, the use
of metal to the blanket module is still required to avoid the helium flow into the
fusion plasma. For the increase of helium pressure inside of blanket module,
additional pumping is required.

CONCLUSION

As the plasma facing material, silicon-carbide has advantages for both fuel hydrogen retention and chemical erosion, compared to graphitic materials such as CFC. Although the retained deuterium amount was comparable to that of CFC, the annealing temperature required to reduce the fuel hydrogen retention can be approximately 150 K lower than that of CFC. In addition, the chemical erosion was extremely smaller than that of CFC. Therefore, the use of silicon-carbide for plasma facing component in a fusion reactor is quite attractive.

In the use of SiC/SiC composite for blanket, one of major concerns is helium gas permeation. The permeability was measured for several SiC/SiC composite materials made by different methods. The SiC/SiC composite made by (LPS and HP) method using nano-powder of _-SiC as matrix showed an extremely low permeability, 4×10^{-11} m^2/s. In addition, the permeability of the bulk material of SiC produced by the similar method was two orders of magnitude smaller than that of LPS2. The blanket design by using such the SiC/SiC composite and the bulk SiC may become possible.

ACKNOWLEDGEMENT

This work was partly supported by Core Research for Evolutional Science and Technology (CREST) as a part of R&D of Composite Material for Advanced Energy Systems' Research Project.

REFERENCES

[1] For example, see T. Hino, Y. Hirohata, Y. Yamauchi and S. Sengoku, "Low Activated Materials as Plasma Facing Components", 18th IAEA Fusion Energy Conference, IAEA-CN-77/FTP1/08, Sorrento, Italy, (2000).

[2] M.F. Stamp, S.K. Erents, W. Fundamenski, G.F. Matthew and R.D. Monk, "Chemical Erosion Yields and Photon Efficiency Measurements in the JET Gas Box Divertor", J. Nucl. Mater., **290-293**, 321-325(2001).

[3] S. Higashijima, H. Kubo, T. Sugie, T. Nakano, S. Konoshima, H. Tamai, K. Shimizu, A. Sakasai, N. Asakura, S. Sakurai and K. Itami, "Impurity Behavior Before and During the x-point MARFE in JT-60U", J. Nucl. Mater., **290-293**, 623-627(2001).

[4] Y. Yamauchi, Y. Hirohata and T. Hino, "Hydrogen and Helium Retention Properties of B$_4$C and SiC Converted Graphite", Fusion Eng. and Design, **39-40**, 427-432(1998).

[5] T. Hino, "Japanese Universities' activities for PFC Development and PMI Studies", Fusion Eng. and Design, **39-40**, 439-444(1998).

[6] T. Hino, J. Jinushi, Y. Yamauchi, M. Hashiba, Y. Hirohata, Y. Katoh and A. Kohyama, "SiC/SiC Composite as Plasma Facing Material", Applied

Electromagnetics and Mechanics, edited by T. Takagi and M. Uesaka, JSAEM, pp.157-158, (2001).

[7]M. Hashiba, T. Jinushi, Y. Yamauchi, Y. Hirohata, T. Hino, Y. Katoh and A. Kohyama, "Gas Permeability of SiC/SiC Composite as Blanket Material of Fusion Reactor", J. Vac. Soc. Jpn., 45, 145-148(2002). [in Japanese]

[8]A. Kohyama, Y. Katoh, S.M. Dong, T. Hino and Y. Hirohata, "SiC/SiC Composite for Fusion by NITE Process and Its Performance", Submitted to 19th IAEA Fusion Energy Conference, Lion, France (2002).

EVALUATION OF INDUCED ACTIVITY OF SIC COMPOSITES IN FUSION NEUTRON IRRADIATION ENVIRONMENT

T. Noda, M. Fujita, H. Araki, W. Yang, H. Suzuki, and Q. Hu
National Institute for Materials Science, CREST-ACE
1-2-1 Sengen, Tsukuba, Ibaraki 305-0047, Japan

A. Kohyama
Institute of Advanced Energy, Kyoto University, CREST-ACE
Gokasho, Uji, Kyoto 611-0011, Japan

ABSTRACT

The effect of nuclear data and impurities on the evaluation of induced activity for SiC_f/SiC composites prepared with Hi-Nicalon, Hi-Nicalon Type-S and Tyranno-SA fibers was examined. The first wall of SiC_f/SiC composite for He gas cooled and water cooled blankets of fusion devices was assumed to be irradiated for 10 MW.y/m². The simulation calculation of induced activity was made based on FENDL/A2.0, EAF-99, JENDL3.2 and ENDF/B-VI. No large difference in the activity of ^{26}Al produced in SiC between nuclear data sources was observed. The main impurities in the composites are Al, Fe and Ni contributing to the dose rate of SiC_f/SiC composites. Although the SiC_f(SA)/SiC contains fair amount of aluminum, the dose rate of all composites is expected to decrease by about seven orders of magnitude after cooling for several tens of years.

INTRODUCTION

SiC composite is considered as a candidate low activation material for the blanket first wall of fusion reactors and is being developed under international collaboration [1]. Recently SiC_f/SiC composite using new SiC fibers such as Hi Nicalon Type-S and Tyranno-SA fibers with irradiation resistance and high thermal conductivity have been produced [1]. However, SiC composite generally contains fair amounts of impurities, which may affect the induced radioactivity. Moreover, the accumulation of ^{26}Al from ^{28}Si is considered as an important issue for SiC from the viewpoint of waste disposal.

The impurities in commercial SiC composites have been evaluated and among several ways of processing SiC composites, chemical vapor infiltration (CVI) is a conceivable process to produce ceramic composites with a high purity [2]. In a previous paper [3], an evaluation of induced activity of SiC/SiC composites prepared with CVI was made and it was found that the main impurities contributing to the dose rate are several ppm of Fe and Ni for the cooling within 100 years while the activity of ^{26}Al, a long-half lived nuclide, cannot be ignored for the longer cooling. In the evaluation of induced activity of SiC, it was pointed out that the nuclear data source is the most important factor and it is necessary to check the effect of the nuclear data source, especially on the estimate of the ^{26}Al production according to ^{28}Si(n,np+d)^{27}Al(n,2n)^{26}Al reactions [4].

In the present paper, the effect of nuclear data and impurities on the induced activity of several SiC_f/SiC composites prepared with Hi Nicalon Type-S and Tyranno-SA fibers prepared with CVI was examined. The induced activity was evaluated by simulation calculations for the first wall of He gas cooled and water cooled blankets.

PREPARATION AND IMPURITY MEASUREMENT OF SIC$_F$/SIC COMPOSITES

Hi-Nicalon SiC fiber (SiC_f(Hi)), Hi-Nicalon Type-S fiber (SiC_f(S)), and Tyranno-SA (SiC_f (SA)) were used as preforms. SiC_f(Hi) has an excess of carbon and a density of around 2.7 g/cm^2 that is slightly lower than the theoretical density of SiC, 3.2 g/cm^2. The compositions of SiC$_f$(S) and SiC$_f$(SA) are nearly stoichiometric. Especially SiC$_f$(SA) has a high thermal conductivity, though it contains aluminum at 2 % and oxygen at 0.3% as additives. The uni-directional (UD) fiber preforms with a disk shape, 10mm ϕ x 3mm, were infiltrated with SiC using thermal decomposition of trichloro-methyl-silane(MTS) [2]. Composites formed were identified with X-ray diffraction. Impurity measurements were made with neutron radio-activation analysis (NAA) [2] using JRR-3 and JRR-4 reactors. For the determination of the Al content, inductively-coupled-plasma atomic emission spectroscopy (ICP-AES) was applied [5], since it is difficult to quantitatively analyze Al in SiC by NAA.

SIMULATION CALCULATION OF INDUCED ACTIVITY

Transmutation and induced radioactivity calculations were performed using the IRAC code [10]. Nuclear data in 42 energy groups covering the range from 0 to 15 MeV, were taken from FENDL/A-2.0, JENDL-3.2 and ENDF/B-VI. EAF-99 was also considered. Neutron spectra were calculated using ANISN for the first wall of the blanket of a one-dimensional fusion reactor model [3].

Table 1 shows the compositions of the blanket used for the calculation. Two models of He gas cooled and water cooled structures were considered for the calculation. The main structures are composed of carbon armor of 2 cm, first wall of 1.5cm, blanket/shield of around 50 cm, SUS316 vacuum vessel, super-conducting magnet and liquid helium vessel [3]. In the present calculation, the first wall of the blanket is assumed to be composed of SiC composites. For the

Table 1 One dimensional model of the blanket for neutron spectrum calculation

Blanket	Thickness(cm)	Composition (vol ratio)
He blanket		
Blanket	19.5	$0.25SiC+0.14Li_2ZrO_3+0.56Be+0.05He$
1st Sheild	15.0	$0.56SiC+0.24B_4C+0.2He$
2nd Shield	15.0	$0.665SiC+0.285B_4C+0.05He$
Water blanket		
Blanket	19.5	$0.4725Be+0.1575Li_2O+0.05SiC+0.05H_2O+0.1He$
1st shield	15.0	$0.9SS+0.1H_2O$
2nd Shield	15.0	$0.95SS+0.3H_2O$

He cooled system, the blanket materials are Li_2ZrO_3, Be and He gas. For the water blanket, SiC, Li_2O, Be and H_2O are the main constituents.

Fig.1 shows the neutron spectra at the SiC first wall of the blanket under the neutron wall loading of 1 MW/m². The neutron spectra at the first wall position were quite affected by the composition of blanket/shield materials. Clear differences in the spectrum between He and water blankets were observed. The flux of thermal neutrons of the He cooled system is lower by about two orders of magnitude than the water blanket, while fluxes of high energy neutrons are rather higher. The difference in the spectrum is considered to be mainly due to the different neutron scattering cross sections between H_2O and He.

Using the neutron spectrum of fig.1, the activation calculation for the SiC_f/SiC composites was made at the first wall of the blanket with a total neutron flux of 3.84×10^{14} n.cm^{-2}.s^{-1} for 10 years irradiation.

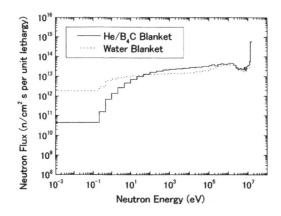

Fig.1 Neutron spectra at the first wall of the Blanket.

RESULTS AND DISCUSSION
IMPURITIES IN SiC COMPOSITES

All specimens prepared had a density higher than 90 % and the ratio of SiC fiber to SiC matrix was around 50%. The SiC matrix was identified to be β-type SiC by X-ray diffraction. As reported in a previous paper, about 35 elements are

Advanced SiC/SiC Ceramic Composites

included in the composites as a result of the NAA and chemical analyzes. **In table 2**, impurity concentrations in the composites prepared with CVI are shown. The SiC$_f$(SA)/SiC contains Al at 1 %, Zr at 28 ppm and Ba at 1.3 ppm in addition to Fe and Ni which are main metallic impurities in the SiC$_f$(Hi)/SiC and SiC$_f$(S)/SiC composites. The concentrations of other metallic impurities are below ppm for all composites.

Element	HiSiC/SiC	Type-S/SiC	Tyranno-SA/SiC
Al	**11	**15	**1.0%
Ti	<1.9	-	-
V	<0.3	<4	0.34
Mn	0.8	<0.019	<0.04
Na	1.70	0.31	0.16
K	0.82	0.17	0.054
Sc	0.0021	0.0013	0.0014
Cr	2.1	0.46	0.84
Fe	4.5(*2.2)	4.6	3.3
Co	0.029	0.13	0.0051
Ni	0.8	6.0	2.2
Cu	2.1	1.1	0.5
As	0.083	0.015	0.016
Br	0.07	0.03	0.0088
Sb	0.015	0.005	0.0069
La	0.021	0.01	0.0043
Ce	0.053	<0.03	-
Ta	0.0047	0.019	<0.0007
W	0.1	0.055	0.015
Au	0.0003	0.00065	0.000089
Hg	<0.006	<0.002	<0.004
Ga	<0.002	<0.003	-
Mo	0.22(*0.6)	0.1	0.13
Nb	(*0.8)	-	-
Ba	<0.4	0.28	1.3
Zn	0.37	0.085	<0.03
Zr	<1	<0.3	28
Ag	<0.0095	<0.004	<0.004
Nd	<0.07	<0.2	-
Eu	0.0015	0.0012	0.0037
Hf	0.160	<0.0006	0.52
Pt	<2	<3	<2
O(%)	0.38	0.418	0.15
N(%)	0.01	0.052	0.05
Cl(%)	0.02	0.056	0.06

*: GDMS, **: Chemical analysis, -: not measured

Table 2 Main impurities in the SiC/SiC composites prepared with CVI (mass ppm).

EVALUATION OF INDUCED ACTIVITY
EFFECT OF NUCLEAR DATA ON THE INDUCED ACTIVITY OF SiC

Fig.2 shows the decays of dose rate of pure SiC after the shutdown of the reactor. The SiC was assumed to be irradiated in the He blanket for 10MW a/m^2. The calculation was made using nuclear data of FENDL/A2.0, JENDL3.2 and ENDF/B-VI. JENDL Activation File 96 was also used for the ^{28}Si(n,d)^{27}Al calculation since JENDL3.2 does not cover this reaction. The activity decreases by about seven orders of magnitude after cooling for one day. However, the activity of around 10^{-5}-10^{-4} Sv/h is kept for a million years due to the radioactive nuclide ^{26}Al. The long-lived activity of SiC is controlled by ^{26}Al($t_{1/2}$=7.2x10^{5}y) which is mainly produced by the following two-step reactions:

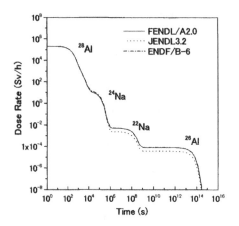

Fig.2 Effect of nuclear data on the decay behaviors of pure SiC after irradiated for 10MW a/m^2.

$$^{28}\text{Si(n,d)}^{27}\text{Al(n,2n)}^{26}\text{Al} \tag{1}$$

and

$$^{28}\text{Si(n,np)}^{27}\text{Al(n,2n)}^{26}\text{Al} \tag{2}$$

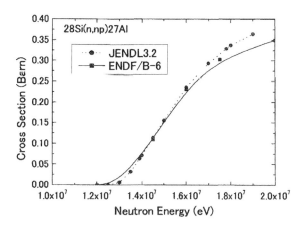

Fig.3 Nuclear cross-section of ^{28}Si(n,np)^{27}Al reaction.

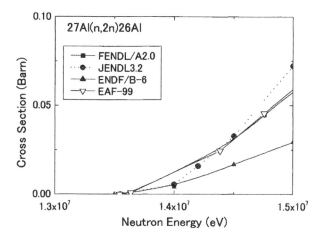

Fig.4 Nuclear cross-sectio of ^{27}Al(n,2n)^{26}Al reaction.

Figs.3 and **4** respectively show the different nuclear data for ^{28}Si(n,np)^{27}Al and ^{27}Al(n,2n)^{26}Al. The comparison between four nuclear data sources are shown in **table 3**.

FENDL/A2.0 adopted the data of JENDL3.2 for ^{28}Si(n,np)^{27}Al. EAF-99 data for these reactions are the same for FENDL/A2.0. As seen in these figures, there is not a large difference in cross-section among these nuclear data. However, the threshold energy of these reactions is near 13MeV. As the result, the small difference in cross-section seriously affects the predicted production of ^{26}Al as seen in fig.2. The active level of ^{26}Al for JENDL3.2 is lower than those for FENDL/A2.0 and ENDF/B-VI.

Table 3 Nuclear data sources for main reactions related with ^{26}Al formation.

Reaction	JENDL3.2	FENDL/A-2.0	ENDF/B-VI	EAF-99
^{28}Si(n,np)^{27}Al	O	JENDL3.2	O	JENDL3.2
^{28}Si(n,d) ^{27}Al	JENDL ActivationFile 96	ADL-3	O	ADL-3
^{30}Si(n,α)^{27}Mg	O	JENDL3.1	O	FEI
^{27}Al(n,2n)^{26}Al	O	ADL3.1	O	ADL-3.1

EFFECT OF NEUTRON SPECTRUM ON THE INDUCED ACTIVITY OF SiC/SiC

The simulation calculation was also made for water blanket. However, the effect of different neutron spectra was hardly observed for the decay behavior of dose rate of pure SiC, since ^{22}Na and ^{26}Al productions were mainly due to neutrons with an energy higher than 1MeV and there is no difference in flux at around 14 MeV as shown in fig.1.

EFFECT OF IMPURITIES ON THE INDUCED ACTIVITY OF SiC/SiC

Fig. 5 shows the decay of dose rate of several SiC composites after shutdown of the reactor with a helium blanket for 10MW a/m^2 assumed irradiation. The calculation was made using FENDL/A2.0 for the compositions indicated in table 3. Since SiC$_f$(SA)/SiC contains a fair amount of Al, the activity levels of ^{24}Na and ^{26}Al are higher than those of SiC$_f$(Hi)/SiC and SiC$_f$(S)/SiC. The activity decreases by about five orders of magnitude after cooling for about one day. However, the activity of around 1 Sv/h is kept for several tens of years by ^{91}Y(t$_{1/2}$ = 58.5 day), ^{54}Mn(t$_{1/2}$=312.5 day) and ^{60}Co(t$_{1/2}$=5.27 y), which are respectively produced from Zr, Fe and Ni contained as impurities. Furthermore, the γ-ray intensity of around 1 mSv/h for SiC$_f$(SA)/SiC is maintained for about one million year due to the radioactive nuclide ^{26}Al. On the other hand, the several ppm of Al in the SiC$_f$(Hi)/SiC and SiC$_f$(S)/SiC composites does not increase the ^{26}Al level of pure SiC. The level of around 1 mSv/h for both SiC$_f$(Hi)/SiC and SiC$_f$(S)/SiC, meets a low-level waste disposal limit of 2 mSv/h [7], though it is difficult to treat SiC$_f$(SA)/SiC as a low level waste because of its high Al concentration.

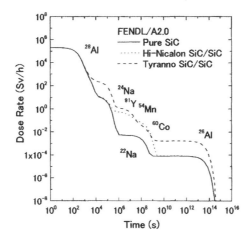

Fig.5 Decay behaviors of dose rate for various SiC/SiC composites after irradiated for 10MW a/m^2.

Furthermore, all composites satisfy the limit after cooling for several tens of years, if the dose rate of 10 mSv/h is assumed as a maximum limit of remote handling recycling level [7].

CONCLUSION

The effect of nuclear data and impurities on the evaluation of induced activity for SiC_f/SiC composites prepared with CVI process was examined. A first wall of SiC composites for He and water blankets under 10 $MW.y/m^2$ irradiation condition was assumed. The following conclusions are drawn:

There is not a large difference in the activity of ^{26}Al produced in pure SiC between FENDL/A2.0, JENDL3.2 and ENDF/B-VI .

Neutron spectrum differences between He gas blanket and water blanket have almost no affect on the dose rate of SiC.

The main impurities in the composites are Al, Zr, Fe and Ni contributing to the dose rate of SiC_f/SiC composites. Especially the activity of ^{26}Al produced in $SiC_f(SA)/SiC$ with aluminum of 1 % is by one order magnitude higher than those of in $SiC_f(Hi)/SiC$ and $SiC_f(S)/SiC$ composites. $SiC_f(Hi)/SiC$ and $SiC_f(S)/SiC$ composites meet the limit of low level waste disposal after cooling for several tens of years. Furthermore, all composites including $SiC_f(SA)/SiC$ satisfy the limit of remote handling recycling level of 10 mSv/h.

ACKNOWLEDGEMENT

This work was supported by the CREST, Japan Science and Technology Corporation.

REFERENCES

[1] R.H.Jones, L.L.Snead, A.Kohyama and P.Fenici, " Recent advances in the development of SiC/SiC as a fusion structural material", *Fusion Eng. Des,.* **41** 15-24 (1998).

[2] T.Noda, H.Araki and H.Suzuki, "Processing of high purity SiC composites by chemical vapor infiltration(CVI)", *J.Nucl.Mat.,* **212-215** 823-829 (1994)823.

[3]T.Noda, M.Fujita, H.Araki and A.Kohyama, "Impurities and evaluation of induced activity of CVI SiCf/SiC composites", *Fusion Eng. Des.*, **51-52** 99-103 (2000).

[4]I. C. Gomes, D.L.Smith and E.T.Cheng, "Status of cross-section data for gas production from vanadium and 26Al from silicon carbide in a D-T fusion reactor", *Fusion Technol,.* **34** 706-713 (1998).

[5]H.Yamaguchi, H.Haraguchi, and H. Okochi, " Determination of aluminum and calcium in silicon carbide by ICP-AES after matrix isolation", *Bunseki Kagaku,* **40** 271-276 (1991).

[6]T.Noda, H.Araki, F.Abe, and M.Okada, Induced activity of first wall materials of fusion reactors", *Trans NRIM,* **30** 185-215(1988).

[7]P.Roco and M.Zucchetti, "Advanced management concepts for fusion waste", **258-263**1773-1777 (1998).

KEYWORD AND AUTHOR INDEX

Printed and bound by CPI Group (UK) Ltd, Croydon, CR0 4YY

Printed and bound by CPI Group (UK) Ltd, Croydon, CR0 4YY

16/04/2025

14658452-0002